An Amateur's Guide to Observing and Imaging the Heavens

An Amateur's Guide to Observing and Imaging the Heavens is a highly comprehensive guidebook that bridges the gap between beginners' and hobbyists' books and the many specialised and subject-specific texts for more advanced amateur astronomers. Written by an experienced astronomer and educator, the book is a one-stop reference providing extensive information and advice about observing and imaging equipment, with detailed examples showing how best to use them. In addition to providing in-depth knowledge about every type of astronomical telescope and highlighting the strengths and weaknesses of each, the author offers advice on making visual observations of the Sun, Moon, planets, stars and galaxies. All types of modern astronomical imaging are covered, with step-by-step details given on the use of DSLRs and webcams for solar, lunar and planetary imaging and the use of DSLRs and cooled CCD cameras for deep-sky imaging.

IAN MORISON spent his professional career as a radio astronomer at the Jodrell Bank Observatory. The International Astronomical Union has recognised his work by naming an asteroid in his honour. He is patron of the Macclesfield Astronomical Society, which he also helped found, and a council member and past president of the Society for Popular Astronomy, United Kingdom. In 2007 he was appointed professor of astronomy at Gresham College, the oldest chair of astronomy in the world. He is the author of numerous articles for the astronomical press and of a university astronomy textbook, and writes a monthly online sky guide and audio podcast for the Jodrell Bank Observatory.

An Amateur's Guide to Observing and Imaging the Heavens

Ian Morison
University of Manchester and Gresham College

CAMBRIDGE
UNIVERSITY PRESS

CAMBRIDGE
UNIVERSITY PRESS

32 Avenue of the Americas, New York, NY 10013-2473, USA

Cambridge University Press is part of the University of Cambridge.

It furthers the University's mission by disseminating knowledge in the pursuit of education, learning and research at the highest international levels of excellence.

www.cambridge.org
Information on this title: www.cambridge.org/9781107619609

© Ian Morison 2014

First published 2014

Printed in the United States of America

A catalogue record for this publication is available from the British Library.

Library of Congress Cataloguing in Publication data
Morison, Ian, 1943–
An amateur's guide to observing and imaging the heavens / Ian Morison, University of Manchester and Gresham College.
    pages   cm
Includes bibliographical references and index.
ISBN 978-1-107-61960-9 (pbk.)
1. Astronomy – Observers' manuals.   2. Astronomy – Amateurs' manuals.
3. Astronomical photography – Amateurs' manuals.   I. Title.
QB64.M674   2014
522–dc23   2013037543

ISBN 978-1-107-61960-9 Paperback

This book is dedicated to the many amateur astronomers worldwide whose knowledge I have attempted to distil in these pages.

Contents

Color plates appear between pages 98 and 99.

Preface

Although I have been a radio astronomer all my working life, I have also greatly enjoyed observing the heavens. At the age of 12, I first observed the craters on the Moon and the moons of Jupiter with a simple telescope made from cardboard tubes and lenses given to me by my optician. I also made crystal and valve radios, and my friends and I set up our own telephone network across our village using former army telephones. Both of these activities were to have a major bearing on my later life.

As I write, I have my father's thin, red-bound copy of Fred Hoyle's book *The Nature of the Universe* on the desk beside me. It was this book that inspired me to become an astronomer.

I was able to study a little astronomy at Oxford but, continuing my interest in radios, was also in the signals unit of the Officers' Training Corps. As I was revising for my finals I spotted an advertisement for a new course in radio astronomy at the Jodrell Bank Observatory. Because I was interested in both astronomy and radios this seemed a good idea and I began to study there in 1965.

My PhD supervisor had been giving evening classes in astronomy at a local college and due to illness asked me if I would take them over from him. The university loaned me two telescopes, a brass 3.5-inch refractor and a 6-inch Newtonian, for use with the classes and I was allowed to keep these throughout the year, enabling me to begin observing the planets, stars and galaxies more seriously. Not long after, I was able to acquire a 10-inch Newtonian in a massive fork mount to see a little more.

In 1990, I helped found the Macclesfield Astronomical Society and am now its patron. I have enjoyed the company and help of its members over the succeeding years as, for example, when they helped me convert my Newtonian into a truss Dobsonian.

Around this time I began to assist with astronomy weekends held by the Society for Popular Astronomy, one of two national amateur astronomy societies in the United Kingdom, and was asked to become its president in 2000. When my term of office ended I remained on its council and, more recently, became its instrument advisor, helping members with their choice and use of telescopes.

With the giving of evening and weekend courses it seemed sensible to acquire additional telescopes of different types to demonstrate and use with the students, and so, over the years, their numbers increased. Reviewing telescopes and mounts for the three UK amateur astronomy magazines allowed me to sample even more, and I was sometimes tempted to purchase them!

Some years ago I co-wrote two books to join the many available for beginners to the hobby. There are also a very large number of books for advanced amateurs covering in detail the many aspects of the hobby. There seemed to be quite a wide gap between the two, and it is this gap which this book aims to fill – covering in some depth the types of telescopes with their strengths and weaknesses, as well as the wide variety of accessories and imaging equipment that can be used with them.

I do not believe that one can write about anything without having actual experience of its use, and so the writing of this book gave me a wonderful excuse to purchase items of equipment that I did not already own. I have really enjoyed trying out new things and hope that the book will encourage others to do so. For example, I had never even thought about doing any spectroscopy but found it a fascinating branch of the hobby – as I hope Chapter 17 will show.

Imaging is now a major part of amateur astronomy, and I have covered every type of imaging, giving many examples with step-by-step details of how the various imaging processes are carried out. In no way do I claim to be an expert imager – the superb images seen in books or on the Web may well be obtained by sophisticated equipment and take many hours of observing and processing – but what I have tried to do is to show how one can get quite reasonable results fairly quickly and so become encouraged to strive to do even better.

I most sincerely hope that this book will help you to increase your knowledge and develop your abilities in what I regard as the most fascinating of all pastimes.

Acknowledgements

As the dedication of this book implies, my first acknowledgement goes to the many amateur astronomers who have shared their wisdom both in their writing of books – many of which are referred to in this volume – and in their contributions to astronomy forums and Web sites. I have spent many enjoyable hours gleaning what I could from them and have tried to pass on what I have learnt.

My friends in the Macclesfield Astronomical Society have been a great source of help and encouragement, and I offer grateful thanks in particular to Stephen Wilcox and Roy Sturmy for the loan of equipment and to Andrew Greenwood, Christopher Hill and Paul Cannon for the use of images to help illustrate the book. Peter Shah, Damian Peach, Greg Peipol and Dr Fritz Hemmerich have also kindly allowed me to use their superb images. My thanks also go to them.

Specifically regarding the chapter on spectroscopy, I have to thank Tom Field for providing helpful comments and advice, Ken Elliot for loaning me his company's CCDSPEC spectrograph and William Wiethoff for the use of a superb spectrum of the quasar 3C 273.

I thank Rich Williams of the Sierra Stars Observatory Network in the United States, who kindly gave me imaging time on two of the Network's telescopes so that I could investigate remote observing.

I would also like to thank Vince Higgs, Sara Werden and the team at Cambridge University Press who have steered this book through to publication and Jayashree Prabhu and her team at Newgen Knowledge Works who have prepared the book for printing. Particular thanks go to Mary Becker who has carried out a superb task in the copyediting of what is a challenging and technical text.

Finally, but not least, I must thank my wife for supporting me as I spent far too many hours at the computer and for putting up with the fact that, far too often, mounts and telescopes were spread across the lounge and dining room ready for use and one complete bedroom was taken over for their storage.

Prologue: A Tale of Two Scopes

When Mars was closest to the Earth in August 2003, the Macclesfield Astronomical Society held a star party at Jodrell Bank Observatory with quite a number of telescopes set up to observe it. As the evening progressed a consensus arose that two scopes were giving particularly good images: my own FS102 4-inch Takahashi Fluorite Refractor (at around £3500, or $5000, with its mount) and an 8-inch Newtonian on a simple Dobsonian mount newly bought for just £200 ($300). I personally preferred the view through the f6 Newtonian but others thought that the f8 FS102 gave a slightly better image, so we will call it a draw. It is worth discussing why these performed so well and, just as importantly, why perhaps the others did not.

The majority of scopes had been set up on a large concrete patio outside our visitor centre, but the FS102 and Dobsonian were on grass and not observing over the patio. This, I believe, was the major reason these two scopes had performed so well. During the day (remember it was August) the concrete would have absorbed heat, which was then released during the evening, causing localised air turbulence through which the scopes mounted on the patio were viewing Mars. It is not, therefore, surprising that the two mounted on grass performed better. One of the world's top solar telescopes, the Big Bear Solar Observatory, rises out of a lake so that it is almost totally surrounded by water in order to minimise any local thermal effects. One of my friends went to Egypt to observe the transit of Venus in 2004. It was very hot in the holiday complex, and he said that it would have been nice to observe from the shallow end of the swimming pool. I suspect that, had he done so, he would have had steadier images too! An obvious piece of related advice is that when observing the planets, particularly in winter, one should not observe them over rooftops, as the turbulence caused by the escaping heat will severely degrade the image. Peter Shah, one of the country's leading astro-imagers, whose beautiful image of M31 is shown in Plate 15.8, has recently lagged the concrete pier on which his telescope is mounted to improve its imaging quality!

The second reason I suspect that the two scopes performed so well is that they are simple. The FS102 has a two-element objective, one element of which is made of fluorite to give it a virtually colour-free image. One key requirement for planetary imaging is high contrast, and refractors excel at this with well-baffled tubes and multi-coated lens elements that scatter very little light. It happens that fluorite elements essentially absorb and scatter no light, and this is one reason the FS series of fluorite Takahashi scopes (sadly no longer made) are said to have the highest contrast of any telescope. The Dobsonian was brand new, and thus its mirrors would have been in very good condition.

The optical performance of a colour-free refractor is essentially perfect and limited solely by the aperture. This determines the resolution and so determines the finest detail that can be seen. A Newtonian has, however, two features that degrade the optical performance: the spider that supports the secondary mirror (which causes the rather pretty diffraction spikes seen with brighter star images) and the secondary mirror itself (which also degrades the image somewhat). Both effects spread light a little away from where it should be and so reduce what I call the micro-contrast of the image, as will be discussed in detail in Chapter 1. But wait. The Dobsonian had twice the aperture so, if theoretically perfect, would have twice the resolution of the FS102. The diffraction effects of the secondary depend somewhat on the 'seeing' – the steadiness of the atmosphere – but, even at worst, would still equal that of the 4-inch refractor. Newtonians suffer from 'coma' – stars begin to look like little comets away from the centre of the field of view, but of course when one is observing planets in the centre of the field this would have no effect. Finally the 8-inch Dobsonian will collect four times as much light as the 4-inch FS102 and that must help too, so it's not too surprising that the two gave comparable performances.

The Dobsonian came with two simple eyepieces made with just three elements rather than the four-, five- or even seven-element designs probably used with the more expensive telescopes. In one sense, the fewer elements the better, as there are fewer surfaces for light to scatter from and less glass to absorb it. A three-element eyepiece will not have as wide a field of view as more complex designs and will tend to show some false colour towards the field edges, but this is no problem when one is observing in the centre of the field and so a simple triplet eyepiece may actually perform better for planetary observing.

I have tried to explain why the two scopes could perform equally well and one reason the others may not have done so well – their location. But there could have been two further reasons, particularly with the more complex 'catadioptric' tele-scopes – those that use both lenses and mirrors to form an image. These tend to need a longer time for the tube interior to cool down to ambient temperature, and also the image quality falls off rapidly if the telescope is not well collimated. As discussed in Chapter 6, the ultimate test for this is to see the 'shadow' of the secondary mirror at the dead centre of an out-of-focus stellar image.

The lessons that come out of this tale are, firstly, think about where you site your telescope; secondly, make sure your scope is cooled down and is well collimated; and lastly (and by far most importantly), a simple telescope with simple eyepieces can outperform many a far more expensive one. You do not need to spend large amounts of money to achieve exquisite views of the heavens!

1
Telescope and Observing Fundamentals

This chapter will discuss the fundamentals of telescopes and observing which are independent of the telescope type or, as in the case of the contrast of a telescope image, dependent on aspects of the telescope design. Later sections in the chapter will discuss the effects of the atmosphere on image quality due to the 'atmospheric seeing' and the faintness of stars that can be seen due to its 'transparency'. The final sections will give details as to how the stars are charted and named on the celestial sphere and how time, relating to both the Earth (Universal Time) and the stars (sidereal time) are determined.

There is one problem that can cause some confusion: the mixing of two units of length: millimetres and inches. Quite a number of US and Russian telescopes have their apertures defined in inches, and the two common focusers have diameters of 1¼ and 2 inches. However, the focal lengths of telescopes and eyepieces are always specified in millimetres, as are the apertures of more recent US, Japanese and European telescopes. In this book I have used the unit which is appropriate and have not tried to convert inches into millimetres when, for example, referring to a 9.25-inch Schmidt-Cassegrain telescope. In calculations and where no specific telescope is referred to, millimetres are always used.

1.1 Telescope Basics

Focal Ratio

A telescope tube assembly will have an objective of a given diameter, D, and have a focal length F. The ratio of the two, F / D, is called the focal ratio, f (Figure 1.1). Typical focal ratios range from 4 to 15. It is easier to design an optical system with a larger focal ratio, so telescopes whose focal ratios are towards the lower end may need more complex – and hence expensive – optical designs or, as in the case of a Newtonian telescope with a short focal ratio, additional corrector lenses.

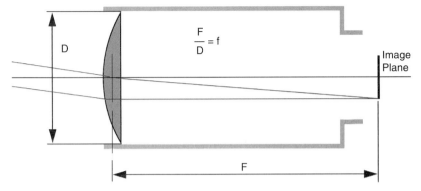

Figure 1.1 The definition of f, the focal ratio of a telescope.

Telescope Magnification

The magnification is given by F/f_e, where f_e is the focal length of the eyepiece. For example, when a 13-mm eyepiece is used with my 820-mm refractor, the magnification is $820/13 = 63$. It is often thought that the prime purpose of a telescope is to give very high magnifications, but the highest useful magnification is limited theoretically by the objective diameter (which defines its resolution) and practically by the 'seeing' (Section 1.5) at the time. This relates to the turbulence in the atmosphere above the telescope and usually limits the effective resolution to 2–3 arc seconds. (An arc second is 1/3,600th of a degree.)

The eye has a resolution of about 1 arc minute (60 arc seconds) in daylight. But this is reduced when the pupil is fully dilated for night viewing to perhaps 2 arc minutes. So, supposing that the effective resolution of the observed image is 2 arc seconds, a magnification of just 60 will bring this up to 2 arc minutes. However, for many objects, spreading the light out over a greater area on the retina can help, so higher magnifications are useful, but it is very rare – perhaps under superb seeing conditions – that magnifications of more than 200 will enable one to see more. There is a rule of thumb that the magnification used with a telescope should not exceed ×50 per inch of aperture (or twice the aperture in millimetres) so, for example, a 6-inch (150-mm) telescope would not be expected to work well at magnifications greater than 300.

The Exit Pupil Diameter and Its Effects

There can be a problem when using very high or very low magnifications. If you hold up a telescope or pair of binoculars towards the sky or a bright wall, the eye lens of the eyepiece will show an illuminated circle. This circle is called the 'exit pupil' and is the column of light that leaves the telescope to enter your eye (Figure 1.2). Its diameter is simply given by the diameter of the aperture divided by the magnification. There are problems if this is too small, such as using a very high magnification with a small-aperture telescope when 'floaters' within the eye can become very obvious or when

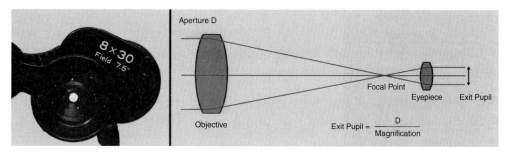

Figure 1.2 The definition of the exit pupil of a telescope or binocular.

it is too big, as then, if the exit pupil is greater than the diameter of the iris, not all the light collected by the telescope will enter the eye and the effective diameter of the telescope will be reduced. Young people have a fully dilated iris of ~7 mm but, as we age, this reduces to ~5 mm. My 80-mm refractor has a focal length of 550 mm so, if I use my 40-mm Paragon eyepiece, I will get a magnification of 13.75 and hence an exit pupil of 5.8 mm. As a result, not all of the light collected by the telescope will enter my ~5-mm pupil and the effective diameter of my telescope is reduced to 70 mm. To get the very best out of my telescope I would need to use a 35-mm eyepiece, which would give a magnification of 15.7 and an exit pupil of 5.1 mm.

In the same way as the resolution of a camera lens tends to peak and then fall off as the aperture is reduced, so the eye will give its maximum resolution when not all the aperture of the eye's lens is in use. Studies indicate that this is when the effective aperture of the eye is about 1.5–2 mm in diameter. The eye's effective aperture when used with a telescope is *not* the diameter of the dilated pupil, but the diameter of the telescope's exit pupil. You may well find that, when you are observing a planet, a particular eyepiece might give you the clearest image. It is then quite likely that the telescope/eyepiece combination is giving an exit pupil around this diameter.

Field of View

Finding the field of view of an eyepiece when used with a given telescope is a little more complex. If one looks through an eyepiece at the open sky, one sees a white circular aperture whose size is limited by what is called the 'field stop' of the eyepiece. This aperture defines what is called the 'apparent' field of view of the eyepiece. For simple three-element eyepieces this might be just 40 degrees but, in the case of complex wide-field eyepieces, can exceed 100 degrees. If you know the apparent field of view, then simply dividing this by the magnification will give an approximate actual field of view. The eyepiece with a 13-mm focal length referred to earlier has an apparent field of view of 68 degrees so, given the magnification of 63, will have a true field of view of just slightly more than 1 degree when used with a scope of 820-mm focal length – very nice to view the Moon surrounded and contrasted against a black sky. A second way is to measure the diameter of the field stop with a ruler; 18 mm in this case. The true field is then given by multiplying the ratio of the field stop divided by the focal length (18/820 = 0.22) by 57.3 (the number of degrees in a radian) to give 1.26 degrees. I suspect that this is a more accurate result. A third way is to observe a star close to the celestial equator and time how long it takes to cross from one side of the field of view to the other when the telescope is not tracking. This time in seconds is divided by 86,164 (the number of seconds in a sidereal day) and multiplied by 360 to give the actual field of view in degrees.

Light-Gathering Power

Larger-diameter objectives collect more light, so enabling one to see fainter objects. To explore this further we need to understand the concept of apparent magnitude. The Greek astronomer Hipparchus placed the stars into six magnitude groups: the brightest in magnitude 1 and the faintest in magnitude 6. Quantitative measurements have since shown that the first-magnitude stars were about 100 times brighter than the sixth-magnitude stars, and this was made a definition. The magnitude scale is logarithmic, so that a fifth-magnitude star will be some factor (say, Z) brighter than a sixth. In the same way, a fourth-magnitude star will be *the same factor Z* times brighter than a fifth-magnitude star. Thus a fourth-magnitude star will be $Z \times Z$ times brighter than a sixth-magnitude star, and so a first-magnitude star will be $Z \times Z \times Z \times Z \times Z$ times brighter than a sixth-magnitude star. But this brightness ratio has been defined as 100, so Z must be the fifth root of 100 = 2.52.

It was found that some stars were brighter than this – given 'zeroth' magnitude – and some even brighter still when the magnitude scale becomes negative, with, for example, Sirius at magnitude −1.5. The planets can be even brighter, with Venus reaching magnitude −4.7.

Let us take a real example to see how this determines what we might be able to see with a pair of 40-mm-aperture binoculars. With our dark-adapted eyes observing at site with no light pollution and a transparent sky, we might be able to see a sixth-magnitude star. If we assume that our eye has a pupil of diameter 6 mm we can see that the binoculars have a diameter 6.6 times greater. They will thus collect $(6.6)^2$, or ~44, times more light.

Now an increase in collecting area of 2.52 times would allow one to see 1 magnitude fainter stars, an increase of in area of 6.35 (2.52 × 2.52) times, 2 magnitudes fainter. Continuing, a 16 times increase in aperture should show stars 3 magnitudes fainter, a 40 times increase 4 magnitudes fainter and (by definition) a 100 times increase 5 magnitudes fainter. So, with our binoculars collecting 44 times more light, we should be able to see stars slightly fainter than 6 + 4 magnitudes – slightly more than 10th magnitude.

It is worth noticing that I specified a site with little light pollution and transparent skies – everyone will have noticed that from a given site when the atmosphere is very clear we can see far fainter stars. The transparency of the atmosphere has a double effect, as not only will the atmosphere, if somewhat 'hazy', reduce the light that we receive from the stars, but it will also scatter back far more light pollution, making it even harder to see the stars. This is discussed in greater detail in Section 1.5.

Books often give a table of the 'limiting magnitude' that a scope of a given aperture can observe. I would rather give a table of what I call the 'magnitude gain' of a telescope, which I think is far more useful and relates directly to the observing conditions. Note the magnitude of the faintest star that you can observe in the direction in which you wish to observe (which takes into account your eyesight, the sky transparency, light pollution and elevation) and simply add the appropriate telescope gain to find the faintest-magnitude star that you should be able to observe in that direction.

75 mm	4.5	90 mm	4.9	102 mm	5.2	114 mm	5.9
130 mm	6.3	150 mm	6.7	180 mm	7.1	200 mm	7.4

Under perfect conditions towards the zenith, a young eye might be able to spot a star with a magnitude of 6.5. If this is added to the magnitudes just given, one gets what is called the 'limiting magnitude' of a telescope, which, in the case of a 150-mm scope, would be ~13.2. Faint stars, which would not be seen using a low magnification under light-polluted skies, will become more apparent if the magnification is increased by using eyepieces of shorter focal length as, with increased magnification, the brightness of the sky background is reduced but that of the stars is not. This increases the contrast between them and the sky and enables stars perhaps 1–2 magnitudes fainter to be seen.

The Resolution of a Telescope

The detail in an image viewed by a telescope is theoretically limited by its resolution, which increases with telescope aperture. (But, as mentioned earlier, this is usually limited by the atmosphere.) The resolution of a scope of given aperture can be measured experimentally by, for example, observing when the two stars of a close double can just be split. This approach gave rise to the empirical 'Dawes limit' proposed by W. R. Dawes and gives the resolution, in arc seconds, as $R = 4.56/D$, where D is in inches, or $R = 116/D$, where D is in millimetres. For a scope of aperture 100 mm this gives 1.16 arc seconds.

The image of a star under perfect observing conditions and when one is using a telescope such as a refractor which has an unobstructed aperture is in the form of a central disk – called the 'Airy disk' – which contains 84% of the light surrounded by a number of concentric rings of decreasing intensity. The whole is called the 'Airy pattern' (Figure 1.3). This is the result of the diffraction of light as it passes through the telescope aperture.

The 'Rayleigh criterion' states that a telescope can resolve two stars when the peak of one star's diffraction pattern falls into the first minima of the other (Figure 1.4). This gives a somewhat lower resolution limit than that defined by Dawes, as in the Raleigh criterion there is a drop of ~26% in brightness between the two peaks whereas in the case of the Dawes limit the drop is only 5%. The angular separation between the centre of the Airy disk and the first minima of the Airy pattern is given, in radians (1 radian = 57.3 degrees), by 1.22 λ/D, where λ is the wavelength of the

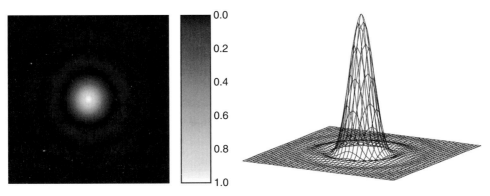

Figure 1.3 The Airy pattern shown (*left*) as a greyscale pattern and (*right*) as a 3D plot. (Image: Wikimedia Commons)

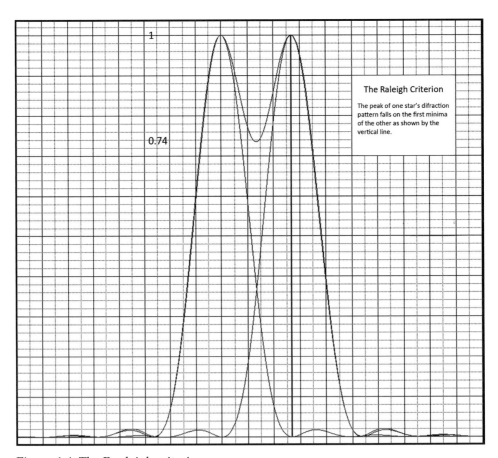

Figure 1.4 The Rayleigh criterion.

Table 1.1 *The theoretical resolution in green light for a number of telescope apertures*

Aperture (mm)	Resolution (arc seconds)
102	1.35
150	0.92
200	0.69
300	0.46

light. (Both λ and D must be in the same units.) Using the wavelength of green light of 5.5×10^{-7} m, this gives a resolution, in arc seconds, of $138/D$, where D is measured in millimetres. Thus, with a lens of aperture 100 mm, one gets a theoretical resolution of ~1.4 arc seconds. The resolution of some typical aperture telescopes is given in Table 1.1.

The form of the Airy pattern was first computed by George Biddell Airy and is a complex calculation but, interestingly, the approximate result can be simply derived from quantum theory. The Heisenberg uncertainty principle states that the more accurately one knows the position of an object (in this case a photon) the less well defined is its motion. If light passes through a slit of width D – which thus defines its position along one axis – the uncertainty principle shows that the light will have an angular spread along that direction given, in radians, by λ/D. However, in the case of a circular aperture, the light path is constrained in two axes, so increasing the angular spread and the size of the Airy disk will be correspondingly larger, hence the factor 1.22.

In the case of visual observing, the resolution is almost always limited by the atmosphere, but if the atmospheric seeing is excellent and very short exposures are made (in the form of a video sequence) and the best of these images are processed as described in Chapter 11, then the theoretical resolution *can* be obtained, as witnessed by the wonderful images of Jupiter taken by Damian Peach and others.

Curvature of Field

Ideally the image plane produced by a telescope would be flat. In practice, unless some corrective optics are included, the image plane is curved, with the outer parts of the image in focus slightly closer to the lens or mirror than the centre. The shorter the focal length of the telescope, the greater the effect. In visual observing, the eye is able to accommodate some curvature of field, and this is not too much of a problem, but when imaging is carried out with a large CCD array, the outer parts of the field may be out of focus. It is possible to purchase lens attachments, called field flatteners, that correct for this and some of the latest telescope designs, such as the

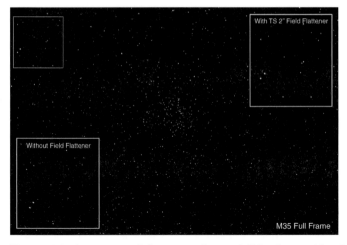

Figure 1.5 An image of the open cluster M35 taken with a Nikon D7000 DSLR with APSC sensor, with and without Telescop-Service 2-inch field flattener.

Celestron 'Edge HD' Schmidt-Cassegrains, incorporate them into the optical tube assembly. Figure 1.5 (and Plate 1.5 in colour) shows an image of the open cluster M35 taken with an 80-mm refractor and a Nikon D7000 DSLR with an APSC-size sensor. Insets show the field corner taken with, and without, the use of a Teleskop-Service 2-inch field flattener.

1.2 The Contrast in a Telescope Image

Image contrast is perhaps one of the key properties of a telescope and one which is particularly important when one is viewing the Moon and planets. It is a subject that is not too well understood, with erroneous statements often appearing in the astronomical press. The following two sections will, I hope, enable you to understand the various elements that come into play to determine what is termed the 'contrast' of an astronomical image. The approach that I believe gives the best understanding of the subject splits the discussion into two parts: firstly, that of the overall contrast of the image and, secondly, what I term the 'micro-contrast' of an image. I am not aware of any other author taking this approach, but I honestly believe that this is by far the best way of considering this very important aspect of telescope design.

The Overall Contrast of an Image

So what is meant by the overall contrast of an image? In an ideal world, the light that is recorded by a CCD camera of a particular feature in an image will have come only from the light emitted or reflected by that feature alone. This will rarely be the case. The lunar image at the left of Figure 1.6 is obviously of low contrast – the

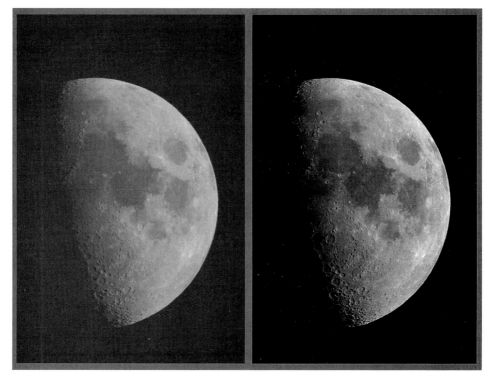

Figure 1.6 A low-contrast (*left*) and high-contrast (*right*) image of the Moon.

blacks are grey, not black. In this case, the reason is that it was taken in twilight and so sky light was falling uniformly across the image. (By taking a sky image with the same exposure just to the left of the Moon and subtracting this from the lunar image removed much of this unwanted light and produced the image on the right, which has much higher contrast – quite a good tip! This is covered in Chapter 18.) However, after dark, there may well be light pollution spreading light into the image and, even with no light pollution, there will still be some 'air glow' reducing the contrast a little.

The overall image contrast (so that blacks are not as black as they should be) can also be reduced by factors in the design of the telescope being used. One often reads that reflecting telescopes give lower-contrast images than refractors, but one major reason for this statement no longer holds quite so true. A telescope mirror having a simple aluminium coating will reflect ~86% of the light falling upon it. As reflecting telescopes will have two mirrors, only ~75% of the light entering the telescope will reach the eye or camera. This will reduce the effective size of the telescope somewhat but not, in itself, reduce the contrast of the image. But what of the remaining 25%? Some of this light will be absorbed within the mirror coating, but a significant portion will be scattered and can fall anywhere within the image, so reducing its overall contrast. (If you shine a red laser beam at a mirror surface so that the reflected beam is away from your eyes, you will easily see where the beam meets the surface, thus showing that light is being scattered.) A major reason refractors give images with higher overall contrast than reflectors is that objective lenses may scatter only ~2% of the light passing through them.

This is why I believe that the high-reflectivity coatings that are now applied to many astronomical mirror surfaces are so important. With ~95% reflectivity, not only will they give somewhat brighter images but they will also greatly reduce the amount of scattered light, so improving the overall contrast. A high-reflectivity coating is well worth having even if at an additional cost: not only will the telescope perform better but a second advantage is that the mirror surface allows far less moisture to penetrate and is likely to last perhaps 25 years before it has to be re-coated. I have a 10-year-old Newtonian whose mirror was one of the first to be given a high-reflectivity coating and it still looks like new.

A second reason for reducing the overall contrast is that light scatters off the interior of the optical tube assembly. This is also why refractors can provide such high-contrast images, as a series of knife edge baffles reducing in size can be located within the optical tube to trap any scattered light. Matt-black flock coatings may also be used in both types of telescope to reduce this, and high-specification reflecting telescopes may also be equipped with a series of baffles mounted within the tube. If a reflecting telescope has a glass correcting element mounted at the front of the tube assembly, it is a good idea to use a dew shield (which extends the tube assembly outwards), not just to reduce the building up of dew on its surface but to prevent extraneous light from falling upon it. The best of these are also equipped with internal baffles. Even with a Newtonian, when extraneous light is a nuisance, an outward extension to the tube will help.

The overall design of the telescope will affect the overall contrast as well. It is impossible to beat a well-designed refractor, but Newtonian telescopes, where one observes across the tube assembly to the far wall, are almost as good. This also applies to the more complex Schmidt-Newtonians and Maksutov-Newtonians. My 150-mm Maksutov-Newtonian has a set of baffles immediately across from the focuser to prevent any light scattered off the tube walls from entering the field of view (Figure 1.7). Few standard Newtonians seem to be so equipped, and the application of some flocking opposite the focuser could well make a useful improvement.

The telescope designs that have the greatest problem with overall contrast are those where the light path exits through the primary mirror, as one is then looking up towards the sky. Such telescopes incorporate an internal baffle tube so that the incident light into the telescope is hidden by the secondary mirror and its support. This does involve some design compromises, as extending the baffling to increase the overall contrast may well restrict the light falling on the outer parts of the image – called 'vignetting'. Even so, extraneous light can still enter the baffle tube and be scattered into the image. Again, a dew shield will greatly help. Increasing the size of the circular secondary mirror support will also help, but this then has an impact on the second cause of reduced contrast within an image, which I term 'micro-contrast', as discussed in the next section.

Figure 1.7 A Maksutov-Newtonian showing baffles opposite the focuser.

The Micro-Contrast of an Image

As already described, the overall contrast will be reduced by *scattered* light either from parts of the optical tube assembly or from the surface of a mirror or within (to a far lesser extent) the objective lens. The micro-contrast, however, is determined by the effects of light that is *diffracted* by parts of the optical tube assembly. In this case, light is moved from its rightful place only by angular distances measured in arc seconds or arc minutes and is thus particularly important in the case of observing planetary disks where the angular scale of the observed object is similar in scale and the features on the surface may have low contrast as well.

The Effects of a Central Obstruction on the Airy Pattern

The first effect of diffraction is that caused by the fact that the telescope will have an aperture of a given size. The result is that a point source of light such as a star gives rise to a disk of light (whose size is determined by the aperture of the telescope) surrounded by concentric rings forming the Airy pattern, as described earlier. If the aperture is unobstructed, as with a refractor, 84% of the light falls within the central disk and thus 16% lies in the rings – with the majority within the first ring, whose diameter is about twice that of the central disk. This changes when the aperture is blocked by a central obstruction such as the secondary mirror in a reflecting telescope. As the obstruction increases as a percentage of the aperture, more light is transferred from the central disk into the rings. Table 1.2 shows how much light remains in the central disk as the percentage of the aperture obstructed by the secondary mirror increases. At the same time, the angular diameter of the central disk actually reduces slightly, so possibly helping to resolve double-star systems. The actual energy in the central disk

Table 1.2 *The energy within the central Airy disk as a function of the relative size of the central obstruction*

Central obstruction (%)	0	10	20	30	40	50
Energy in central disk (%)	84	82	76	68	58	48

relative to the maximum possible (84%) is called the 'Strehl ratio', discussed more fully in Section 1.4.

It is generally reckoned that an obstruction of up to 15% has virtually no observable effect. The size of the secondary in a Newtonian or Maksutov-Newtonian (M-N) is reduced as the focal ratio is increased, and there are scopes of this type specially optimised for planetary viewing. An f9 Newtonian and some M-N scopes have secondary mirrors giving an obstruction close to 15%. The very small secondary in an M-N telescope optimised for planetary imaging is shown in Figure 1.7.

A way of improving the effect of the secondary mirror is to reduce its size so that only the very central part of the field of view is fully illuminated. That is, if you place your eye (without an eyepiece) at the centre of the focuser you will see all of the mirror, but as you move away from this central position the area of the mirror that you see will be reduced. Thus less of the mirror will be illuminating these points in the field of view, so causing vignetting away from the centre of the field. This is, of course, no problem when you are viewing the planets provided that they are kept in the centre of the field of view. It is possible to have two changeable secondary mirrors, a small one for planetary observing and a larger one for deep-sky use to more fully illuminate the whole field of view

A second diffraction effect is found when the secondary mirror has to be supported by a spider, as in Newtonian telescopes. This produces a thin cross, which is centred on stellar objects. In fact, the result is quite pretty, and it has been known for astroimagers using refractors to stretch two strings across the objective to give the same effect as a spider! It is not obvious on extended objects such as a planetary disk, but the effect is still present and will reduce the micro-contrast of the image. When a curved spider is used the spikes disappear, as the light is spread around more uniformly, but light will still be scattered away from its rightful place.

If one has an aperture telescope of, say, 250 mm or more in aperture which has spider or secondary obstruction, a useful trick that can be used for planetary observing is to employ an off-axis mask, as shown in Figure 1.8. This gives an unobstructed aperture and thus a cleaner image. If the seeing is not good, the reduced aperture may actually give better images of the planets, as the column of light passing through the atmosphere is smaller. It is often said that a telescope with an objective size of 130 or 150 mm may well give the best planetary images, as this aperture matches the size of the atmospheric 'cells' moving over the telescope.

Finally, it is often said that Schmidt-Cassegrains (S-C) are not good for planetary observing, as they have a large central obstruction. Figure 1.9 shows how the effective resolution of a 200-mm telescope falls off as the central obstruction increases in size.

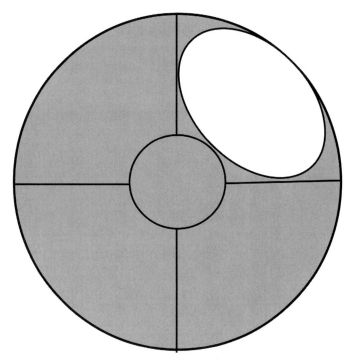

Figure 1.8 An elliptical off-axis mask to give a reflecting telescope a smaller, but unobstructed aperture.

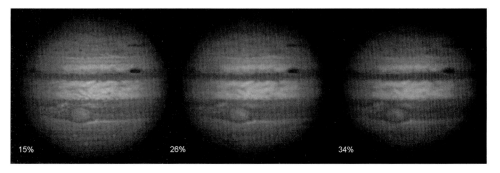

Figure 1.9 The effect of a secondary obstruction on the resolution of a telescope.

Unless the seeing is near perfect, the effective size of the central disk is given by the diameter of the first ring in the diffraction pattern. This is about twice the diameter of the central disk of an unobstructed aperture, so the effect is to halve the nominal resolution. True, but your S-C is likely to be at least 8 inches in diameter and so will still have the same resolution as a 4-inch refractor – and no one says that *they* are no good for planetary observing. But the image will be far brighter and this may allow greater magnification to be used. When the seeing is superb, the fact that the diameter of the central disk reduces slightly as the central obstruction becomes larger may actually increase the resolution of a telescope, particularly when close double stars are being observed.

1.3   Rich-Field Telescopes

As the name implies, a rich-field telescope is one that will show the observer the maximum possible number of stars within the field of view when he or she is looking, say, towards the Milky Way. Using such a telescope to sweep along the Milky Way on a dark moonless night reveals one of the most beautiful sights that can be seen in the heavens.

At first sight you might think that if you buy a bigger telescope you will be able to see more stars. It is not, however, quite as simple as that. Yes, a telescope with a bigger aperture will enable you to see fainter stars but, in general, larger-aperture telescopes (having longer focal lengths) will have smaller fields of view, so limiting the number of stars that can be encompassed within them. In fact, it's not trivial to work out what might be best, and it requires knowledge of the number of stars of a given brightness per unit area of the milky way.

It turns out that we tend to see the most stars when the exit pupil of the telescope/ eyepiece combination equals that of the dark-adapted eye. In young people this is approximately 7 mm but sadly, as previously mentioned, as we age this drops down to 5 or 5.5 mm. As described earlier, the exit pupil is the telescope aperture divided by the magnification, so if we use 5 mm as our pupil size, then a 50-mm pair of binoculars would need to have a magnification of at least 10 times (hence the common 10 × 50 specification). With the classic 80-mm refractor that is now so common, one would need a minimum magnification of 16, with a 102-mm telescope a magnification of 20, with a 150-mm telescope 30, with a 200-mm telescope 40 and with a 300-mm telescope 60. (Younger people with larger pupils can use lower powers.) The low-power eyepieces used to give the minimum magnification tend to come equipped with 2-inch barrels, so enabling the use of a large-diameter field stop to give a wide field of view. This will be further discussed in the section on eyepieces in Chapter 7.

A telescope using a low-power eyepiece to give a wide and rich field is not just for observing the milky way. The combination of my 80-mm refractor and a low-power eyepiece gives a wonderful view of the Pleiades Cluster in Taurus (Figure 1.10), easily able to encompass the whole. The Double Cluster in Perseus is another wonderful object, just seen with the unaided eye, but an 80- or 100-mm refractor used at low power will show the two clusters beautifully. A 200-mm telescope will allow you to see the orange-coloured red giant stars sprinkled across the field. One hardly need mention the region around M42, the Great Nebula in Orion – one of the most wonderful sights in the heavens. The constellation Auriga, passing high overhead during part of the year, lies along the milky way and includes three Messier open clusters. A real test, requiring a moonless night and transparent sky, is to observe the Rosette Nebula in Monocerous.

One shouldn't limit oneself to our own galaxy. The Great Nebula in Andromeda, our nearest giant galaxy in space and lying at a distance of 2.5 million light years, spans a little more than 3 degrees in length. An 80-mm telescope with a low-power wide-field eyepiece can easily encompass it. If you have a 2-inch focuser, then you

Figure 1.10 The Pleiades Cluster, M45, viewed through an 80-mm refractor and low-power eyepiece.

owe it to yourself to buy a suitable eyepiece so that you can enjoy these wonderful rich-field views for yourself.

1.4 Image Quality

One factor in determining the image quality of a telescope is the precision with which the mirrors or lenses have been made. Refractors where the false colour (chromatic aberration) has been eliminated are often said to give exquisite 'pin sharp', images, whereas those produced by reflecting telescopes, particularly Schmidt-Cassegrains, are often said to be somewhat 'mushy'. There are several reasons this might be so:

1. A refractor is usually perfectly collimated – that is, optically aligned – and tends never to need collimation, whereas reflectors tend to lose their collimation and (as shown in Chapter 6) need to be regularly collimated to give of their best.
2. The central obstruction of a reflecting telescope will reduce the image quality somewhat compared with an identical aperture refractor.
3. The overall contrast of refractors will tend to be higher.
4. A refractor is likely to have an optical system of higher accuracy and so to be closer to optical perfection. The reason is very simple. It is actually four times easier to

make a lens with a given optical precision than a mirror. Suppose a mirror has a 'perfect' surface except for a small part which is ⅛ wavelength below its surroundings. As the light is reflected, the error in the light path length will be double this and so be ¼ wavelength. The error in the path length for any deviation from a perfect surface is doubled. Let us now consider the surface of a lens which has a depression of ⅛ wavelength. Had it been filled with glass, the difference in optical path length would have been increased by the fact that the glass has a refractive index of ~1.5 – but this is only ~50% more than the path though air. So the actual path length error will only be 1/16 wavelength – one-quarter that of the equivalent error in a mirror!

5. The expensive refractors which are prized for their exquisite images tend to have more time expended on them by a master optician than is generally the case with mirrors and so will tend to be closer to optical perfection.

This is not to say that reflecting telescopes cannot be superb, and one can purchase mirrors, such as those made by the Zambuto Optical Company, which are very close to optical perfection and give images comparable to the very best apochromat refractors. Happily, the vast majority of telescope optics are now excellent and, at a Scottish star party in January 2013, the image of the central region of the Orion Nebula as seen through a Celestron 11-inch Schmidt-Cassegrain was the best I have ever encountered – the stars making up the Trapezium showing as perfect pinpricks of light. A well-collimated Newtonian or Schmidt-Cassegrain whose mirrors have high-reflectivity coatings can be a superb telescope.

How Is the Quality of a Mirror or Lens Assessed?

There are a number of ways by which a manufacturer of mirrors or lenses may specify the quality of the optics.

Strehl Ratio

This is perhaps the most obvious and best-defined way of specifying optical quality and one which is coming into more common use. As previously described, a refractor, as it has an unobstructed aperture, would ideally place 84% of the light from a star (when observed under perfect seeing) into the central disk of the Airy pattern. If the objective were perfect it would be said to have a Strehl ratio of 1. No objective is perfect, although some of the very best reach Strehl ratios greater than .95 under test conditions. Some manufacturers may be willing to quote minimum Strehl ratios for their objectives – a value such as .95 or .96 – but it is interesting to note that the very 'high end' manufacturers, whose lenses often equal or exceed a .98 Strehl ratio, may not do so for fear of competition among their purchasers. The important point to note is this: it is the atmospheric seeing that will virtually always limit the image quality of a good telescope, and a Strehl ratio of .95 or above can essentially be regarded as perfect.

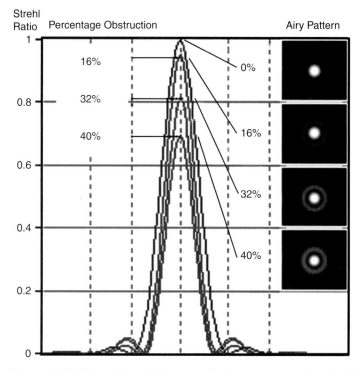

Figure 1.11 The effect of the central obstruction on the Strehl ratio and Airy pattern.

When a telescope has a central obstruction, the light within the central disk is reduced and so the Strehl ratio is bound to be less and, if the actual Strehl ratio of the telescope were given (likely to be nearer .8 rather than .9), its value would not be representative of the inherent quality of the mirror. In this case, the Strehl ratio quoted is for the mirror assuming that there is no central obstruction. Values greater than .9 may be then be measured and quoted. Figure 1.11 shows how the Strehl ratio decreases as the size of the central obstruction increases.

The Rayleigh Criterion for Optical Quality

Lord Rayleigh carried out tests on telescopes whose optics had been made to differing quality standards and empirically showed that, if the optics give a wavefront error which is less than ¼ wavelength of green light, the result is visually indistinguishable from a perfect optical system under normal seeing conditions.

As described earlier, a mirror has to be made to a surface accuracy of ±⅛ wavelength for this to be achieved. The quality of a mirror is thus often determined by quoting a 'peak-to-valley' (P/V) difference in terms of wavelength. If we assume that some parts of the mirror are ⅛ wavelength above and some parts are ⅛ wavelength below the mean surface, then the mirror will have an overall P/V value of ¼ wavelength. Many manufacturers will state that their mirrors meet this criterion.

Where low- or medium-magnification observations are made, such as when star clusters and galaxies are observed, a telescope made to this accuracy will be essentially perfect, and most observers would not be able to tell the difference between, say, a mirror polished to ¼ wave accuracy and one figured to 1/10 wave. However, when high magnifications are used to view the planets under excellent seeing conditions, experienced observers will be able to tell the difference. Some manufacturers, such as Orion Optics in the UK, will thus provide mirrors with a range of accuracy. The company's 'standard' grade offer gives a 1/4 P/V wavefront, 'professional' grade 1/6, 'research' grade 1/8 and 'ultra' grade a 1/10 P/V wavefront. A 1/4 P/V wavefront error corresponds to a Strehl ratio of .82, whilst a 1/10 P/V wavefront error will provide a Strehl ratio of .97. (As a Strehl ratio of greater than .95 is essentially perfect, mirrors of greater accuracy are not really required.)

A note of warning: Rayleigh specified his criteria relating to green light, where the eye is most sensitive. Some manufacturers may specify the wavefront error in the red light of a helium-neon laser at 6,320 angstroms rather than green light at around 5,000 angstroms. This is a ~26% increase in wavelength and so will overstate the optical quality.

A P/V value can understate the accuracy of a mirror: if just a small part of the mirror is in error by this amount but the majority of the surface is more accurate, the mirror will perform better than might be expected. A mirror can be better evaluated by measuring the surface profile at a very large number of points across the surface to provide a far better feel of its overall accuracy, as described later.

Root-Mean-Squared Surface Accuracy

Mirrors can now be tested using interferometric methods, and these yield another way of defining the surface accuracy. The root-mean-squared (RMS) accuracy of a mirror provides a statistical measure of the departure of the surface from the ideal shape. The RMS value for a mirror will be considerably less than the P/V value and may thus make the mirror appear better than it actually is. For example, a mirror with an RMS value of 1/35 (0.028) wave corresponds to a 1/10 wave P/V value.

The RMS value can be converted into the Strehl ratio of the mirror. The approximate formula is

$$\text{Strehl ratio} = 1 - (2\pi \times \text{RMS})^2$$

Putting in, say, 0.028 for the RMS value, multiplying by 2π and squaring gives 0.032, resulting in a Strehl ratio of .97.

Diffraction Limited

Manufacturers often state that their optics are 'diffraction limited'. This implies that it is the Airy diffraction pattern, determined by the aperture of the mirror or lens, which limits the image quality, not the quality of the mirror. It is not a well-defined term. It is often said that this corresponds to a P/V wavefront error of ¼ wave, but others, I suspect correctly, believe that the correction has to be better than a 1/6.6 wavefront error for this to have an element of truth.

Mirror Smoothness

This is something that few manufacturers will specify but has a significant impact on the micro-contrast of an image. A pattern of ripples in the surface which may be less than 1/30 wavelength in amplitude – and so will not figure in the overall mirror specification – will nevertheless reduce the contrast of the features on a planetary disk. This is one area where hand-finished mirrors made by the world's top opticians such as Carl Zambuto and Robert F. Royce outperform those from large-scale manufacturers. The Zambuto Optical Company Web site gives considerable insight into this aspect of mirror performance.

1.5   The Effects of the Atmosphere

Sadly, unless we have access to the Hubble Space Telescope, we have to observe the heavens through our atmosphere. This has two effects on the image we observe, caused by the atmospheric 'seeing' and 'transparency'.

Atmospheric Seeing

This is the name given to the blurring and twinkling of astronomical objects caused by turbulence in the Earth's atmosphere. There are formal definitions based on the diameter of what is called the 'point spread function' or 'seeing disk', which corresponds to the fuzzy blob seen in a long-exposure photograph of a star. At high-altitude observatories such as that at Mauna Kea or La Palma, the seeing disk may be as little as 0.4 arc second in diameter but from ground level could be as much as 4 arc seconds, though there are locations such as in Florida or Barbados where the seeing can be surprisingly good. This is attested by the wonderful images of Jupiter taken by Damian Peach from Barbados using a technique called 'lucky imaging', which will be fully described in Chapter 11 on lunar and planetary imaging.

Amateur astronomers use the 'Pickering scale', devised by William H. Pickering, to describe the quality of the seeing during an observing session. This has a scale of from 1 to 10, ranging from 'very poor' to 'excellent'. An excellent visual animation of how a star image will appear under the different seeing conditions can be found at www.damianpeach.com/pickering.htm. It is only in the highest-four seeing categories that the first ring of the Airy disk can be seen but, even when the seeing is not that good, there can be brief moments of clarity when a still patch of air comes between the planet and observer. A simpler five-category scale can also be used (see Figure 1.12):

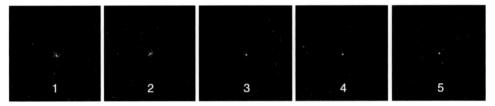

Figure 1.12 A five-point seeing scale as described in the text.

1. Bad   Boiling image without any sign of diffraction pattern.
2. Poor   Broken-up central disk. Missing or partly missing diffraction rings.
3. Average   Central disk deformations. Broken diffraction rings.
4. Good   Light undulations across the diffraction rings.
5. Excellent   Perfect, near motionless diffraction pattern.

The 'seeing' is caused by small-scale fluctuations in the atmosphere of both temperature and density, giving rise to turbulence that results in a blurred, moving or scintillating image. This latter aspect manifests itself in the 'twinkling' of stars, which is thus a reliable indicator of how good the seeing is. The fact that the planets are less prone to scintillation indicates that the angular sizes of the atmospheric cells are of the order of arc seconds across or less.

There are three levels in the atmosphere where turbulence occurs:

1. *Near ground (0–100 m).* This is due to convection currents either from the heat of nearby houses or from the cooling of ground, concrete or tarmac that has been heated during the day. An unvarying terrain such as large areas of grass or water tends to suffer least, as it loses its heat more slowly and evenly. One of the world's best solar observatories, the Big Bear Solar Observatory, is located on an artificial island in a lake which provides a cooling effect on the atmosphere surrounding the building and eliminates ground convection currents that would normally cause optical aberrations. Unless the telescope itself has reached ambient temperature, it can cause turbulence in the surrounding air. If mounted on a steel pillar within an observatory, this should be lagged. Run-off roof observatories tend to have better characteristics than a dome. (Some domes have louvers in their sides to allow air to flow over the mirror, and smaller telescopes are now using 'clamshell' domes that collapse back to the ground.)
2. *Central troposphere (100 m to 2 km).* The turbulence at these altitudes is determined largely by the topography and population areas upwind of the observing site. The air downwind of a city or range of hills will contain turbulent eddies which destroy the image quality. It is thus best to have an observing site with the sea or a flat arable landscape for a considerable distance upwind. These help to produce a laminar airflow, resulting in stable images. It is thus not surprising that the world's great optical observatories are located on islands far out to sea or with an ocean upwind of the observing site.

3. *High troposphere (6–12 km)*. Jet streams in the upper atmosphere can often produce images which appear stable but are devoid of fine detail. Their positions vary with the time of year. They can bring trains of weather systems across a country and thus also limit the number of clear nights when one could make observations. Jet stream forecasts for northern Europe and parts of North America can be found at www.netweather.tv/index.cgi?action=jetstream;sess=.

The Canadian Weather Office provides a wonderful facility for observers in North America, which is a map showing the seeing quality on the five-point scale given earlier for up to 48 hours ahead. This can be found at www.weatheroffice.gc.ca/astro/index_e.html.

Seeing will be poor when a cold front has passed over, replacing warm air with cold. This gives rise to local convection currents as the ground and buildings lose heat. Conversely, warm air tends to be more stable, particularly when a high-pressure area is present and mist or fog has formed. At these times, though the 'transparency' (discussed later) will be poor, the seeing can be excellent. The formation of cumulus clouds in the afternoon indicates convection in the lower atmosphere, so that seeing will tend to be poor for some time after sunset and, as a general rule, will tend to be better in the hours before dawn. High-altitude cirrus clouds and light winds often indicate that a night of good seeing is in prospect. The ultimate test is, of course, to observe the image of a bright star and compare it with the videos on Damian Peach's Web site given earlier. If the image of the star is defocused then the patterns of the atmosphere's turbulence passing across the telescope can be easily seen.

In the United States, the states where the seeing tends to be best are those in the south-east, with southern Florida enjoying good seeing throughout much of the year due to a stable high-pressure region and a smooth airflow from the Gulf of Mexico. Arizona and New Mexico fair well too. The more northerly states towards the Canadian border tend to have poorer seeing due to the presence of the polar jet stream.

In the United Kingdom, seeing conditions are often good in the south and south-east, particularly near the coast, but further north the seeing is often poor. (I can attest to this!) The seeing is usually better during the summer months than in the winter. This is not always the case, as in 2012 the jet stream settled over the UK, which resulted in one of the wettest summers for many years, giving few opportunities for observing and poor seeing when one could.

Atmospheric Transparency

The fact that our atmosphere absorbs most of the infrared and ultraviolet radiation that passes through it is fortunate for us but, happily, the ozone layer that protects us from the ultraviolet absorbs only 1 or 2% of the light visible to our eyes. However, even clean air scatters light by a process called 'Rayleigh scattering'. This affects blue light far more than red, which is why our daytime sky is blue and why the sun seen close to the horizon is red, as the blue light has been scattered away from the line of sight. Without Rayleigh scattering, our daytime skies would be black.

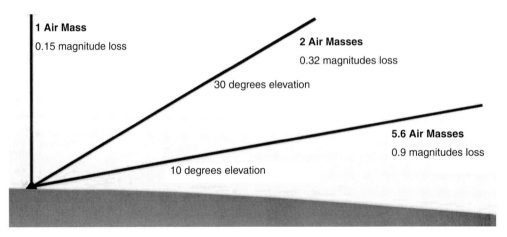

Figure 1.13 The effect of air mass on extinction.

At the blue-green wavelengths where our eyes are most sensitive at night, Raleigh scattering reduces a star's brightness by 0.14 magnitude at the zenith, with ozone adding another 0.016 magnitude: this is called the 'zenithal extinction'. The extinction obviously increases as a star is observed closer to the horizon as what is called the 'air mass' increases from 1 at the zenith to 2 air masses at 30 degrees elevation and 5.6 air masses at 10 degrees elevation (see Figure 1.13). A star will then appear 0.32 and 0.9 magnitude fainter respectively. It is obviously best to observe objects as high in the sky as possible.

Sadly, the air is never totally clean, with dust, humidity and emissions from power plants, aircraft and motor vehicles combining to form what are termed 'aerosols'. The result is that, even from a totally dark-sky site, a star's brightness will be reduced. For example, on a moderately poor night a star which is observed at an elevation of 30 degrees above the horizon will lose 0.8 magnitude and thus only 48% of its light will reach our eyes. (If you could observe from the top of Mauna Kea, the star's brightness would be reduced by only 0.3 magnitude and 76% of the light would be detected, a reason for situating the world's largest telescopes atop high mountains.) I have had the opportunity to observe the Moon's dark side, lit only by 'earthshine', from a height of 9,000 ft on Mauna Kea, and it appeared almost as bright as the full Moon as seen from sea level.

The aerosols in the atmosphere have a second effect: that of reflecting light from the ground back to Earth – light pollution. The greater the amount of aerosols in the atmosphere, the greater the amount of light that is reflected back, so helping to mask the already attenuated light from the star or deep-sky object.

It so happens that on the nights when the transparency is very poor, the seeing is usually very good, and a calm night in autumn can be an excellent time for observing and imaging the planets. Conversely, when the transparency is very good, the seeing can be really bad but then, should the Moon not intrude, deep-sky objects that do not need a high magnification to observe are the better targets.

The Canadian Weather Office provides a second facility for observers in North America, which is a map showing the sky transparency on a five-point scale from

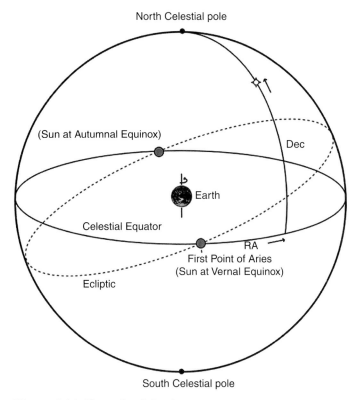

Figure 1.14 The celestial sphere.

'excellent' down to 'very poor' for up to 48 hours ahead. This can be found at www. weatheroffice.gc.ca/astro/index_e.html.

1.6 Charting the Stars

The terms and ideas in this section will become important when telescope mounts are discussed in Chapter 8.

Star Positions in Right Ascension and Declination

The modern stellar coordinate system is analogous to the way in which we define positions on the surface of the Earth and uses the orientation of the Earth in space as its basis. Firstly, the Earth's rotation axis is extended up and down to the points where it reaches our imaginary celestial sphere. The point where the axis meets the sphere directly above the North Pole is called the 'North Celestial Pole' and that below the South Pole is the 'South Celestial Pole'. If the Earth's equator is extended outwards it will cut the celestial sphere into two – into the Northern and Southern Hemispheres – forming the 'celestial equator' (see Figure 1.14).

There is one path around the celestial sphere that is of great importance: that of our Sun. If the Earth's rotation axis were at right angles to the plane of its orbit around the

Sun, the Sun's path would trace out the celestial equator but, as the axis of the Earth's rotation is inclined to its orbital plane by an angle of 23.5 degrees, the path of the Sun is a great circle, called the 'ecliptic', which is inclined by 23.5 degrees to the celestial equator. The Sun spends half the year in the southern half of the celestial sphere and the other half in the northern. Its path thus crosses the celestial equator twice every year: once at the 'vernal equinox', on March 20 or 21, as it comes into the Northern Hemisphere and 6 months later, when, at the 'autumnal equinox' on September 22 or 23, it passes back into the southern skies.

Just as a location on the Earth's surface has a 'latitude', defined as its angular distance from the equator towards the poles, so a star has a 'declination' ('Dec' for short), given as an angle which is either positive (in the Northern Hemisphere) or negative (in the Southern Hemisphere). The 'Pole Star' in the northern sky is close to the North Celestial Pole near to +90° (or 90° north) in declination, and the region at the South Celestial Pole (where there is no bright star) is at −90 ° declination.

The second coordinate proves to be rather more difficult. On the Earth we define the position of a location round the Earth by its 'longitude'. But there has to be some arbitrary zero of longitude. It was sensible that the zero of longitude, called the 'prime meridian', should pass through a major observatory and that honour finally fell to the Royal Greenwich Observatory in London.

As mentioned earlier, the path of the Sun gives two defined points along the celestial equator that might sensibly be used as the zero of 'right ascension' – the points where the ecliptic crosses the celestial equator at the vernal and autumnal equinoxes. The point where the Sun moves into the Northern Hemisphere was chosen and was given the name 'the first point of Aries', as this was the constellation in which it lay. Star positions are measured eastwards around the celestial sphere from the first point in Aries to give what is called the star's 'right ascension' (RA).

However, RA is not measured in degrees but in time, with 24 hours equivalent to 360 degrees. So the celestial sphere is split into 24 segments, each of 1 hour and the equivalent of 15 degrees around the celestial equator.

Angular Measure

A great circle measures 360 degrees in angular extent.
Each degree is divided into 60 arc minutes.
Each arc minute is divided into 60 arc seconds.
There are then 3,600 arc seconds in 1 degree.
('Arc seconds' and 'arc minutes' can also be written as 'seconds of arc' and 'minutes of arc' respectively.)

Precession

Should you locate the point where the Sun crosses the ecliptic at the vernal equinox on a star chart (with position RA = 0:00 hours, Dec = 0.0 degrees), you might be surprised to find that it is not in Aries, but in the adjacent constellation Pisces. This

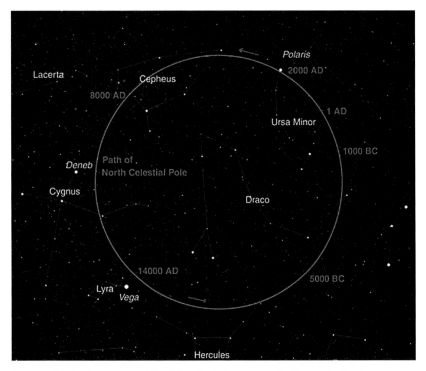

Figure 1.15 The path of the North Celestial Pole through the heavens.

is the result of the precession of the Earth's rotation axis – in just the same way that the axis of rotation of a spinning top or gyroscope is seen to precess. The precession rate is slow – one rotation every ~26,000 years – but its effect over the centuries is to change the positions of stars as measured with the coordinate system described earlier, which is fixed to the Earth. Consequently, a star chart is valid for only one specific date. Current star charts show the positions of stars as they were at the start of the millennium and will state 'Epoch 2000' in their titles. One result of precession is that the 'Pole Star', Polaris, is close to the North Celestial Pole only at this partic-ular time in the precession cycle (Figure 1.15). In ~13,000 years, the bright star Vega will be near the North Celestial Pole instead (though by no means as close). It also means that constellations currently not observable from the UK such as the 'Southern Cross' will become visible for a time low above the southern horizon. (Try putting the year 15,000 into the free planetarium program 'Stellarium', and look due south on a January evening from the south of England.)

1.7 Constellations and Stellar Nomenclature

A constellation is one of 88 regions of the celestial sphere defined by the International Astronomical Union in 1922 and based on the 48 constellations listed by Ptolemy in his second-century *Almagest*, with further constellations added as the Southern Hemisphere sky was charted. Most constellations include a number of bright stars forming patterns whose shapes represent humans (Orion, the hunter), animals (Leo, the lion), birds (Cygnus, the swan) or objects (Lyra, the lyre). Many bright stars have names which have been handed down through the centuries, but only a few tend to be widely used, such as for the stars Betelgeuse and Rigel in the constellation Orion and Regulus in Leo.

In 1603 Bayer devised a system in which the brighter stars in a constellation were identified by small letters of the Greek alphabet; α, β, γ, δ and so on. Usually, the brightest star in a constellation was designated α and the second-brightest β, continuing in sequence to fainter stars, but sometimes, as in Ursa Major, their position within the constellation, rather than their brightness, was used. As an example, the star Vega in the constellation Lyra is also called α Lyrae, using the genitive form of the constellation name. (Often an abbreviation of the constellation name is used – for example, one may refer to ϵ Lyr, the famous 'double-double' star not far from Vega. In addition, the Greek letter is sometimes spelt out.)

In 1725 Flamstead catalogued the fainter stars in a constellation by giving them a number which increased in right ascension across the constellation. As an example, 51 Pegasi (or 51 Peg) is a star just to the west (right) of the Square of Pegasus, around which the first planet orbiting a star was discovered.

In about 135 BC the Greek astronomer Hipparchus compiled a catalogue of at least 850 stars. A satellite, named *Hipparcos* in his honour, was launched in 1989 equipped with a 29-cm Schmidt telescope designed to measure star positions to very high accuracy. Resulting from its observations, the Hipparcos Catalogue, a high-precision catalogue of more than 100,000 stars, was published in 1997. Thus the star Rigel, β Orionis, is also known as HIP 24436, and 51 Pegasi as HIP 113357. These catalogue numbers are used to identify faint stars in, for example, the Stellarium planetarium program. Quite a number of other star catalogues exist, with Rigel also known as HD 34085 in the Henry Draper Catalogue of 225,300 stars down to magnitude 9 and as SAO 131907 in the Smithsonian Astrophysical Observatory Star Catalogue published by the Smithsonian Astrophysical Observatory in 1966 and containing 258,997 stars.

1.8 Time

Local Solar Time

For centuries, the time of day was directly linked to the Sun's passage across the sky, with 24 hours being the time between one transit of the Sun across the meridian and that on the following day. This time standard is called 'local solar time' and is the time indicated on a sundial. The time such clocks would show would thus vary across the United Kingdom, as noon is later in the west. It is surprising the difference this makes. In total, the UK stretches 9.55 degrees in longitude from Lowestoft in the east to Mangor Beg in County Fermanagh, Northern Ireland, in the west. As 15 degrees is equivalent to 1 hour, this is a time difference of slightly more than 38 minutes.

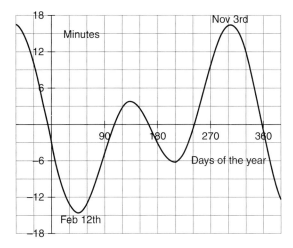

Figure 1.16 The 'equation of time' – the difference between GMT and local solar time at Greenwich Observatory. (Image: Wikimedia Commons)

Greenwich Mean Time

As the railways progressed across the UK, this difference became an embarrassment and so London or 'Greenwich' time was applied across the whole of Great Britain. A further problem had become apparent as clocks became more accurate: because the Earth's orbit is elliptical, the length of the day varies slightly. Thus 24 hours, as measured by clocks, was defined to be the *average* length of the day over one year. This time standard became known as 'Greenwich Mean Time' (GMT).

The Equation of Time

This has the consequence that, during the year, our clocks get in and out of step with the Sun. The difference between GMT and the local solar time at Greenwich is called the 'equation of time' (Figure 1.16). The result is that the Sun is not always due south at noon, even in London, and the Sun can transit (cross the meridian) up to 16 minutes 33 seconds before noon as measured by a clock giving GMT and up to 14 minutes 6 seconds afterwards. This means that sunrise and sunset are not usually symmetrically centred on midday, and this does give a noticeable effect around Christmas time. Though the shortest day is on December 21, the winter solstice, in London the earliest sunset is around December 10 and the latest sunrise does not occur until January 2, so the mornings continue to get darker for a couple of weeks after December 21 whilst, by the beginning of January, the evenings are appreciably longer.

Universal Time

Greenwich Mean Time was formally replaced by Universal Time (UT) in 1928 (though the title has not yet come into common or legal usage in the UK) but was essentially the same as GMT until 1967, when the definition of the second was changed. Prior to this, 1 second was defined as 1/86,400th of a mean day as determined by the rotation of the Earth. The rotation rate of the Earth was thus our fundamental time standard. The problem with this definition is that, due to the tidal forces of the Moon, the Earth's rotation rate is gradually slowing and, as a consequence, the length of time defined by the second was increasing. For a short while the orbital period of the Earth around the Sun was used to determine the second, but in 1967 an entirely new definition of the second was advanced based on the period of oscillation of atoms in an atomic clock:

The second is the duration of 9,192,631,770 periods of the radiation corresponding to the transition between the two hyperfine levels of the ground state of the caesium 133 atom.

Thus our clocks are now related to an atomic time standard which uses caesium beam frequency standards to determine the length of the second.

But this has not stopped the Earth's rotation from slowing down, and so very gradually the synchronization between the Sun's position in the sky and our clocks will be lost. To overcome this, when the difference between the time measured by the atomic clocks and the Sun (as determined by the Earth's rotation rate) differs by around a second, a leap second is inserted to bring solar and atomic time back in step. This is usually done at midnight on New Year's Eve or June 30. Since the time definition was changed, 22 leap seconds have had to be added, about one every 18 months, but there were none between 1998 and 2005, showing that the slowdown is not particularly regular. Leap seconds are somewhat of a nuisance for systems such as the Global Positioning System (GPS) Network and there is pressure to do away with them, which is, not surprisingly, opposed by astronomers. If no correction were made and the average slowdown over the past 39 years of 0.56 second per year continued, then in 1,000 years UT and solar time would have drifted apart by ~9 minutes.

Sidereal Time

If one started an electronic stopwatch running on UT as the star Rigel, in Orion, was seen to cross the meridian (due south) and stopped it the following night when Rigel again crossed the meridian, it would be found to read 23 hours, 56 minutes and 4.09 seconds, not 24 hours. This period is called the 'sidereal day' and is the length of the day as measured with respect to the apparent rotation of the stars.

Why does the sidereal day have this value? Imagine that the Earth was not rotating around its axis and we could observe from the dark side of the Earth facing away from the Sun. At some point in time we would see Rigel due south. As the Earth moved around the Sun, Rigel would be seen to move towards the west and, 3 months later, would set from view. Six months later after setting in the west, it would be seen to rise in the east, and precisely one year later we would see it due south again. So, in the absence of the Earth's rotation, Rigel would appear to make one rotation of the

Earth in one year and so the sidereal day would be one Earth year. But, in reality, during this time, the Earth will have made ~365 rotations so, in relation to the star Rigel (or any other star), the Earth makes a total of ~365 + 1 rotations in one year and hence there are ~366 sidereal days in one year. The sidereal day is thus a little shorter and is approximately 365/366th of an Earth day.

The difference would be ~1/366th of a day, or 1,440/366 minutes, giving 3.93 minutes, or 3 minutes 55.8 seconds. The length of the sidereal day on this simplified calculation is thus approximately 23 hours, 56 minutes and 4.2 seconds, very close to the actual value.

2
Refractors

In the nineteenth and early twentieth centuries, every 'gentleman astronomer' would have had a 3½-inch (90-mm) brass refractor with a focal length of 42 inches (1,080 mm) and so a focal ratio of 12. Indeed, I have one myself, though I would not claim to be a gentleman! They may have had a 6-inch (150-mm) Newtonian telescope as well. But as Newtonians with larger apertures became available and as more emphasis was put on deep-sky observing, refractors went out of fashion. Over the past 30 years, however, they have had a renaissance as improved glasses and computer-aided design have made available refractors that can give exquisite images of the planets, whilst those of shorter focal lengths give wonderful wide-field views of open clusters and the Milky Way. I really do feel that every amateur astronomer should have one.

The Dutchman Hans Lippershey, a spectacle maker, is generally credited with the design of the simple refracting telescope, which uses two lenses to create a magnified image of a distant object, although it is unclear if he actually invented it. He applied for a patent in 1608 but did not receive one, as there were several claims made by other spectacle makers. In Italy, Galileo Galilei heard of the device and carried out experiments to find the optimum design of the singlet objective lens. His empirical design of a biconvex lens with differing radii of curvature was almost exactly that which would be designed now by ray tracing methods. He used his telescopes, whose magnifications ranged up to 30, to observe the Moon and planets and discovered what are now known as the four Galilean moons of Jupiter. He observed that Venus could show almost full phases during part of its orbit and realised that this could happen only if Venus passed beyond the Sun, thus showing that Venus orbits the Sun, not the Earth, and so proving the Copernican model of the solar system.

A singlet objective lens gives an image which suffers badly from 'chromatic aberration' – in that the light at different wavelengths of the spectrum comes to a focus at differing distances from the objective. As the eye is most sensitive in the green part of the spectrum, this is the wavelength range which is brought to a focus. As a result both the blue and red parts of the spectrum will be out of focus and, together, will result in the focused image being surrounded by a purple fringe. Isaac Newton did not believe that this problem could be overcome and this led him to invent the reflecting telescope, which does not suffer from chromatic aberration.

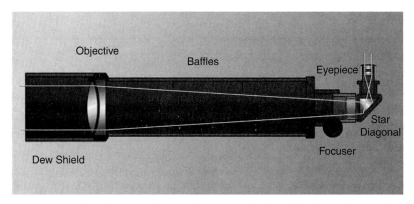

Figure 2.1 The classic refractor design with doublet lens. (Image: Starizona)

In 1733 an English barrister named Chester Moore Hall invented what is termed an 'achromatic doublet' as seen in Figure 2.1, which used two glass elements, one of flint glass and one of crown glass. To prevent others from discovering how his objective lens was made he ordered the two types of glass from different retailers. Unfortunately, both retailers sub-contracted their orders to a jobbing optician named George Bass, who thus uncovered the secret behind Hall's invention. Some years later, an optician, John Dolland, bought a reading glass from Bass and learnt from him how an achromatic doublet might be made. In 1758 Dolland read a famous paper to the Royal Society on achromatic doublets and was given the patent rights to his design. He never acknowledged the work of Hall, whose case was argued by Jesse Ramsden. (Ramsden was the inventor of a commonly used telescope eyepiece.)

The achromatic doublet is composed of two lenses made from glasses having different amounts of dispersion. Typically a convex lens made of crown glass is mated to a concave lens made of flint glass so that their chromatic aberration is cancelled out. The converging, positive power of the convex lens is not quite equalled by the diverging, negative power of the concave lens, so together they form a weak positive lens. I have calculated the focal lengths of both a singlet lens and an achromatic doublet for blue, green-yellow and red light where the overall focal length for the green-yellow wavelength of 5,893 angstroms is 967 mm. In the case of the singlet lens, the blue (4,861 angstroms) and red (6,563 angstroms) wavelengths have focal lengths of 954 and 970 mm respectively – a spread in focal length of 16 mm! However, the achromat gives focal lengths of 962.3 and 964.6 mm for the red and blue light, both less than the focal length of 967 mm for green-yellow light and with a reduced overall spread of 4.7 mm.

This was for a generic pair of glasses. If the two types of glass are chosen with care, particularly if one has a very low dispersion or, even better, a lens made of calcium fluorite crystal, the correction can be amazingly good and the resulting doublet lenses are effectively colour free when in focus. Such doublet lenses are often called 'ED doublets', where ED stands for 'extra low dispersion'. Pure fluorite lenses are now rare, due partly to cost but also to the fact that some glasses, sometimes called 'FD' or

'SD glasses', produced in recent years have properties that are very similar to those of fluorite and can be used with an appropriate mating glass element to make lenses which are essentially as good.

2.1 Glass Types Used in Refractors

In the descriptions of specific telescopes that follow, mention will often be made of the glass types used to manufacture their lenses. There is usually one ED, FD or SF glass used either as one element of a doublet or as the central element in a triplet. A key property of an ED glass is its 'Abbe number', which determines how little dispersion is introduced as the lens refracts light. In summary, the types are as follows:

1. Fluorite crystal, CaF_2, with an Abbe number of 94.99: very expensive and hard to work with and now found only in relatively small diameter lenses.
2. FPL-53 'Super ED' glass made by Ohara in Japan with an Abbe number of 94.94: less expensive and easier to work with than fluorite crystal with an Abbe number very similar to that of fluorite. Chemically more stable than fluorite. Also referred to as synthetic fluorite, or 'SF', glass. It is easier to design a colour-free lens using FPL-53 than glass with lower Abbe numbers, but it is somewhat harder to figure and does not maintain its shape as well with changes in temperature. Triplet lenses using fluorite or FPL-53 can give essentially perfect correction for chromatic aberration both in focus and outside of focus.
3. FPL-51 ED glass made by Ohara in Japan with an Abbe number of 81.54: less expensive and easier to work with than FPL-52 and chemically very stable. It maintains its shape better than FPL-53 with changes in temperature, and this is an asset when apertures greater than about 115 mm are considered. The colour correction will not be quite as good as with smaller lenses made using FPL-53 and may show a little false colour in out-of-focus images but essentially none in in-focus images – which is what really matters.
4. FCD1 ED glass made by Hoya in Japan with an Abbe number of 81.6: very similar to FPL-51.
5. H-FK61 glass made by CDGM in China with an Abbe number of 81.6: very similar to FPL-51.

It should be pointed out that the mating elements used to match the characteristics of the ED element are just as important in controlling chromatic aberration as is the accuracy with which the lenses are figured, coated and assembled. One will often find telescopes being sold under different brand names that look, apart from the company's logo, identical. The mechanical components are sourced from various machining companies around the world, and several manufacturers may well use the same components. However, the optical design and types of glass used to make the doublet or triplet lens may well be different, so not all similar-looking telescopes may perform the same.

Achromats – Good and Bad

Telescopes in which generic glasses are used to make the standard achromatic doublet have been given a rather bad name of late. This is unfair. If a telescope having a focal ratio of f12 to f15 is made using a simple achromat, the chromatic aberration is minimal and, as the curvature of the lenses used to form the doublet will be less than that for smaller focal ratios, the other lens aberrations, notably spherical aberration, will be less. The problem is that simple achromats having focal ratios of less than f7 have come onto the market, and these do show significant false colour and are not to be recommended except for use as a guide scope. Some people are more tolerant of the false colour than others, and filters can be obtained to limit the overall visible band to lessen its effects.

An achromat objective will bring two colours of the spectrum to the same focal point and should be corrected for both spherical aberration and coma at one focal point – preferably within the green-yellow part of the spectrum, where the eye is most sensitive. Ernst Abbe of the Zeiss Company defined the specification of a more advanced lens that he termed an 'apochromat', which will bring three colours to the same focal plane. Assuming that the design is done well, an apochromat can reduce the observed colour fringing to near zero. At the same time an apochromat should also be corrected for spherical aberration and coma at two widely separated visual wavelengths rather than one. This specification is very difficult to meet and may not even give the best optical performance. It is unlikely that any of the apochromat telescopes now available meet all parts of his definition.

The Rise of Apochromats

The era of apochromat telescopes for use by amateur astronomers really began in 1981. An amateur telescope maker named Roland Christen had come across a batch of an abnormal dispersion flint glass that had been ordered by NASA but not used. Christen was able buy this glass and designed some triplet lenses using it for one element. His prototype 5-inch f12 refractor produced the best images of Jupiter of any telescope at the Riverside Telescope Making Conference and earned him the prize for the most innovative optical design. Not long after, he gave up his well-paid job, set up a company named Astro-Physics, and went into the business of making telescopes full time. The refractors made by his company, as in Figure 2.2, are some of the most prized telescopes now available, with a long waiting time before one can acquire a new one.

A second notable optical designer, Thomas Beck of TMB Optics, also designed a series of very high quality refractors, with their triplet objectives being made by the Russian optical company LZOS and mounted in tube assemblies made by APM in Germany. Sadly, Beck died unexpectedly in late 2007; his family is continuing his company, but supplies of the scopes he designed are limited at present. APM continues to provide superb apochromat refractors with lenses made by LZOS.

In Japan, Vixen and Takahashi designed doublet objectives where one element was made from a pure fluorite crystal. The Takahashi FS series has become a legend,

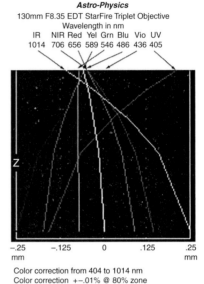

Astro-Physics
130mm F8.35 EDT StarFire Triplet Objective
Wavelength in nm

| IR | NIR | Red | Yel | Grn | Blu | Vio | UV |
| 1014 | 706 | 656 | 589 | 546 | 486 | 436 | 405 |

| −.25 | −.125 | 0 | .125 | .25 |
| mm | | | | mm |

Color correction from 404 to 1014 nm
Color correction +−.01% @ 80% zone
Aspherized @ 546nm OPD = .025, .99 Strehl

Figure 2.2 An Astro-Physics 130-mm, f5.3 Starfire EDF. The plot shows the superb colour correction and a Strehl ratio of .99 (essentially perfect) of the earlier 130-mm, f8.35 refractor.

but only the telescope of 60-mm aperture is still available, due partly to cost and for environmental reasons. Fluorite has the lowest dispersion of any lens type so, when mated with a suitable glass element, superb colour correction can be made with just two elements. One cause of reduced contrast in an image is scattering of light within the objective lens. Fluorite crystal scatters virtually no light, so these telescopes are believed to show the highest contrast of any telescope. A Takahashi FS102 f8 telescope is one of my most prized possessions.

There is one other optical design that is used in high-quality refractors: that of four elements in two doublets, one acting as the objective and the second, nearer to the eyepiece, acting to correct the aberrations inherent in the objective doublet. This is called a Petzval arrangement. TeleVue, perhaps better known for its eyepieces, makes the NP101is and NP127is, whilst Takahashi produces the FSQ85-EDX and the FSQ106-EDX. These are designed to give very wide flat fields of view and, with large focusers, can be used with the largest CCD cameras now available, so are perfect for astro-imaging. Not surprisingly, these four-element refractors are probably the most expensive per inch of aperture of any telescope that can be bought!

One US company of note is the Telescope Engineering Company, or TEC, whose first apochromatic refractors came to the market in 2002. Perhaps the most famous of its line is the APO140, f7, ED oiled triplet (Figure 2.3). In this objective, the three elements, rather than being air spaced, are separated by thin films of oil. This has some real advantages over an air-spaced triplet: the oil filling any minute imperfections in the internal surfaces, which may not even need to be finely polished. (Applying water

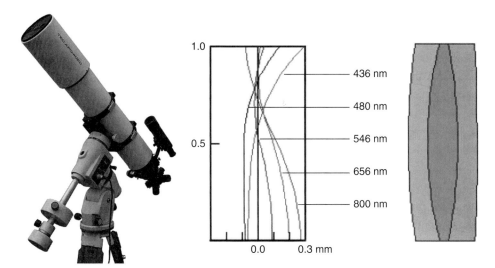

Figure 2.3 The renowned TEC 140-mm-aperture apochromat with oil-spaced objective.

to a piece of roughened glass, perhaps found on a shoreline, will make it appear smooth.) The middle element is made from Ohara FLP-53, a very high dispersion fluorite glass, virtually as good as a pure fluorite element, and, as the objective has only two air/glass surfaces at front and rear, there is less internal scattering, thus giving high-contrast images. A further advantage of an oiled-spaced objective is that the central element in the lens can cool more quickly to the ambient temperature and follow any reduction in temperature during the night rather better than an air-spaced triplet. This is a telescope that many amateurs aspire to own.

2.2 The Asian Invasion

Many high-quality apochromatic telescopes are now made in China and Taiwan and imported by a number of companies which badge them under their own names. One such is William Optics, which markets an excellent range of doublet and triplet apochromatic refractors. My own 'get up and go scope' is an 80-mm-aperture Zenithstar II ED APO made using a doublet objective, one element of which is made from Ohara FLP-51 glass. This has a longer than average focal ratio of f6.8, which enables this less costly optical design to give excellent performance with essentially no false colour visible when in focus. (Though some is visible out of focus, does this matter?) Incidentally this is made by the Long Perng Company, which I suspect is behind many telescopes of this type. The address of the company – 6, Alley 73, Lane 244, Hsin-Shu Road, Hsin Chuang, Taipei 242, Taiwan – rather implies a little backstreet workshop, but the firm was founded in 1974 and has quite impressive premises! A very detailed review of this telescope can be found on the author's 'Night Sky' Web site under 'Related Links'. To give superb image quality with the lower focal ratio of 5.9 (and hence provide a wider field of view) the GTF81 five-element APO uses

Figure 2.4 The Sky-Watcher Equinox 100 PRO, f9 fluorite ED apochromatic refractor.

a three-element objective with one element made from Ohara FLP-53 glass coupled with a two-element ED field flattener. Along with some of the company's other scopes, it uses a digital read-out rack-and-pinion focuser. The range includes triplets of 123-, 132- and 158-mm aperture – all highly regarded.

Another well-known range of refractors is made by the Chinese company Synta and sold under the 'Sky-Watcher' brand name. (Synta now owns Celestron, so presumably some of their telescopes will share similar components.) The Equinox PRO series (see Figure 2.4) includes refractors of 80-, 100- and 120-mm apertures. These are called 'fluorite ED APOs', as they incorporate a doublet objective with one element made from Ohara FLP-53 glass, which performs essentially the same as a pure fluorite crystal. I had a chance to review the 120-mm, f7.5 for a UK astronomy magazine and was very impressed with the quality of its optics; I would certainly have bought it had I not already owned the Takahashi FS102.

There is a fundamental problem in that the term 'apochromat' or 'apo' is now applied to virtually all telescopes where the chromatic aberration is significantly less than that of an achromat of the same focal ratio. Sometimes the term 'semi-apo' is used, usually for a doublet with a high-dispersion glass as one of its elements to minimise, but not totally eliminate, chromatic aberration. Now even telescopes like these tend to be termed 'apos'. A true apochromat objective is likely to require the use of three elements rather than two. This naturally increases the cost, and the objectives will take longer to reach ambient temperature when brought out into the cold.

As the effects of chromatic aberration are reduced to near zero, other aberrations play a more important role. One, in particular, is called 'spherochromatism' – the variation of spherical aberration with wavelength. The optical designer has to balance all these out to try to produce an objective lens of high quality at an affordable price. One reason apochromatic lenses tend to have three elements is that the curvature of the individual lenses is less; this makes spherical aberration and spherochromatism less of a problem and increases the tolerances in its alignment, so making fabrication easier.

The main thrust in the design of refractors in recent years has been to both increase their apertures and reduce their focal ratios. This is partly driven by astro-imaging enthusiasts, as larger apertures will allow shorter exposures, and shorter focal lengths give wider fields of view. Another problem is that refractors of large aperture and longer focal ratio (so giving an increased focal length) tend to get unwieldy and require bigger and hence more expensive mounts to support them.

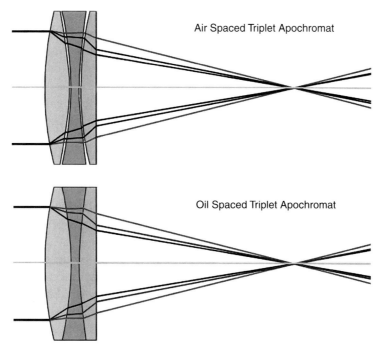

Figure 2.5 Air- and oil-spaced apochromats (not to scale).

Few telescopes are now designed so that optical quality is the top priority. One such was the Thomas Back–designed TMB/APM 130-mm aperture, f9.25 super-planetary APO (now replaced by the APM 130-mm aperture telescope of 1,200-mm focal length) where the increased focal ratio over the more usual f7 ratio allows exquisite views of the Moon and planets. Unless it is important to have the widest possible field of view, a refractor with a focal ratio greater than f6.5 may well perform better and is likely to be less expensive than one of f/6 or less.

Oil-Spaced or Air-Spaced Triplets?

Astro-Physics has used a number of oil-spaced objectives in the past and its current 140-mm, f7.5 Starfire EDF incorporates one, whereas its 175-mm, f/8 Starfire EDF uses an air-spaced triplet (see Figure 2.5). The advantages of an oil-spaced triplet have been described in relation to the TEC 140-mm refractor, but what advantages might an air-spaced triplet have? In an oil-spaced triplet, the two radii of curvature of the central element are determined by the inner radii of curvature of the outer elements. By separating the three lenses, the designer can alter the radii of curvature of this central element and this gives two more degrees of freedom with which to optimise the design, so perhaps being able to reduce the spherical and spherochromatic aberrations of the lens. The modern multi-coating that can be applied to the lens surfaces helps to minimize any internal reflections or scattering, so that the difference in this respect compared with an oil-spaced objective is now less than in the past. The one problem is that it is more difficult for the central element to cool, and thus more time will be needed to cool down to ambient outside temperatures.

2.3 Curvature of Field, Field Flatteners and Focal Reducers

In an ideal telescope, the field of view would be perfectly sharp at all points, implying that the focus point for all points within the field of view is a flat plane. In practice, however, the focal surface will tend to be the surface of a sphere centred on the nodal point of the objective. Thus the focus for points farther from the axis of the telescope will lie closer to the lens. When observing visually, the eye is often able to accommodate this and some eyepieces may even apply some correction so that the whole field of view can appear sharp. Refractors of longer focal length will suffer less from this problem. It can, however, be a real problem for astro-imagers, as their sensors are perfectly flat. The smaller the sensor and the longer the focal length the less of a problem this is, and my FS102 of 816-mm focal length gives excellent stellar images across the size of an APSC-size sensor. To achieve wide-field images, refractors of shorter focal length are required, and for many it may be necessary to add a field flattener prior to the DSLR or CCD camera. Field flatteners are often made specifically for an individual refractor, but some will work well with, for example, all 80-mm-aperture refractors with focal ratios close to f6. One such example is the Sky-Watcher 2-inch field flattener (meaning that it fits into 2-inch focuser barrels), which is optimised for focal ratios between f5.5 and f6. It has a T-ring adaptor and so, with the appropriate bayonet, can directly couple with a DSLR camera. A second, the Teleskop-Service TS 2-inch field flattener, will work for an even wider range of telescopes having focal ratios of between f5 and f8. Field flatteners often reduce the effective focal ratio – well liked by astro-imagers, as this both increases the field of view and reduces exposure times.

Plate 1.5 is an example of the effect of using a TS 2-inch field flattener with my 80-mm, f6.8 refractor. The image of M35 was a stack of twenty 8-second exposures with the sky glow removed, as described in Chapter 18. The two insets show how the quality of the image in the corners of the APSC frame was markedly improved when the field flattener preceded the Nikon D7000 DSLR. This field flattener is also used with my 127-mm, f7 refractor with equally good results. The f6.3 focal reducers used with Schmidt-Cassegrain telescopes have also been used with some success with refractors, and so I have made an adapter to use a Celestron model with my 127-mm, f7 refractor to give an effective focal ratio of f4.4 with an increase in the field of view. Not surprisingly, there is some vignetting towards the corners of the frame.

2.4 Some General Points

Apochromatic or semi-apochromatic refractors tend not to come with a mount and are called 'optical tube assemblies' (OTAs), but suitable mounts will be discussed later. The cost depends partly on the quality of construction of the OTA but largely on the type of objective lens used. But let's look at a couple of general points before considering the types of objectives that are available.

Focuser and Eyepieces

The scope *must* have a focuser capable of mounting the eyepieces having a 2-inch barrel, as will be explained in Chapter 7. This is the only way to get the very wide-field views that refractors are so suited for. Of course, there is nothing to prevent the use of eyepieces of short focal length to give high magnifications, and these scopes can give really nice views of star clusters and planets as well as the brighter deep-sky objects.

The focuser barrel diameter is also important if the telescope is to be used for wide-field astro-imaging: too narrow and it will vignette the outer parts of the field of view of a 'full-frame' sensor. So, as well as the 'standard' barrel size of 2.5 inches, focusers having larger diameters such as 3.5 and even 4 inches may be provided as standard or available as an option.

Focal Ratio

Refractors tend to have focal lengths ranging from f5 to f8. The shorter the focal ratio for a given aperture, the wider the potential field of view, but the harder it is for the lens designer to correct for chromatic (and other) aberrations. To give the same quality image, an 80-mm-aperture telescope with a 400-mm focal length (f5) will require a higher-quality objective than for one with a 560-mm focal length (f7). Or, to put it another way, for a given type of lens, a telescope with a longer focal length will give better images.

There is a further significant, but rather subtle point that I believe should point you to an OTA of longer focal length if you want to observe wide-field views visually. I am not convinced that you necessarily win by using a focal length at the shorter end of the focal ratio range. It relates to the amount of light that can enter your eye. The 'exit pupil' of the telescope should not exceed the dark-adapted diameter of your pupil; otherwise much of the light collected by the telescope will be lost! Using, say, an eyepiece of focal length 35 mm with a telescope having an 80-mm aperture and 400-mm focal length gives an exit pupil of 7 mm, whereas with an OTA of 545-mm focal length (f6.8) the exit pupil is 5.1 mm. Unless you are young, it is unlikely that your pupil, when dark adapted, will be much greater than ~5 mm, so much of the light (up to half) collected by your scope will not enter your eye. So divide the focal length of your scope by the proposed focal length of the eyepiece to get the magnification (i.e., 500/35 = 14.3); then divide the aperture by this to get the exit pupil (80/14.3 = 5.6 mm). If this is significantly greater than your dark-adapted pupil, you need to rethink!

2.5 A Refractor Survey

The final part of this chapter will look at the wide range of aperture sizes, focal ratios and objective constructions that are available. It will re-emphasise many of the points discussed in general in the preceding sections. It is impossible to include every telescope manufacturer, but I hope the following survey will lead you in the right direction.

60- to 66-mm Refractors

These compact telescopes can be used, say, on expeditions to view eclipses where weight considerations come into play. An example is the Takahashi FS-60CB with a fluorite doublet, which can be serve as a travel scope but is also, when used with a dedicated field flattener, perfect for wide-field astro-imaging. Altair Astro offers a 60-mm triplet refractor equipped with a very nice helical rack-and-pinion focuser. One might also consider a 60-mm spotting scope as an alternative – I took one, equipped with a Baader Solar Film filter, to observe a total eclipse of the Sun in Turkey.

70- to 80-mm Refractors

Telescopes of this size are very popular, with variants made by many manufacturers. The cost is determined largely by the types of objective lenses being employed.

Simple achromats use a two-element objective with one element of flint glass and one of crown glass. There will be some false colour around bright objects such as the limb of the Moon or the planets Venus and Jupiter, but this is often over-emphasised. But it is important to have a focal length at the longer end of the range if you want to use the achromat for visual observing. Achromatic refractors having focal ratios of ~f7–f10, which could be reasonably good for visual use, seem very hard to come by. There are quite a number of f5 scopes, such as the Sky-Watcher Startravel 80T and Orion Short Tube 80-A, that are well suited for use as guide scopes, but I could not honestly recommend them for visual use. I would far rather that one raided the piggy bank to move up to the next level of objective.

Semi-apochromatic refractors use a low-dispersion glass as one of the two elements. This is usually FPL-51 or an equivalent. Given a focal ratio of f6 or f6.8, they can produce images with virtually no false colour when in focus and provide superb images for a very reasonable cost. Excellent refractors of this specification are the f6.8 Williams Optics Zenithstar II ED and the f6 Antares Sentinel 80-mm semi-apo.

Apochromatic refractors use objectives made of either three elements or two elements, where one element is either fluorite crystal or Ohara FPL-53 glass. These give images which for a given focal length will be better than the semi-apo types. But, in my view, an f6.8 semi-apo will be *very* close in image quality to an f5 apo. The f6.2 Sky-Watcher Equinox-80 PRO OTA and the f7.5 Sky-Watcher Evostar-80ED DS-PRO are excellent examples using a doublet lens where one element is made of FPL-53 (Figure 2.6). Perhaps the ultimate in 80-mm refractors is the f5.9 Williams Optics Grand Tourismo 81 FD, which uses an FPL-53 element in an air-spaced triplet objective. A point to note, however, is that a triplet design will take somewhat longer to cool down to ambient temperature, as the inner element is insulated from the outside air by the outer two elements.

Astro-Imaging Refractors

A number of refractors in this aperture range are designed specifically for astro-imaging. One such is the Takahashi FSQ-85ED, which uses four elements in a Petzval

Figure 2.6 A Sky-Watcher 80-mm refractor with two-element objective, one element using FPL-53 fluorite glass. It is equipped with a 2-inch star diagonal and 50-mm finder scope.

configuration to give a broad, flat field. William Optics produces a 75-mm, f5.9 ED doublet made from Ohara glass, which has a dedicated field flattener/reducer available to give a reduced focal ratio of f4.8.

Telescope Mounts

None of these scopes are that heavy and can even be used on a sturdy tripod. They can obviously be mounted on small equatorial mounts, but I would be inclined to use an Alt/Az mount. My 80-mm refractor is used in conjunction with an iOptron Minitower: it is exceedingly quick to set up and its 'go-to' ability and tracking are excellent. This mount is perhaps more than is required for an 80-mm refractor, and iOptron now produces a smaller version called the CubePro, which is about half the price and is also highly regarded.

A second low-cost 'go-to' mount suitable for a 60-mm or 80-mm refractor is the Sky-Watcher 'Allview', which has nearly 43,000 objects in the controller's catalogues. It can carry telescopes up to 5 kg in weight. An interesting feature is that, once aligned, it can be moved around the sky by hand without losing alignment or positional information, so that it can be slewed quickly from one object to another.

The larger-aperture refractors to be discussed later can be used with both Alt/Az and equatorial mounts, with the latter better suited to astro-imaging. In their larger sizes they are more unwieldy (to be technical, they have a greater 'moment of inertia') than compact telescopes such as the Schmidt-Cassegrain and so, for a given weight, a more substantial mount will be required.

90- to 120-mm Apochromat Refractors

There is a wonderful variety of telescopes available in this aperture range employing either doublets or triplets with ED glass or, in one case, fluorite as one of the elements.

Takahashi produces a range of four refractors: the SKY 90, which has a fluorite crystal doublet; the TSA 102, an ED triplet which is the worthy successor to the legendary fluorite FS102; the FSQ-106ED using four elements in a Petzval configuration to provide a flat, very wide-field image plane, optimised for wide-field deep-sky imaging; and the TSA 120, which incorporates an air-spaced, f7.5 triplet objective.

TeleVue Optics manufactures two four-element Petzval telescopes: the NP-101 and NP-101is, the latter having an enlarged focuser, making it more suitable for imaging.

Less well known, but highly regarded refractors have lenses designed and made in Hungary which are incorporated into telescopes sold by CFF Telescopes. In this aperture range they offer a 102-mm-aperture, f6.3 refractor using an oil-spaced ED triplet. The Strehl ratio of their lenses is guaranteed to exceed .96 in green light and is typically .98. The handcrafted telescopes often require some time for delivery, as they are made in small batches.

Williams Optics provides a new FLT 110 design which incorporates an f7, air-spaced triplet, designed by Thomas Back (good) and equipped with a 3-inch rack-and-pinion focuser capable of holding heavy imaging equipment. It can be provided with a dedicated field flattener.

The Japanese company Vixen produces the ED103S and ED115S f8 refractors using triplet ED objectives, whilst their AX103S incorporates an additional field flattener element for astro-imaging.

Synta, which, as mentioned earlier, owns Celestron, markets a wide range of refractors under the Sky-Watcher brand in this aperture range, with each aperture available with a range of objectives and tube design. The Esprit range includes the 120-mm, f7, ED air-spaced triplet refractor.

APM offers a 107-mm, f6.5 triplet apo refractor utilising Ohara FPL-53 glass for the central element, giving very high optical quality images with field flatteners and reducers available for astro-imaging.

Synta markets some excellent Sky-Watcher ED doublet telescopes of 100- and 120-mm aperture. In the UK, Altair Astro offers 102- and 110-mm semi-apo refractors using ED doublets based on an Ohara FPL-51 ED element as well as 102 and 115 f7 Altair Wave ED triplets which have dedicated field flatteners for imaging use.

127- to 140-mm Apochromat Refractors

Unless one is able to purchase a second-hand Takahashi FS127 (almost impossible, as owners do not want to sell them), all refractors in this aperture range use triplet objectives, either oil or air spaced.

The premium (and hence expensive) refractors are made by Astro-Physics with its 130-mm, f6.3 and 140-mm, f7.5 triplets; APM with its 130-mm, f6 and 130-mm, f9 LZOS manufactured lenses; Takahashi with its f7.7, TOA 130 range; and TEC with its 140-mm, f7 telescope. TeleVue Optics has its NP-127is, using four elements in a two-group Petzval arrangement to give a broad, flat field designed for wide-field imaging. The Astro-Physics and TEC objectives are oil spaced, with the others air spaced. All will provide superlative performance, and it will always be seeing that limits what is visible, not the telescope.

CFF Telescopes, already mentioned, offers telescopes having 127-mm objectives with focal lengths of 890 and 1,200 mm (f7 and f9.5 respectively), as well as a 140-

mm telescope of focal length 1,200 mm (f8.6). All these use fluorite-based (FPL-53?) oil-spaced objectives which have one aspheric surface and are guaranteed to have a Strehl ratio of greater than .96. These are equipped with Moonlight or Starlight Instruments high-quality focusers. As with Astro-Physics, their handcrafted telescopes have to be ordered and may take some time to be delivered.

William Optics offers a new Chinese-manufactured FLT 132 DDG triplet APO refractor which is equipped with a digital read-out, 3-inch rack-and-pinion focuser. This has an air-spaced triplet with FPL-53 glass as its central element.

Very similar telescopes appear under a wealth of brand names. For example, the optical tubes and rings of the Meade Series 6000 130-mm f7, the German Teleskop-Service's TS Optics Photoline Triplet APO 130/915 mm and the British Altair Astro's Altair Wave Series 130-mm, f7 ED Triplet APO look identical. However, this does not mean that the telescopes are the same, as the objectives and focusers may well be different: the Altair Wave uses Japanese Ohara FPL-51 glass for its central element, whilst the Meade uses a Chinese FK61 glass of very similar specification for its central element. The Altair Wave lenses are individually numbered, and an interferogram may be provided to attest to their optical excellence. There are also a number of 127-mm f7.5 telescopes such as the well-regarded Explore Scientific version, which uses Hoya FCD1 glass as the central element of the triplet.

150 mm and Above

Few manufacturers stray into such a rarefied atmosphere. Takahashi has on offer a TOA 150 utilising both SD and ED glass in its air-spaced triplet objective. CFF Telescopes uses GPU Optical oil-spaced triplets in their 160-mm, f6.5 and 180-mm, f7 aperture telescopes and APM, using LZOS Russian lenses, has a range of 152-mm, f7.9 and 175-mm, f8 refractors. Can your house be re-mortgaged?

My Own Refractors

I do like refractors! These include two achromats: a brass 90-mm, f12 telescope from another era and a Helios 150-mm f8. This was bought largely to acquire the EQ5 mount that came with it but is optically not that bad. Yes, it shows a fair bit of chromatic aberration, but this is minimised by use of a Baader Fringe Killer that filters out the blue and red ends of the spectrum. It works very well doing monochrome imaging of the Moon in H-alpha light – of course, the fact that it is a simple achromat is irrelevant, and it produced an excellent image of the 2004 transit of Venus when used with a filter made from Baader Solar Film.

I then bought a Takahashi FS102 f8 telescope with a fluorite doublet. This telescope gives exquisite images and has been used for both visual observing and, more recently, astro-imaging, with examples appearing in this book. Having been shown by one of the world's top astro-imagers how an 80-mm refractor could be used for wide-field astro-imaging, I acquired a William Optics Zenithstar II f6.8 ED doublet. Some examples of its use in astro-imaging are also given. I have also found that it is an excellent telescope to take on holiday, used just with a camera tripod, and to demonstrate at winter star parties when it provides wonderful views of the Pleiades Cluster.

A World-Class Refractor

For some years I had wanted to purchase a larger-aperture apochromat refractor, and the writing of this book gave me an excuse to purchase one. A TEC 140-mm telescope would have been nice but rather too big for my available mounts, so I was looking for one with an aperture of 127 or 130 mm. I was finally able to acquire a CFF Telescopes 127-mm-aperture, f7 refractor incorporating an oil-spaced triplet objective made in Hungary (Figure 2.7). A Web site relating to the company, GPU Optical, that preceded CFF Telescopes gives great insight into the design of triplet lenses and the reasons that an oil-spaced design was chosen. As already mentioned, an oil-spaced triplet has two fewer optical degrees of freedom than an air-spaced triplet, so that the optical correction might not be quite as good. With a focal ratio of f7, the two objective types can be virtually equal in quality, but one surface of the CFF objective is made aspherical to equalise or even improve the optical performance. In Chapter 1, the Strehl ratio of an objective is defined. A value of 1 is perfection, and a value greater than .95 will mean that it is virtually always the atmosphere, not the lens, that will determine the image quality. CCF Telescopes' individually numbered lenses (mine is no. 0010) are guaranteed to have a Strehl ratio of .96 and typically reach .98. I believe that this company, whose origins date to 2007, is close in concept and equal in quality to Astro-Physics, making a small number of individually handcrafted telescopes. Though mine was brought directly from the company, a small dealer network is gradually being set up across Europe.

The CFF 127-mm-aperture, f7 tube assembly can be equipped with either a 2.5-inch Moonlight or 3-inch Starlight 'Feather Touch' dual-speed focuser. The latter, with which my refractor was equipped, uses a helical rack-and-pinion drive and is reputedly one of the best focusers available today – adding significantly to the overall cost. The interior of the tube assembly is flocked and, in addition, a number of knife edge baffles are set along its length. When one looks in through the lens, the interior appears totally black – nothing can be seen. The tube is covered with a white 'crackle' finish with absolutely no flaws. The weight, including tube rings, is 9.9 kg, and it comes in a very high quality flight case, though there is no fixed foam packing.

Tests observing the limb of the Moon for false colour showed that none appeared to be present either in focus or on either side of focus. (Note: it is important not to use the top or bottom edges of the Moon for this test, as refraction in the atmosphere will add some blue or red colouration.) A severe test is to photograph a TV aerial or the branches of a tree outlined against a bright white sky. Again, no false colour was observed and the image had very high contrast.

The contrast obtained with the CFF 127 in observations of the Moon appeared equal to that of my Takahashi FS102 – a telescope that due to its fluorite doublet is reckoned to have the highest overall contrast of any telescope. Visual observations of

Figure 2.7 The author's 127-mm, f7 refractor, manufactured by CFF Telescopes, mounted on a Losmandy GM8 equatorial mount. It is equipped with an aspherical oil-spaced triplet lens and Starlight Instruments 3-inch dual-speed, helical rack-and-pinion focuser.

Jupiter and the Orion Nebula equalled those achieved with the Takahashi in quality, and the fact that the 127-mm objective was collecting ~50% more light than the 102-mm Takahashi meant that I was able to see a little more. I honestly believe that, as the heading of this section implies, I have been able to acquire a world-class telescope which is as near optical perfection as is possible to achieve.

3
Binoculars and Spotting Scopes

Why Use Binoculars?

In a nutshell, binoculars will give you the best view possible of objects, such as the Andromeda Galaxy and the Pleiades Cluster, that are simply too large to be seen in the field of view of most telescopes and act as a 'rich-field telescope' giving wonderful views of the Milky Way (see Figure 3.1). They are usually more compact and weigh less than a telescope system and so can be taken abroad when the luggage allowance precludes a telescope. They are great things to have in any case! The standard parameters of a pair of binoculars are pretty obvious, but there are a few less obvious features that can be quite important, as we will see.

3.1 Binocular Specifications

Magnification

This is the first number given in the basic specification of a pair of binoculars. Typical numbers are 8 and 10. One might think that the greater the magnification the better – but this is generally not the case. The greater the magnification, the smaller the field of view will be (as described later) but, perhaps even more important, the more the image will appear to jump about. Unless the binoculars are to be mounted on a tripod or are image stabilized, a magnification greater than 10 is not to be recommended.

Objective Size

This is simply the diameter of the objective (front) lens measured in millimetres and is the second number in the specification – for example, a pair of 8 × 40 binoculars has a magnification of eight times and an objective diameter of 40 mm. The larger the diameter, the more light that is collected and so, in principle, dimmer objects will be seen. Thus large objective lenses will be an asset for astronomical use, but as a consequence the size and weight of the binoculars will increase. The most common objective sizes are 20, 35–42 and 50 mm. The light collected increases as the square of the diameter, so

46

Figure 3.1 The author's Helios 8 × 40 Porro Prism binoculars with rubber eyecups and Leica 8 × 42 Roof Prism binoculars with pull out eyecups.

a pair of 50-mm binoculars will collect more than six times as much light as a 20-mm pair – and so allow one to see objects that are about 2 magnitudes fainter.

Exit Pupil

Together, the objective size and magnification determine the diameter of the binoculars' exit pupil. This is the diameter of the cylinder of light that leaves the eyepiece and enters the eye. (One can see the exit pupil by pointing the binoculars at a white wall or sky and observing the brightly lit circle appearing on the eye lens of the eyepiece. This is shown in Chapter 1, Figure 1.2.) The exit pupil diameter is given by simply dividing the size of the objective by the magnification. Let's take two examples: pairs of 8 × 20 and 8 × 40 binoculars. The first will have an exit diameter of 20 mm/8 = 2.5 mm and the second 40 mm/8 = 5 mm. However, the light that can enter the eye is determined by the diameter of your iris, which in daytime is about 2.5 mm – so all the light from the 8 × 20 pair will enter the eye in daylight, but only a quarter of the light collected by the 8 × 40 will do so! There is thus no point in using large-diameter objectives in broad daylight, and a lightweight pair of 8 × 20 binoculars cannot really be beaten for daylight use. However, of course, as twilight and darkness fall, the iris opens up as the first stage of dark adaptation. For most, the pupil diameter will increase to about 5 mm, and so all the light from the 8 × 40 (or 10 × 50) binoculars can enter the eye. However, a pair of 7 × 50 binoculars with an exit pupil of ~7 mm may be suitable only for younger astronomers whose pupils will be able to dilate up to 7 mm.

It should be pointed out that when 8 × 20 binoculars are used, it is important that what is called the 'inter-ocular distance' – the distance between the eyepieces – be correct so that the exit pupils and pupils exactly line up. When one is using, say, 8 × 40s with their 5-mm exit pupils, there is a fair margin for error, so they are perhaps a little easier to use; of course, for astronomical use it is worth having a pair of binoculars with larger objectives, and it does not matter at all if some light is wasted when they are used in daylight. Perhaps, for all-round use, 8 × 40s provide the best compromise, as they will not be too large or heavy and do not magnify too much, so can give a steady image.

Field of View

This depends on, firstly, the magnification – the greater the magnification the smaller the field of view – and, secondly, the design of the eyepiece. The binocular specification may say 'wide field' or 'extra wide field'. Then, for a given magnification, you will see a larger field of view than with a pair having a normal eyepiece. This depends on the complexity, and hence cost, of the eyepieces used in the binoculars. For almost all uses, including astronomical, it is better to go for a wide field of view, but do expect to pay a little more. The field of view can be defined in two ways. The first is simply by giving the field of view in degrees – which is what astronomers would like to know. The second gives the linear distance that could be encompassed at a given distance – usually given number of feet at 1,000 yards. In this case, convert the feet into yards, divide this by 1,000 and multiply by 57.3 to get the field of view in degrees.

Eye Relief

This is almost never discussed but can be important if, as is commonplace, you wear glasses. With many binoculars, if you have your glasses on and even if the eye caps are folded back, you will find that you cannot see the whole of the nominal field of view. (You know when you are seeing the whole field of view when the edge of the field is a sharp transition from image to black.) Obviously you can take your glasses off, but that can be a nuisance and, if you suffer from significant astigmatism, may not give you as good an image as you could see with the help of your glasses. The solution is to get binoculars with good eye relief (16–20 mm). This tends to be the case with more expensive binoculars but is often not mentioned in the specifications. So, if you are trying out a pair of binoculars, check that there is a well-defined field of view. Some binoculars have rubber surrounds that are extended when glasses are not used but can be folded back when glasses are used, whilst others have push-in, pull-out eye cups.

Lens Coatings

The size of the objective determines how much light enters the binoculars. Not all will pass into your eye. Some will be reflected at the various glass/air interfaces in the binoculars – prisms as well as lenses. This light is scattered and will reduce the contrast of the image as the darker parts of the image become apparently lighter. Good binoculars will have *all* the lenses and prisms multi-coated ('fully multi-coated' in the specification) to minimize reflections, thus reducing the scattered light and so giving much better images. Naturally this increases their cost.

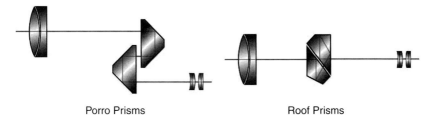

Porro Prisms   Roof Prisms

Figure 3.2 The two types of prism configurations used in binoculars. (Image: Starizona)

Prisms

In the conventional design of binoculars, Porro prisms are used (see Figure 3.2). These can be made of different glass types. The BAK-4 glass prisms give better performance than the cheaper BAK-7 types. If you look at light exiting the eyepiece (at arm's length) when pointed at a white wall, the exit pupil must be round, not square. If square, the prisms are too small! More recently, roof prisms have begun to be used to give very compact designs, but to work well they must be manufactured with very high precision – hence good-quality binoculars using this prism type are more expensive than Porro prism types. To give the clearest and sharpest images, the roof prisms need to include a 'phase coating', which cheaper pairs may not have. It is rare for this to be mentioned in the specification.

Other Features

It is quite nice if binoculars are waterproof or at least splash proof and, to prevent internal misting of the lenses, filled with nitrogen (which implies that they must be airtight). Some have rubber coatings to make them more robust.

3.2   Image-Stabilized Binoculars

The larger the objective lenses and the greater the magnification, the fainter the stars that can be seen. Firstly, they will collect more light, but secondly, higher magnifications reduce the background sky brightness, so enabling the stars to stand out better. Thus the larger pairs, such as those designated 10×50, can be pretty good at picking out many of the fainter Messier objects. The problem is that binoculars of this size and magnification tend to be hard to hold steadily (with 8 being about the maximum useful magnification when handheld), so tripods or monopods must be used to steady them. Image-stabilized binoculars get round this problem by including sensors and optical elements to counteract the movements of our hands and so steady the image (see Figure 3.3).

The image stabilization allows even higher magnifications to be used, further reducing the sky brightness, so rather than 8×42 and 10×50 they tend to have magnifications of 10 or 12 and 15 or 18 respectively. There is, however, a more subtle effect

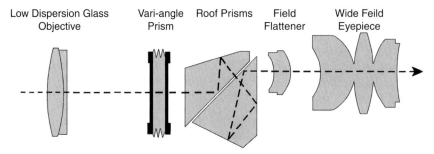

| Low Dispersion Glass | Vari-angle | Roof Prisms | Field | Wide Feild |
| Objective | Prism | | Flattener | Eyepiece |

Figure 3.3 The optical path of Canon image-stabilized binoculars passes through a Vari-angle Prism, which is controlled by electronic gyroscopes.

Figure 3.4 Canon 10 × 30, 10 × 42 L and 15 × 50 image-stabilized binoculars.

that enables significantly fainter objects to be seen. With a handheld, non-stabilized pair, the light from a star will be falling on many adjacent rods or cones in your retina as the image wanders about. This makes it difficult for the brain to detect faint objects. When stabilised, that light remains concentrated on a very small area of the retina and is then far easier to detect. In addition, it is thought that our brain can integrate for a few seconds, the net result being that stars at least half a magnitude fainter will be seen than otherwise. The effect of switching the stabilisation on when you are viewing the Pleiades Cluster is quite amazing!

The Japanese company Canon was one of the first to introduce image-stabilised binoculars (Figure 3.4). These are widely regarded as some of the very best for astronomical use, and I have been lucky to own a second-hand pair of the original 15 × 45s. Electronic gyroscopes detect the binoculars' movements and control a fluid-filled adjustable prism to stabilise the image. The objective lenses of Canon's largest pairs have been increased to 50 mm in diameter, and the stabilisation control button now uses a single press to switch the stabilisation on or off rather than keeping it continuously pressed. The first-rate optics use ultra-low-dispersion glass, and the eyepieces have 15 mm of eye relief, so that most people who wear glasses can still see the full field of view without removing them.

Of course, the greater the magnification, the smaller will be the field of view, but the 15 × 50s incorporate wide-field eyepieces having a 67-degree apparent field of view, which translates to an excellent 4.5-degree true field of view – nicely encompassing the Andromeda Galaxy. The Moon is pretty impressive too – viewing it with these binoculars is almost like using a small telescope – as is the Double Cluster in Perseus. It is even possible to spot the three brightest stars in the Trapezium at the heart of the Orion Nebula. All four Galilean satellites of Jupiter can be easily seen and Saturn appears elongated, giving a hint of its ring system with its brightest satellite, Titan, close by.

Under truly dark skies it is really rather surprising what can be seen. When I used my 15 × 45s in Galloway, one of the darkest regions in the UK, the Dumbbell Nebula in Vulpecular stood out beautifully, as did the galaxy pair M81 and M82 in Ursa Major – so many of the Messier objects can be easily seen. In New Zealand and with Baader Solar filters attached, I was able to observe the 2012 transit of Venus with ease and, later, had wonderful views of the southern Milky Way and the Magellanic Clouds along with the enigmatic object Omega Centauri and the globular cluster 47 Tucanae.

More recently, Canon has brought out a pair of 10 × 42L IS binoculars. They have been given an 'L' designation – reserved for Canon's highest-quality lenses – and incorporate two ultra-low-dispersion elements in each optical train. Incorporating wide-field eyepieces with 16 mm of eye relief, the true field of view is 6.5 degrees, so they will show twice the area of sky at one time as the 15 × 50s. They are lighter too – 1,130 rather than the 1,180 grams for the 15 × 50s. Both are powered by two AA batteries, which last for a couple of hours or so. The optical quality of the 10 × 42s is superb, and they can focus as close as 9 ft – as opposed to ~19 ft for the larger binoculars. The Canon range also includes a lower-cost 10 × 30 pair having a 6-degree field of view and weighing just 620 grams. The optical quality of this smaller pair is very good, but not quite up to the standard of the more expensive ones. It is said that they will equal 8 × 42 non-stabilized binoculars when used for stargazing and are, perhaps, the best choice for astronomers on a budget.

I have referred only to Canon IS binoculars, as the only real alternatives are the (virtually identical) 14 × 40s made by Nikon and Fujinon, but these seem to be less readily available and are, perhaps, better suited to terrestrial and marine, rather than astronomical, use due to the characteristics of the stabilizing mechanism.

It is said that your best telescope is the one you use the most. In my case, it is my binoculars that are always with me on my travels and help keep me in touch with the heavens.

3.3 Giant Binoculars and Their Mountings

Giant binoculars are those whose objectives have apertures greater than 50 mm. In order to keep the exit pupil down to 5–7 mm, the magnification has to increase in line with the aperture. Those that have exit pupils greater than ~5.5 mm are really only suitable for younger observers. This is the case for 9 × 63 pairs, which have an exit pupil of 7 mm. A very popular size for astronomical use has objectives of 70-mm aperture, such as the 15 × 70 Celestron SkyMaster and Helio Apollo binoculars, which have an exit pupil of 4.66 mm. It is possible to handhold these for a short while, but larger binoculars with 80- or 100-mm objectives will need to be supported. Celestron provides 20 × 80 and 25 × 100 pairs at relatively low cost and, at similar cost, Helios can provide 22 × 85 and 25 × 100 pairs. Revelation provides both 15 × 70 and 20 × 80

Figure 3.5 Celestron 20 × 80 (*top*) and Meade 25 × 100 (*bottom*) binoculars with an Orion parallelogram binocular mount.

pairs with fully multi-coated optics and Bak4 prisms. At significantly greater cost, Helios also produces the Quantum 7 25 × 100, which is equipped with triplet semi-apochromatic lenses and incorporates oversized BAK-4 prisms and a fully multi-coated optical system.

Binocular Mountings

Particularly with the very large binoculars just described, some support must be provided. The obvious solution is to use a tripod, and some binoculars have a tripod bush provided or clamps can be obtained to mount them. Except for relatively low elevation objects, tripods do not actually work very well, and it is generally reckoned that monopods work better. The observer sits in a beach chair with the monopod between the legs. More sophisticated solutions use a parallelogram structure with a counterweight, as shown in Figure 3.5.

3.4 Binocular Observing

Let's first discuss some fundamentals. Due to the size of the objectives compared with the size of the pupils of our eyes, we would expect to see fainter objects than with the unaided eye. And we will. Rather than state the limiting magnitude of a pair of binoculars – that is, the faintest star that could be observed under perfect conditions – I prefer to give what I call a 'magnitude gain'. For a pair of 8 × 40 binoculars this would be about 4 magnitudes, so if in a certain direction one could see a star of magnitude 5, then the binoculars would see stars down to magnitude 9. Theoretically, 10 × 50 binoculars will go down a further half magnitude, but their increased magnification also helps to darken the sky background, so enabling even fainter stars to be seen. But one must remember that unless the binoculars are mounted on a tripod or image-stabilized binoculars are used, the star images will 'dance about' and their light will be spread over a number of rods or cones in the retina and the theoretical increase in sensitivity will not be achieved.

Two further points. Just as for observing with your eyes, binoculars will show you more when your eyes have become fully dark adapted, and the faintest objects will be seen best using averted vision. Also, as far as I know, no one has yet made a pair of 'go-to' binoculars, so their use forces one to learn the sky – that can't be bad!

So What Can Be Observed?

The Sun.

The obvious warning must be given to prevent permanent damage to your eyes! Never, ever observe the Sun directly with any optical instrument unless it is fitted with appropriate solar filters such as those made from Baader Solar filter material well taped across the objectives with no chance of light leakage. It is then possible to observe the bigger sunspots as they rotate around the Sun. This set-up is also good for observing the early stages of a solar eclipse. Without the filters, binoculars are perfect for observing the corona around the *totally* eclipsed Sun, where the streamers may well extend over a few degrees. (Be very careful that the filters are removed only when full totality has been reached and replaced as totality ends.)

The Moon.

The best region to view is along the terminator, where the shadows enhance the detail. You will see the maria, mountain chains and the larger craters.

The Planets.

Planetary viewing is perhaps not best with binoculars, but you could see the Galilean moons of Jupiter and maybe note that Saturn's image is extended due to its rings. Binoculars will easily pick up Uranus and, under good conditions, Neptune. Try to observe their motion through the stars over a few nights.

Double Stars.

The first to try, and visible throughout the year, are the pair Alcor and Mizar, located at the bend in the handle of the Plough in Ursa Major. A harder double, but with a beautiful colour contrast, is the third-magnitude star Albireo, which forms the 'head' of Cygnus, the Swan.

Open Clusters.

Surely the Hyades and Pleiades Clusters in Taurus must be two of the best, but others that are well seen in binoculars include M44, the Beehive Cluster in Cancer and M35 in Gemini. Galileo was able to spot more than 40 stars in M44. A nice cluster is M41, at 4.5 magnitudes and just 4 degrees below Sirius. Another good binocular sight is the pair of clusters that make up the Double Cluster in Perseus, which will be high overhead in late autumn evenings.

Globular Clusters.

These ancient spherical star clusters are rather too small to resolve, but binoculars will enable you to see them as small 'fuzzy' balls. The best in the northern sky is 5.8-magnitude M13 in Hercules, between the right-hand stars of the 'Keystone'. Three others worth looking for are 6.5-magnitude M92, also in Hercules, the fifth-magnitude M22 in Sagittarius – the first globular cluster to be discovered – and seventh-magnitude M15 in Pegasus.

Nebulae.

A must for observers is M42, the Orion Nebula. This is a hazy fourth-magnitude patch in the 'sword' of Orion. It can be seen with the unaided eye, but binoculars give a far better view. There is a tight grouping of bright stars that form the Trapezium at its heart. Their visible light illuminates the dust, whilst their ultraviolet light excites the hydrogen gas to glow the lovely pinkish-red colour that we see in images – but, sadly, not with our eyes. Quite a challenging object, but one I have easily seen in 15 × 45 binoculars, is M27, the Dumbbell Nebula in Vulpecular. This was the first 'planetary nebula' to be discovered and is the remnant of a giant star that exploded at the end of its life, leaving a cloud of dust and gas. Another easily spotted nebula is M8, the Lagoon Nebula in Sagittarius. At fifth magnitude, it lies in one of the most beautiful regions of the Milky Way. This then brings us to what is perhaps one of binoculars' greatest strengths – scanning along the star clouds that make up our view of our own Milky Way galaxy. On a moonless and transparent night well away from the light pollution from towns and cities, it is amazing how many stars are visible!

Galaxies.

Under these same dark and transparent skies, binoculars can be used to observe some of the nearby galaxies. The most obvious is M31, the Andromeda Galaxy, which is the largest galaxy in our local group. With a magnitude of 3.5, its nucleus is visible to the unaided eye, but with binoculars one can get a feel of its full extent, which stretches some 3 × 1 degrees across the sky – so spanning about half a binocular field of view. If you slowly sweep the field of view across the galaxy, you may also get hints of the dust lanes that cut across the faint starlight. A pair of galaxies, M81 and M82 in Ursa Major, with magnitudes of 6.9 and 8.4 respectively, may also be seen under good conditions. Harder to spot, as it is seen nearer to face-on, is M33 in Triangulum. To my eyes its looks like a little piece of tissue paper stuck on the sky.

You can find more information about most of these objects and how to find them in the sky on the Web by searching for the 'Astronomical A list' – a section on the Jodrell Bank Web site that I wrote a few years ago. In 1990 Philip S. Harrington published *Touring the Universe through Binoculars*, a guidebook to virtually all the celestial objects (more than 1,100) that could be seen with binoculars, with the majority visible in 8 × 40 or 7 × 50 pairs. It also includes a binocular guide to the Moon's major features, listing the days in the lunar cycle when they can be best seen, along with chapters on observing the Sun and planets. Still available, both new and second-hand, it could also be a useful guide to what is observable with one of the smaller telescopes such as an 80-mm refractor or 130-mm Newtonian. With Dean Williams, Harrington produced a useful computerised star atlas which shows all 1,100 objects and includes all stars down to 11th magnitude. Prints can be made of any area, to which is appended a list of the visible objects within it. This useful (and free) program, called 'TUBA', can be downloaded from www.philharrington.net/tuba.htm.

A second, very worthwhile book for binocular users is Gary Seronik's *Binocular Highlights: 99 Celestial Sights for Binocular Users*, which is published by Sky and Telescope. Gary's Web site, www.garyseronik.com, is also an excellent source of interesting articles, including a detailed review of the Canon image-stabilized range.

3.5 Spotting Scopes

A viable alternative to small refractors are spotting scopes, which, of course, can also be used for other activities such as birdwatching. They typically come with apertures of 60, 80 and 100 mm. The lower-cost spotting scopes will use achromatic doublets as their objectives but, if they are often going to be used for astronomical use – which tests the optics rather more than terrestrial use – it would be worth paying extra to purchase a scope using an ED doublet. Celestron provides a wide range which includes both the use of ED and fluorite lenses. The Regal 80 F-ED appears to be an excellent value and very similar to my own 80-mm ED spotting scope, which is marketed under the Acuter brand. These are made by Synta (which owns Celestron) and are very similar to the Celestron models. The optical quality of my Acuter 80-mm ED is comparable to that of my William Optics 80-mm refractor. For those who consider optical perfection a must, spotting scopes made by Leica, Zeiss and Swarovski are available – but at some cost.

Many spotting scopes incorporate a 45-degree angle before the eyepiece, which makes them more comfortable for birdwatching, but which (I promise) makes it very difficult to find objects in the night sky. Some Celestron spotting scopes (as shown in Figure 3.6) have a small sighting tube on one side to overcome this problem. Having

Figure 3.6 A Celestron 80-mm spotting scope using an ED doublet objective. It has a small sighting tube to help align on a celestial object.

become very frustrated when using a 45-degree spotting scope without such a sighting tube, I made an angled bracket to hold a red dot finder!

The majority of spotting scopes are provided with a zoom eyepiece. Their optical quality is fine, but you will find that the apparent (and hence real) field of view at the low-magnification end of the zoom range is very restricted, although there is no problem at the higher end of the zoom range. As a result, they will not give as wide a field of view at low power as one might expect. The solution is to buy the fixed-focal-length eyepiece having the longest focal length (and thus the lowest magnification and widest field of view possible) from the range of fixed eyepieces that should be available for purchase for the spotting scope.

4

The Newtonian Telescope and Its Derivatives

As its name implies, the Newtonian telescope was invented by Sir Isaac Newton in 1668. Newton suspected that white light was made up of a spectrum of colours and that the chromatic aberration seen in the singlet objectives used in refracting telescopes was due to the light being split into its constituent colours. He thus reasoned that if an image were made using a concave mirror, chromatic aberration would not occur. He chose to make a spherical mirror out of speculum metal, an alloy of tin and copper, and used a flat elliptical mirror to reflect the converging light path sideways to the eyepiece situated just outside the telescope tube – this being the hallmark of the Newtonian design. The theoretical surface of the concave mirror is parabolic. Usually the mirror blank is first made to have a spherical surface, which is then figured to a parabolic shape. Below about 100 mm in aperture, the difference in the two is sufficiently small for this second step not to be required.

Using a mirror of 33-mm aperture, Newton was able to see the moons of Jupiter and the crescent phases of Venus, but otherwise it is not thought that he carried out any astronomical observations. Over the years the quality of reflecting telescopes improved, reaching a first high point when, in 1781, an amateur astronomer, William Herschel, used a 160-mm, f13 telescope to discover the planet Uranus. His mirror was figured more accurately than those used by professional astronomers of the time, and so his images were clearer, enabling him to see that what had been thought by other observers to be a star had, in fact, a planetary disk.

4.1 Qualities and Drawbacks of Newtonian Telescopes

- They are free of chromatic aberration.
- For a given aperture, they are usually less expensive than other telescope designs of comparable quality.
- Only one surface, that of the primary mirror, has to be ground and figured into a complex shape.
- The eyepiece, being at the top of the tube assembly, is generally easier to access than with other telescope designs.

- The design uses two mirrors in the optical light path. With simple aluminium coatings their reflectivity was of the order of 87%. Thus the brightness would be reduced to 0.87 × 0.87, or ~75%, corresponding to a brightness loss of about 0.3 magnitude. (This is derived using the equation magnitude difference = −2.5 log 10 (brightness ratio).) As discussed in Chapter 1, some of the light lost will be scattered into the field of view, so reducing the overall contrast of the image. However, modern dielectric coatings increase the reflectivity of each mirror to ~97%, reducing the magnitude loss to a minimal amount. The overall contrast will be greatly improved as well.
- The image field suffers from coma away from the optical axis, causing stars to appear like little comets with a flare pointing towards the field centre. This gets greater the farther away from the axis and is inversely proportional to the focal ratio, so Newtonians with low focal ratios such as f4 suffer the most. This is not the problem that it once was, as 'coma correctors' are readily available to overcome this defect.
- The secondary mirror and its supporting 'spider' give rise to diffraction effects which reduce the micro-contrast of the image, as discussed in Chapter 1.
- When they are mounted on an equatorial mount, the eyepiece position, if fixed, can sometimes be hard to get to; however, most telescope tubes can be rotated within their tube rings to overcome this problem.
- When Newtonians are transported, the alignment of the primary and secondary mirrors can be disturbed, giving rise to poor image quality. To align the mirrors, a process called 'collimation' is used, as will be discussed in Chapter 6.

4.2 The Basic Design of a Newtonian

As shown in Figure 4.1, the basic design of a Newtonian is very simple. At the base of the telescope tube a parabolic mirror is supported on a moveable mount so that, by means of three adjustment screws, its optical axis can be aligned along the centre line of the telescope tube. Near the top of the tube a 'spider' supports a secondary mirror which reflects the light sideways to the image plane (the focus of the mirror) just outside the telescope tube, where a focuser supports the eyepiece that is to be used. The secondary mirror is cutting across a converging cone of light and is thus a conic section, which in this case is an ellipse and so the mirror is what is called an 'elliptical flat'. The designer of a Newtonian calculates the minor axis of the mirror, and the major axis is simply the minor axis dimension multiplied by the square root of 2, or 1.141, so an elliptical flat having a minor axis of 50 mm would have a major axis of 71 mm. The secondary mirror forms what is called the 'secondary obstruction', whose diameter is its minor axis.

   The size of the secondary mirror is a compromise. If too small, it will cause the outer parts of the field of view to be vignetted (i.e., darkened as compared with the centre), but the larger it is the greater its diffraction effects, so reducing the micro-contrast of the image. Following a discussion of the vignetting problem, three detailed designs will be described: two of these are optimized for a particular use and one provides a good, all-round telescope. It is not expected that you would aim to design

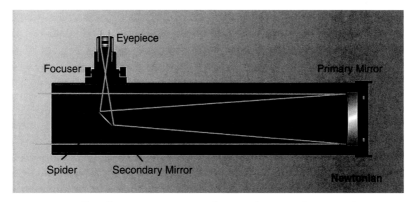

Figure 4.1 The classic Newtonian design. (Image: Starizona)

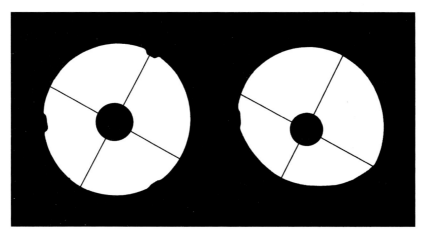

Figure 4.2 The primary mirror of a 200-mm, f6 Newtonian as photographed from the centre of the field of view (*left*) and from the edge of the 1.25-inch focuser (*right*).

a Newtonian yourself – though there is sufficient detail to allow you to do so – but rather that this will make you aware of the design choices that have to be made and so allow you to specify a suitable telescope for your own use.

Vignetting of the Field of View

The size of the secondary mirror will determine what percentage of the field of view will be 100% illuminated. Figure 4.2 shows the view of the primary mirror taken at the centre of the focal (or image) plane and as one moves towards one side. From the centre of the field of view you must be able to see the whole of the mirror – if not, the telescope is badly designed and the effect is equivalent to having a telescope with a smaller primary mirror – but away from the centre you will tend to see less of the mirror, as shown in the figure. This means that less light will fall on that part of the field of view, so causing what is called 'vignetting'.

One might at first think that the obvious thing is to simply use a secondary which will fully illuminate the largest field of view that might be used but, as you will have seen in Chapter 1, the larger the size of the secondary the more light is transferred from the central Airy disk into the surrounding rings, so reducing the micro-contrast in the image. In practice, it does not really matter if the illumination falls to 70% of the maximum at the edge of the field of view when the telescope is used visually, although, for wide-field imaging with a DSLR, a greater percentage might be preferable. The optimum design of a Newtonian thus actually depends on the use for which it is intended. To give you a feel of how the various factors in the design of a Newtonian affect the telescope's performance, I will carry out the design of two Newtonians: one optimised for visual or webcam imaging of the planets, the other for wide-field observing and imaging. A primary mirror of 200 mm will be used for both.

4.3 A Newtonian Optimised for Planetary Observing

The greater the focal ratio of a Newtonian, the smaller the secondary mirror (in the form of an elliptical flat) will be. This will, of course, make the telescope tube quite long if a large-aperture mirror is used. It is generally accepted that if a secondary mirror whose diameter is equal to or less than 15% of the diameter of the primary is used, there is no significant loss of micro-contrast. As this is a major objective in planetary imaging, the design will aim for a secondary that gives a central obstruction of around 15%. Given this requirement, it is possible to calculate the required focal length of the telescope and hence the focal ratio. The logic when working 'backwards' in this way is a little complicated so, instead, let's assume a focal ratio of f9 and see if this meets our requirement. The focal ratio then immediately gives a focal length of 9 × 200 mm, which is 1,800 mm.

The other parameter that is required is the distance of the focal (or image) plane from the optical axis of the telescope – the less this distance is, the smaller the required secondary. The focal plane must obviously be outside the telescope tube but ideally as close to it as possible. The use of a low-profile focuser and relatively simple eyepieces – good for planetary observing – will help reduce this distance when 50 mm might be an achievable value. (There is an expensive ultra-low-profile focuser which could reduce this value to 30–40 mm.) To this distance we must add half the diameter of the primary mirror, 100 mm, together with the distance between the edge of the mirror and the mirror wall and the wall's thickness. To prevent wall currents from affecting the image quality, the internal wall diameter should be of the order of 20 mm greater than the primary diameter and, to provide the maximum overall contrast, a high-quality Newtonian may well have some knife edge baffles inside the tube, which in this case could be 5 mm or so high. Given a thin wall tube, the outer diameter of the optical tube would thus be 220 mm, so the outside of the tube would be 110 mm from the optical axis. It might be noted here that the diameter of the tube (or its internal baffles) can limit the light falling off-axis onto the mirror, and this can thus be a further cause of vignetting away from the centre of the field of view. Given the narrow field of view required for this application this can be ignored, but it will

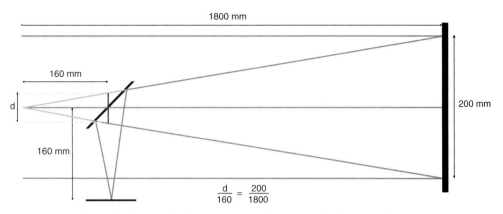

Figure 4.3 The calculation of the minimum diameter of the secondary mirror.

have to be addressed in the following wide-field design. Thus the focal plane will be 110 + 50 mm from the optical axis, giving a total distance of 160 mm. As shown in Figure 4.3, by using similar triangles one can then calculate the required minor axis of a secondary (*d*) that will just fully illuminate the central point in the image plane: $d/160 = 200/1,800$, so that $d = 17.7$ mm.

One then has to decide how wide the field of view that will be fully illuminated should be. As the Newtonian will be used for observing planets with very small angular sizes, this does not need to be more than, say, 10 mm. (As an absolute minimum it must be greater than the diameter of the dark-adjusted eye pupil, which could be up to 7-mm.) This assumes, of course, that one keeps the planet in the centre of the field of view. It turns out that the diameter of the secondary mirror has to have a diameter which is that required to give full illumination at the exact centre of the field of view plus the diameter of the required fully illuminated field. This thus gives a secondary with a minor axis of 27.7 mm. However, it is difficult to ensure that the outer edges of a flat mirror are optically flat, and it is thus wise to use a slightly larger secondary than theoretically necessary, so a flat with a minor axis of 30 mm might be employed. If we divide this by the diameter of the primary mirror and multiply by 100, we will get the percentage of central obstruction, which comes out at 15% – our design goal!

Such a telescope will be somewhat cumbersome and require a sturdy mount, and so one might consider reducing the diameter of the primary mirror to make a more compact telescope. This is not quite as easy as it might appear, as smaller primaries actually need to have greater focal ratios to achieve a 15% obstruction and, though the telescope tube will be thinner, the length will not reduce as much as one might think. Dobsonian mounts, to be discussed later in the chapter, are ideal for large or long telescope tubes, and a US company, Ocean Side Photo and Telescope (OPT), sells a truss tube 8-inch (200-mm), f9 Newtonian of virtually the design just carried out.

In place of an eyepiece one might wish to use a webcam to take video sequences, as will be described in Chapter 11 in the section on planetary imaging. With the focal plane so close to the tube, it might not be possible to bring the CCD sensor to the

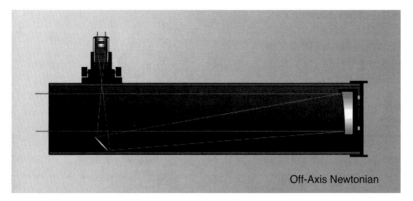

Figure 4.4 An off-axis Newtonian. (Image: Starizona)

image plane. However, to give a good image scale on the sensor, a Barlow lens would probably be used in front of the webcam and this usually solves the problem.

An obvious way of removing the diffraction spikes caused by the secondary spider would be to mount the secondary on an optically flat window. This has the advantage of keeping dust out of the telescope tube but can 'dew up' under humid conditions. It is actually quite hard to make large optical flats and this approach is rarely adopted, partly as it is no more difficult to make corrector plates or meniscus lenses that allow a spherical mirror to be used and so help minimize coma. A spherical mirror is less costly to fabricate than a parabolic mirror and so the overall cost may well be less. With this approach, two types of modified Newtonians have been designed, the Schmidt-Newtonian and Maksutov-Newtonian, both of which will be covered later in this chapter.

Sub-Aperture Masks and Off-Axis Newtonians

There are two ways to optimise a Newtonian for planetary observing which do not give rise to a central obstruction. The most obvious is to use a telescope with a primary mirror of perhaps 300 mm and make an off-axis mask. This could be a circular aperture of ~140 mm or an elliptical aperture of ~140 × 160 mm arranged so that neither the secondary mirror nor its spider lies within its aperture, as shown in Chapter 1, Figure 1.8. This would be a very good way of using a large-aperture Newtonian telescope for planetary observing. The second way is to make a telescope, called an off-axis Newtonian (Figure 4.4), where only part of the primary mirror is used. A large primary would be cut into four quadrants and then each reduced in size to make it circular as if one were making a circular sub-aperture mask for the original mirror. This is a very neat design! However, it requires an exceptional primary mirror around 400 mm in diameter and the skill to cut out the four smaller circular mirrors without stressing the glass, so such designs are not cheap. A range of off-axis Newtonians, including a 150-mm-aperture, f7.5 telescope, are manufactured and sold by Dodgen Optical, LLC. This company also sells the mirrors for use by amateur telescope makers.

4.4   A Newtonian Optimised for Wide-Field Observing and DSLR Imaging

To achieve a wide field, a telescope of short focal length is wanted, so as low a focal ratio as possible will be best. There is the problem of coma, which becomes significant as the focal ratio is reduced – its effect being inversely proportional to the square of the focal ratio. This becomes very significant when the focal ratio drops to f4, and such a Newtonian would have to be used with a coma corrector, such as those produced by TeleVue. Choosing such focal ratio and employing a coma corrector could be said to be cheating as, in effect, the telescope will no longer be a pure Newtonian, but corrector lenses such as focal reducers and field flatteners are now being used with many types of telescope. As coma correctors are often designed for a focal ratio of f5, let us assume the use of a primary mirror of this focal ratio. This immediately gives the focal length of our Newtonian of 200 × 5 mm, that is, 1,000 mm.

For wide-field observing and imaging, the reduction in the micro-contrast of the image due to the use of a large secondary mirror is not important. The effect will be to reduce the theoretical resolution by about a factor of 2, which in this application has no real relevance. What is important is to minimise the vignetting of the field of view when the telescope is to be used with a full-frame DSLR. A drop in brightness down to 70% at the edge of the field is generally regarded as suitable, with image manipulation software such as Photoshop easily able to correct for this if necessary. A full-frame DSLR has a sensor size of 36 × 24 mm. This gives a diagonal of 43 mm, so one might design the telescope to give a fully illuminated field of view of, say, 38 mm. The field stop diameter of a wide-field 2-inch eyepiece is also around 46 mm, so this would work well for visual use too, though the eye is less sensitive to the fall-off in brightness towards the edge of the field.

As previously mentioned, to avoid the telescope tube itself vignetting the outer parts of the field of view, the tube must be somewhat wider than the mirror diameter. Given a field stop diameter of 46 mm, the overall field of view with a 1,000-mm focal length is 46/1,000 × 57.3 degrees. This is 2.6 degrees so, looking towards the mirror from an angle of 1.3 degrees off-axis, the full area of the mirror should be seen unobstructed by the telescope tube. This requires that the tube diameter have a radius which is 1.3/57.3 × 1,000 mm greater than the radius of the primary mirror. That is 23 + 100 mm, so the tube diameter should be 246 mm.

In order for the sensor of a DSLR or CCD camera to reach the image plane, this has to be farther from the telescope tube, and 80 mm, rather than 50 mm, would be a suitable distance. Adding this to the distance of 123 mm of the tube from the optical axis gives 203 mm. Again using similar triangles, the minimum diameter (d) of the secondary mirror is given by $d/203 = 200/1,000$ mm, so $d = 40.6$ mm. To this minimum diameter, which would give full illumination only at the exact centre of the field of view, one must add the diameter of the field that is to be fully illuminated, 38 mm. Allowing a little extra width to avoid the use of the secondary mirror's extreme edges would give a secondary diameter of ~82 mm. The percentage of obstruction would then be 82/200 × 100, or 41%. One should not worry about the loss of theoretical resolution that such an obstruction would give, and neither is the loss of brightness caused by the obstruction important. The effective area of the primary will be $(200)^2 - (82)^2$, which is 33,276 mm^2 rather than 40,000 mm^2 of the primary, corresponding to a loss of just 0.2 magnitude.

4.5 A Newtonian for All-Round Use

A very good compromise in designing a 200-mm Newtonian is to use a focal ratio of f6, and many such telescopes are sold with this basic specification. The focal length will thus be 1,200 mm, and the field of view when a low-power, wide-field, 2-inch eyepiece is used will be ~2 degrees. The secondary mirror will probably be ~50 mm in diameter, and this would give a percentage of obstruction of 25% and provide full illumination over the central ~25-mm-diameter region of the up to 46-mm-diameter field of view. This is a good compromise, but some manufacturers such as Orion Optics in the UK allow the purchaser to choose other secondary diameters should he or she wish to optimise the telescope towards planetary (~36 mm) or wide-field imaging (~60 mm). One could even purchase two flats for use depending on the type of observations planned for a given night!

4.6 Schmidt-Newtonian and Maksutov-Newtonian Telescopes

Like a Newtonian, both of these telescopes employ a mirror at the base of the telescope tube which reflects the incoming light up to an elliptical 'flat' that reflects the light through 90 degrees to form a focus outside the optical tube, where the image is viewed with an eyepiece. This is the 'Newtonian' part of the design. However, instead of a parabolic mirror, a spherical mirror is used. This would, by itself, give disastrous images due to the mirror's spherical aberration. This is corrected by the use of a corrector plate at the front of the optical tube assembly. Not surprisingly, that used in a Schmidt-Newtonian is a thin glass plate similar to that used in Schmidt-Cassegrains, whilst that used in a Maksutov-Newtonian is a much thicker and deeply curved corrector plate as in Maksutov-Cassegrains. (Details of these telescope types are given in Chapter 5.)

Both share three advantages over a classical Newtonian telescope. Firstly, the secondary mirror is supported by the glass corrector plate, so an image will not show the 'diffraction spikes' that are caused by the spider used to support the secondary mirror in a Newtonian. Secondly, the glass corrector plate encloses the tube assembly, so that the primary and secondary mirrors are protected from dust and will thus be less likely to need cleaning. Thirdly, the use of a spherical mirror with an appropriate corrector plate reduces coma and thus gives better off-axis images than are obtained with a Newtonian of equivalent focal length. The corrector plate does, however, give rise to two disadvantages. These telescopes will need longer for the air inside to reach ambient temperature and so enable the telescope to work at its best, and the corrector plate is also prone to dewing up and so a dew tube and perhaps a dew heater will be required.

Both telescope types share three advantages over the respective Schmidt- and Maksutov-Cassegrain versions. Firstly, the light, as in a Newtonian, travels through the tube twice rather than three times, making them slightly less troubled by tube

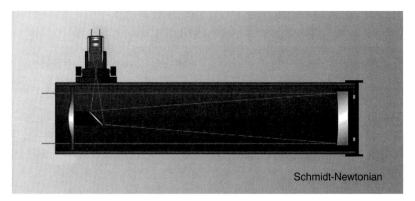

Figure 4.5 A Schmidt-Newtonian. (Image: Starizona)

currents. Secondly, as the eye is looking across the tube assembly rather than up towards the sky as in the Cassegrain versions, it is much easier to baffle any extraneous light that might find its way into the image plane and so give improved image contrast. Thirdly, particularly in the case of the Maksutov-Newtonian, the secondary mirror is far smaller than that in the Cassegrain version, increasing the micro-contrast of an image so important for planetary viewing.

Finally, it has to be said that both types are bulkier and heavier than either Newtonians (as they have the extra weight of the corrector plate) or their Cassegrain versions (as they have greater tube lengths).

The Schmidt-Newtonian

The Schmidt-Newtonian scope replaces the normal parabolic mirror with a spherical mirror (see Figure 4.5). This will introduce spherical aberration, which in this case is eliminated by the addition of a Schmidt corrector plate as used in Schmidt-Cassegrain telescopes.

The objective of the design is to create a telescope with a wide field of view – so requiring a short focal ratio – but unhindered by the significant coma that would be present in a Newtonian of similar focal ratio. While the Schmidt-Newtonian still experiences some coma, it is less (roughly half) than that of a traditional Newtonian.

In the recent past, Meade has manufactured a range of three Schmidt-Newtonians: a 6-inch f5, an 8-inch f4 and a 10-inch f4. They were designed for wide-field visual or imaging use and came with the LXD55 and, later, the improved LDX75 computerised mounts. These were sufficiently rigid for supporting the 6- and 8-inch versions, but perhaps were not really up to the job for the 10-inch version. The tube assemblies are very substantial and relatively heavy for their size though, having short focal lengths, are reasonably compact.

The mirrors, made of Pyrex, were given high-reflectivity coatings, and the corrector plates, which were made of 'water-white glass' for increased light transmission, had high-transmission 'UHTC' coatings. The LXD55 and LXD75 were amongst the

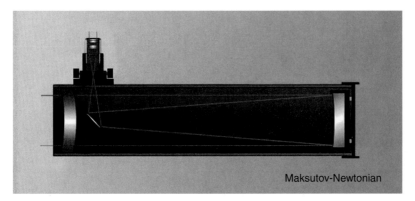

Figure 4.6 A Maksutov-Newtonian. (Image: Starizona)

first computerised mounts available but were not sold without one of the telescopes. Quite a number of amateur astronomers bought the combination and then sold the telescope cheaply – which was how I acquired both the 6-inch (150-mm) and 8-inch (200-mm) optical tube assemblies.

Meade no longer produces these telescopes or mounts, but they are still available on the second-hand market. Their qualities have become more widely appreciated, and they are now becoming quite sought after, being particularly suited for wide-field astro-photography.

Maksutov-Newtonians

As will be described in Chapter 5, in 1941 the Russian optician Dmitri Maksutov designed a Cassegrain-style telescope in which a spherical primary mirror is used along with a full-diameter deeply dished 'corrector plate' to correct for the use of a spherical, rather than a parabolic, mirror and also reduce off-axis aberrations such as coma. The same basic idea has been applied to the Newtonian design to give a Maksutov-Newtonian whose design is shown in Figure 4.6. There are two improvements over Newtonians:

1. The secondary mirror (flat in this case) is supported by the corrector plate, so there will be no diffraction spikes.
2. The use of a spherical primary mirror and corrector plate should give a wider field of view before effects such as coma become apparent and so improve on the performance of a Newtonian of the same aperture and focal ratio.

In the past, the majority of these telescopes were made in Russia by the firm Intes-Micro, which supplies telescopes with apertures ranging from 5 to 10 inches. I own an MN66, which has, as the name implies, a 6-inch (152-mm) aperture and an f6 focal ratio. The company also makes f8 versions of the 6- and 7-inch-aperture scopes, and I have been able to use the MN78, having a 7-inch (178-mm) aperture and f8 focal ratio.

These will have a smaller maximum field of view than the f6 versions but, as with Newtonians, the larger focal length design requires a smaller secondary mirror – and hence central obstruction – so reducing their diffraction effects. This makes them even better for lunar and planetary imaging, where the micro-contrast is most important. The MN78 has a central obstruction which is only 13.4% of the telescope's aperture, as shown in Chapter 1, Figure 1.7. The diffraction effects of such a small obstruction are virtually negligible, and so this telescope should give essentially equal performance to an apochromatic refractor of this aperture – which would be considerably more expensive and also somewhat unwieldy.

Depending on the part of the sky in which one wishes to observe, a Newtonian tube assembly on an equatorial mount has to be rotated to bring the eyepiece to a comfortable observing position. When the tube is rotated (with the tube rings loosened), there is a great tendency for the tube to slide down from its balance position. The MN78 overcomes this problem by having two 7-mm-high ridges extending around the telescope tube that slide in Teflon bearings within the tube rings, so allowing easy rotation of the tube but preventing any longitudinal movement.

At the base of the telescope tube is an extension with a circular panel that can be screwed out to reveal a 12-volt fan. A ring of ventilation holes around the aperture of the telescope allows the air to pass through the tube and so bring the interior of the telescope to ambient temperature – essential before any serious observations can be made. The way to check this is to observe an out-of-focus image of a bright star. Unwanted tube currents show up as a 'bleeding' of the image, rather like flames flickering away from the edge of the circular star image.

As previously described, the designer has to make a compromise as to the size of the secondary mirror, and as this telescope is specifically aimed at planetary observations a small secondary is used with only the central few millimetres of the field of view fully illuminated and the image brightness falling by about one-half at the field edge, which, surprisingly, is not readily apparent to the eye.

The telescope specification states that the primary and secondary mirrors have a reflectivity of 95%. This implies that a multi-coating system has been used, giving two great benefits apart from the increase in image brightness. The first is that the multi-coatings appear to last far longer than standard coatings, perhaps up to 25 years before re-coating is required, but even more significant is the fact that the surface coating is far smoother. This means that far less light is scattered at the mirror surface. This scattered light could fall anywhere across the telescope image, so reducing the overall image contrast.

The optical tube contains knife edge baffles, with a set directly opposite the eyepiece to minimise any light scatter, which would harm the image contrast. They, no doubt, contribute to the superb image contrast of the scope.

The Intes-Micro mirrors are specified to give a minimum of 1/6 wavelength errors, better than the standard ¼ wavelength criterion for good optical performance.

Orion in the United States also sells Intes-Micro-built telescopes under the Argonaut brand, and other manufacturers, such as Synta in China, marketed under its Sky-Watcher brand, have now begun to build and sell Maksutov-Newtonians.

The following section recounting my personal experience with the MN78 will give you a feel of the capabilities of this excellent design of telescope.

4.7   First Light with an Intes-Micro MN78

An hour or so before darkness fell, the scope and mount were set up. I had some misgivings about my ability to handle the lengthy tube assembly, but these proved unfounded. The tube rings were first mounted horizontally on the equatorial head and appropriate counterweights added. The telescope, cradled in my arms, was easily lifted into the rings, which were then secured. Finally the position of the mounting plate was adjusted to balance the telescope tube, finder and eyepiece. As with all Newtonian telescopes, to bring the eyepiece to a suitable position requires the rotation of the telescope tube – which can often slip down in the process. The MN78 has two thin rings surrounding the optical tube which slide within Teflon guides in the tube rings. It rotates easily and cannot slip through the rings – a really useful feature. Sadly, high cloud was beginning to obscure all but the brightest stars and so I observed the star Vega, virtually overhead. The interior of the telescope had totally stabilised and there were no tube currents visible. Racking through the focus position, I could see that the image was totally colour free, both in and out of focus. No false colour was really expected, but as the light passes through a corrector lens there is at least the potential for it to occur. If the telescope is perfectly collimated, then the 'shadow' of the tiny secondary mirror should appear in the precise centre of the out-of-focus star image, which should be circular in shape. Sadly, just as the collimation was observed to be near perfect, the clouds rolled in and further observations had to wait until the next night.

At this point the Crayford-type focuser, which is a little unusual, should be described. The focuser travel was just 18 mm, far less than in most focusers, but it supported a 2-inch eyepiece holder which could be extended by a further 50 mm. This, in turn, could support a 1.25-inch eyepiece adapter, allowing a further 20 mm of extension. The focusing technique was thus to set the focuser travel at its midpoint and then adjust and lock the extension tubes for approximate focus before finally achieving exact focus using the focuser. This is fine with a set of parfocal eyepieces, but a little fiddly if not. One might well ask, why not give a greater range of focuser adjustment? Because the telescope designer wished to use the minimum size of secondary possible, the focal plane is not far from the edge of the telescope tube. If a larger range of adjustment were given, the tube would have to be longer and, when the focuser was racked into its minimum position, would extend into the telescope tube and would intercept some of the light falling on the primary mirror, giving rise to unwanted diffraction effects.

Though my mount is equipped with Wildfire Innovations digital setting circles, I was simply using the supplied 50-mm finder scope. This has both a focusing adjustment and also an adjustment for bringing the cross wires into exact focus (good). Optically it was fine but, as with all straight through finders, when one is trying to find Vega, which was virtually overhead, not that easy to use. As a result I first found, not Vega, but Epsilon Lyrae, the nearby 'double-double' star, in the eyepiece – which would have been my next target. As I increased the magnification, the scope easily separated the two close pairs – a very good start indicating good optical performance of the scope. I then moved to Vega itself in order to 'star-test' the scope. Racking through focus showed a clear, high-contrast set of rings on either side of focus. This showed that the mirror had a smooth surface; a rough mirror, even if the correct shape, gives a blurred set of out-of-focus rings. However, the seeing was not good enough to observe the rings around the in-focus Airy disk. In one sense this was good, as the whole point of the telescope design was to have the minimum light in these rings, so unless conditions were near perfect I shouldn't really have expected to see them!

Close by was the Ring Nebula, M57. It showed as a nice little smoke ring, just as expected. By this time, Jupiter was appearing in the south-east over the roof top of my neighbour's house. Due to heat currents rising from the house, this is *not* a good time to observe a planet! But I could not resist having a first look. The equatorial bands were well in evidence and four moons were visible. Due to the low elevation, refraction was causing false colour above and below the image, and a green filter was used to clean up the image. I tried an OIII narrowband filter which I have used with success with a 12-inch telescope, but there really wasn't sufficient light. Waiting for Jupiter to rise higher in the sky (and away from the rooftops) I moved round to observe the Double Cluster in Perseus, and this was perhaps the most pleasing view of the evening. I was using a TMB 40-mm Paragon eyepiece. This is certainly one of the better 2-inch wide-field eyepieces and has a field stop diameter of 46 mm (essentially the largest possible), so giving the widest field of view, at 1.8 degrees, possible with this telescope. This nicely encompassed the two clusters, which extend for somewhat more than 1 degree. This was the best wide-field image that I had ever observed with this eyepiece, with pinpoint stellar images virtually to the edge of the field. This showed that this telescope/eyepiece combination was essentially perfect. This is not such an obvious statement as it might seem. All telescopes (and particularly refractors of short focal length) show some field curvature, which means that if the centre of the field is precisely in focus, the outer parts may not be. Eyepiece designers may attempt to correct for this to some extent. The Paragon has had little correction applied, which implies that the MN78 has a pretty flat field, and together they make an excellent match!

It was about this time that the corrector plate began to dew up. The moisture was removed with warm air from a hair dryer, but it does indicate that a dew shield is really a must. An excellent, baffled dew shield is available (I have one for my MN66). Not only will it help prevent dew from forming, but it will also prevent one's fingers from straying onto the front surface of the corrector plate.

As Jupiter was now higher in the sky, I observed it again. Io had disappeared, but the remaining satellites did appear to show disks rather than be simply point-like, and the equatorial bands were showing some structure. I used a 13-mm Vixen wide-angle eyepiece, a 9-mm TMB planetary eyepiece and a TeleVue 7-mm Nagler. These gave magnifications of 111, 160 and 206 respectively. The seeing conditions did not, however, really allow the highest of these to be sensibly used.

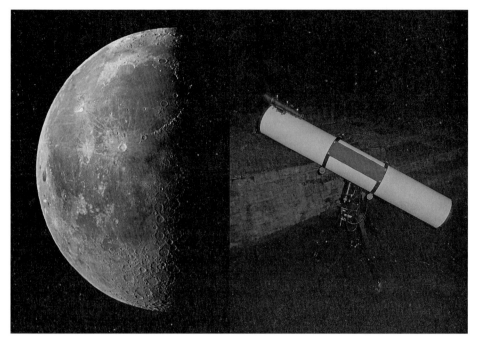

Figure 4.7 The third-quarter Moon and the Intes-Micro MN78 mounted on a Losmandy GM8 with which it was taken.

The night remained clear, and by 5:30 AM the nearly 24-day-old Moon was high overhead, so I took the opportunity to observe and photograph it as sunlight gradually lightened the sky. As is often the case, the seeing had improved markedly during the night, and I was rewarded with some of the crispest and highest-contrast views of the Moon that I have ever seen. Sinus Iridum, the 'Bay of Rainbows', was near the terminator with the tip of the Laplace promontory rising up into the sunlight. To be honest, I am not often up before the dawn to observe the Moon during its later phases and found the very rough area of the surface just to the upper left of the crater Aristarchus interesting, having never observed it before. The low-contrast rays surrounding the crater Kepler showed up well, along with the terraced walls and central peaks of Copernicus. Now was the time for my first attempt at photography using my Nikon D80 10-Mpixel camera.

I usually have problems getting the camera CCD to lie at the focal plane of a telescope, and this was no exception. Using the 2-inch T-mount adaptor, I could not rack the focuser in quite enough to reach focus. There were only 2 mm in it, but by removing the 2-inch extender from the focuser and making a small addition to the T-mount adaptor, the camera sensor was able to reach the image plane. This took a while, so by the time I was able to image the Moon, the sky was getting significantly brighter. However, I was quite pleased with my first lunar shots with the telescope, with the Moon sitting nicely within the 24 × 16 mm CCD area of the camera as shown in Figure 4.7.

I made two other lunar images. That of the third-quarter Moon was made by screwing the lens element of a 2-inch apochromatic Barlow into a T-mount adaptor. Three images were taken to cover the whole Moon and combined using the free 'Microsoft ICE (Image Composite Editor)' program followed by a little sharpening and some 'curve' adjustments, as discussed in Chapter 11. This image has 22,222 pixels across the Moon's disk! There are two significant points to make. The first is that using a 'picker' to measure the brightness of individual pixels in the raw images showed that the overall image contrast was superb – the light level just a few pixels away from the lunar limb and within craters near the limb was less than 1% of maximum brightness in the image. The second is that in the images where the Moon filled the frame, the sharpness extended across the whole CCD frame – proving that the telescope does provide a very flat field and confirming my observations with the Paragon eyepiece.

4.8 The Dobsonian Telescope Mount

In the early 1970s the American John Dobson popularised a very simple type of telescope mount that now bears his name. It is based on what is called a 'lazy Susan'. This is a type of rotating tray, usually circular, placed on top of a table to facilitate the sharing of a number of food dishes by a group of people and is very common in restaurants in China. In the case of a Dobsonian, a base plate, usually made of wood, is placed on the ground and the base of what is called the 'rocker box' is mounted above this, located by a central pivot and supported on the base plate by a mechanism to allow it to rotate smoothly around the central pivot. In the simplest version, the base of the rocker box has a Formica surface which is supported by three Teflon pads mounted on the base plate. The rocker box has triangular cut-outs on each side which have Teflon pads mounted on each side. On the side of the Newtonian tube are mounted two circular disks which are supported by the pads. These might well be made of plastic. The frictional characteristics of the Teflon/Formica and Teflon/Plastic interfaces make it possible to move the telescope very smoothly in both azimuth and altitude (but allow the tube to remain stationary if required) and enable it to be gently pushed to follow an object across the sky. It makes a very stable configuration as the telescope mirror stays very close to the ground.

There are many variations of the basic design, which is often used for large-aperture telescopes. Dobsonians tend to be optimised for the observations of deep-sky objects such as galaxies and nebulae, and their observation does require dark skies. This implies portability, and Newtonians of large aperture (and hence overall length) using single tubes tend to be too big to transport easily. To overcome this problem, in many Dobsonians the telescope is split into two parts: one contains the mirror and the second, the 'cage assembly', contains the secondary mirror and focuser (Figure 4.8). These can often be stored or transported with the cage assembly placed within the lower mirror box to give a very compact package. The two parts are then linked together with a number of struts – most commonly eight struts arranged in what are sometimes called a 'Serrurier truss configuration'. This is not strictly true, as the Serrurier design in a majority of large optical telescopes uses two trusses on either

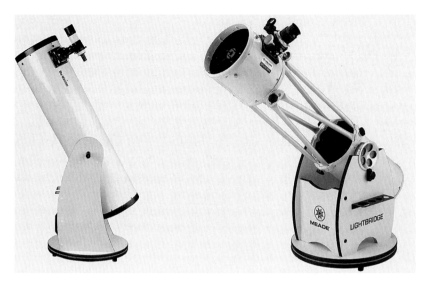

Figure 4.8 A Sky-Watcher Dobsonian with solid tube and a Meade LightBridge with a truss tube.

side of a central support, which keeps the optics on the same optical axis as the telescope is changed in elevation. Dobsonians use only a single truss which, nevertheless, gives the structure great rigidity. The struts are held in place with clamps to allow the telescope to be quickly assembled or disassembled.

Many more modern designs have reduced the overall weight and provide increased portability by reducing the size of the mirror box to little more than a small rotating platform supporting the mirror. The altitude bearing in these designs is given a large radius, which is roughly equal to the radius of the objective mirror. A further advantage is that the telescope's balance becomes less sensitive to changes in the weight in the cage assembly when heavier eyepieces are used.

My first Newtonian telescope, having a 250-mm aperture, was originally mounted in an exceedingly heavy fork mounting which was almost too heavy for me to move. So in the early 1990s I built (with the aid of many friends) a truss-based design. The primary had a relatively large focal ratio of f7, giving a focal length of 1,750 mm. This was my main telescope for nearly 30 years.

4.9 Newtonian-Based Astrographs

In the same way that field flatteners can be used with refractors to provide a wide, flat field suitable for use with large CCD cameras or full-frame DSLRs, so can somewhat more complex optical correctors be used with Newtonian telescopes of low focal ratio. Such systems reduce exposure times and give wide fields of view, so that a Newtonian with corrector lens can make a superb astrograph, that is, a telescope optimised for imaging – in this case wide-field imaging.

In the UK, Orion Optics manufactures a range of 8- to 16-inch AG (AstroGraph) telescopes using a Newtonian design allied to a 'corrected field flattener', a four-

Figure 4.9 Orion Optics (UK) AG16 astrograph (*left*) and Astrosysteme Austria N8 astrograph (*right*). Note the large secondary mirror, which is used to provide an unvignetted wide field of view.

element lens based on a design by Harmer Wynne. The f3.8 mirrors are hand figured to 1/8 wavelength and are provided with sufficient 'back-focus' to allow CCD cameras to come to focus even when, as usual, a filter wheel is in the optical path. Using an 8-inch AG astrograph, the superb astro-photographer Peter Shah has taken the image shown in Plate 15.8 of the galaxy Andromeda Galaxy, M31. Many more of his wonderful images can be seen on his Web site at Astropix.co.uk.

Astrosysteme Austria (ASA) produces a range of 8- to 20-inch Newtonian-based astrographs (N series) using a parabolic primary allied to a corrector lens assembly designed by Phillip Keller (see Figure 4.9). The mirrors, figured to 1/7 wavelength, have a focal ratio of f3.6, so allowing short exposures to be made, and have a useable image plane diameter of 50 mm. The company also produces an 8-inch-series astrograph which uses a hyperbolic mirror and corrector lens assembly to give the even shorter focal ratio of f2.8 and an improved stellar image quality across its 60-mm flat field. The author's image of M33, described in detail in Chapter 16, was taken using ASA's 8-inch Newtonian astrograph and shows superb stellar images right into the corners of the field of view.

5

The Cassegrain Telescope and Its Derivatives

Schmidt-Cassegrains and Maksutovs

The classical Cassegrain telescope utilises a parabolic primary mirror and a small, convex, hyperbolic secondary mirror inside the focus of the primary which directs the light cone through a central aperture in the primary to a focal point beyond. Such a configuration is ideal for large telescopes, as the often-heavy instrumentation is located where the major weight of the telescope lies. As the light path is folded within the telescope tube, the instrument is quite short in relation to its focal length. The convex secondary mirror multiplies the focal length by what is termed the 'secondary magnification', M, which is the focal length of the system divided by the focal length of the primary. One result of a configuration with a high secondary magnification (to yield a compact system) is to give significant curvature of field.

5.1   A Planetary Cassegrain

Relatively few pure Cassegrain telescopes are in use by amateur astronomers; variants of the design, described later, are far more popular. However, one pure Cassegrain has recently come onto the market aimed specifically at planetary observing and imaging which could, perhaps, give these telescopes a little more prominence. The obvious design aim would be to make the secondary mirror as small as possible, and this is achieved by having a primary mirror of higher focal ratio than normal and hence a secondary mirror with a smaller magnification ratio. The telescope would thus be longer than a typical Cassegrain.

Called the Altair-Planeta 250-mm Cassegrain and manufactured for the UK company Altair Astro, the open tube truss design has a focal ratio of f18, giving a focal length of 4,500 mm (Figure 5.1). This means that the central obstruction, at 25.5%, is far smaller than that for typical Cassegrains or Schmidt-Cassegrains, so minimising the loss of micro-contrast as compared with a refractor. The design aims to provide a highly corrected and fully illuminated field. The downside is that the field of view is small – no more than 0.57 degree with a 2-inch eyepiece and 0.34 degree with a 1.25-inch eyepiece. This is, however, perfect for planetary imaging with a webcam, as the

Figure 5.1 Altair-Planeta, 250-mm truss tube Cassegrain. (Image: Altair Astro)

focal ratio of f18 is very close to the theoretical ideal of ~f20. (The reasoning behind this will be discussed in Chapter 11 on planetary imaging.)

5.2 Variants of the Cassegrain Design

Two variants of the classical Cassegrain are used by amateur astronomers: the Dall-Kirkham and Ritchéy-Cretien. The Dall-Kirkham uses a prolate ellipsoid primary allied to a spherical secondary, which is easier to fabricate than the hyperbolic secondary of the classical Cassegrain. It does, however, suffer from greater off-axis coma, making it suitable only when large fields of view are not required or a corrector lens is included in the optical path. Superb examples are made by Takahashi in its Mewlon range of 210-, 250- and 300-mm-aperture telescopes. They use an oversized primary mirror (so that off-axis objects are not vignetted) which is figured to an accuracy of λ/20, giving a wavefront error of λ/10 – superb. These have focal ratios of ~f3 with secondary magnifications of 4, giving overall focal ratios of ~f12. Their central obstructions are around 30%. Imaging is best done when a focal reducer and field-flattening lens are used to give focal ratios of ~f9.2, when fields of 1.2 degrees, 1 degree and 45 arc minutes are obtainable with the 210-, 250- and 300-mm-aperture telescopes respectively.

Optimised Dall-Kirkham telescopes are also available; these include the Orion Optics (UK) ODK f6.8 range of 10- to 16-inch telescopes. They have an integrally fitted corrector lens to provide a wide flat field for visual and photographic use.

Another variant, the Richéy-Cretien design (Figure 5.2), is now use in virtually all large professional telescopes, including the Hubble Space Telescope. They use two hyperbolic mirror surfaces which are chosen to eliminate coma, so allowing a relatively large field of view, and are thus highly suitable for astro-imaging. The spot sizes of stars away from the axis get larger but retain circular shapes, which are more pleasing than comatic images. Until recently, Richéy-Cretien designs have been very expensive – even with small apertures – but modern computer-controlled mirror-grinding technology has made them cost competitive with other designs. They have relatively large secondary mirrors, so reducing their micro-contrast, but are ideal for wide-field, deep-sky observing and astro-photography.

Figure 5.2 Richéy-Cretien optical design (Image: Starizona) and RC Optical Systems 12.5-inch, f9 Richéy-Cretien (Image: RC Optical).

5.3 Catadioptric Telescopes

The term 'catadioptric' simply applies to telescopes that use both mirrors and lenses to form the image. One of the two most common designs, called the Schmidt-Cassegrain, is a development and combination of two other telescope designs. The first is the Cassegrain telescope and the second is the Schmidt camera.

The Schmidt Camera

The object of this well-known design is to produce wide-field images that are used for survey purposes and, for radio astronomers, source identification. To allow for a very wide field of view a spherical mirror is used but, if this were used without a correcting element, the images would be disastrously affected by spherical aberration. Bernard Schmidt solved this problem by designing a thin corrector lens, placed at the centre of curvature of the primary mirror, to correct for the spherical aberration. The camera is thus able to produce superb stellar images over a wide field. The design does, however, give rise to a significantly curved, convex focal plane, so photographic plates having a suitably formed curved surface are used to make the exposures.

Fritz Zwicky was instrumental is acquiring an 18-inch and then the 48-inch Schmidt cameras for the Mount Palomar Observatory. The latter produced the National Geographic Sky Survey which produced two sets of 6-degree square plates, one plate being more sensitive to red light and the other to blue; these have been a major resource in the world's observatories. An identical telescope located at Siding Springs in Australia was able to complete the all-sky survey for the southern skies. A digitised version of the survey is available on the Web, and a description of its use for amateur astronomers is given in Chapter 15 on CCD imaging.

5.4 The Schmidt-Cassegrain Telescope

The Schmidt-Cassegrain telescope, or SCT, has become exceptionally popular amongst amateur astronomers over the past 40 years or so since Celestron introduced the highly popular 'Celestron 8' 8-inch-aperture SCT in 1970 (Figure 5.3). A

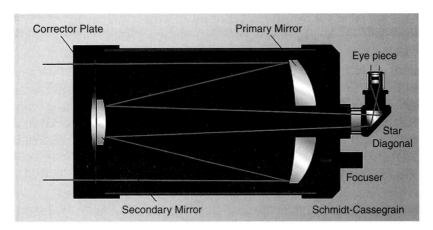

Figure 5.3 The optical design of the Schmidt-Cassegrain. (Image: Starizona)

wide range of SCTs are now made by Celestron and Meade. As their name implies, they use the Cassegrain configuration but have a spherical primary mirror allied to a Schmidt corrector plate to correct for spherical aberration that it would produce. In the configuration used in commercial SCTs, the corrector plate also supports the secondary mirror so there will be no spider diffraction effects. This configuration gives a very compact design – one of the SCTs' much prized assets – but does give rise to quite significant curvature of field. This is no real problem for visual observing but, for imaging use, a focal-ratio-reducing and field-flattening corrector lens is added near the focus. This also reduces the focal ratio from f10 to f6.3, so exposure times will be less.

There is a fundamental problem faced by all Cassegrain designs in that the observer is looking through the primary mirror towards the sky. There is thus the problem of preventing skylight from directly reaching the focal plane and so drastically reducing the overall image contrast. If a large secondary mirror is used this can, by itself, prevent any light from directly reaching the eyepiece, but large secondary mirrors will reduce the micro-contrast of the image. To minimise this effect, the designer will try to keep the central obstruction formed by the secondary to a minimum but, in a typical SCT, this is still rarely less than about 34%. Even with a large secondary, light scattered from the sidewalls of the telescope tube could still enter the eyepiece and so an internal baffle tube is added, which limits the light reaching the field of view to only that reflected by the secondary mirror and thus eliminates light scattered from within the tube assembly. However, an efficient baffle tube will tend to cause vignetting towards the outer edge of the field of view, and therefore the designer has to make an appropriate compromise between the overall contrast of the image and the reduction in illumination of the outer parts of the field of view.

It is vital that the interior of the baffle tube be very well blackened; otherwise, internal reflections from objects such as bright planets or stars within or just outside the field of view can give rise to circular arcs of light. I have, in the past, reviewed one faulty SCT where the first centimetre or so of the central baffle had failed to be blackened and found that unless Jupiter or Saturn was in the exact centre of the field of view a bright circle of light was seen extending from them.

One interesting point is that the corrector plate refracts on-axis light falling close to its edge away from the optical axis. If these rays are to be intercepted by the primary mirror, it must be somewhat larger in diameter than the corrector plate. For a standard 200-mm, f10 design, which has a corrector plate 200 mm in aperture, the primary mirror has to have a diameter of 201 mm – not too significant. However, if off-axial rays which would illuminate the outer parts of the field of view are considered, a primary mirror some 211 mm in diameter would be required if this is not to be an additional cause of vignetting. Looking in detail at the specifications of the Celestron and Meade 8-inch SCTs shows that whilst Meade uses oversize mirrors, Celestron does not. Celestron SCTs will thus suffer a touch more vignetting of the field of view than those made by Meade.

In the majority of SCT designs the primary mirror has a focal ratio of f2 and the secondary has a magnification ratio of 5, giving the overall focal ratio of f10. However, Celestron produces one SCT having an aperture of 9.25 inches (235 mm) with a primary of f2.5 and a secondary having a magnification of 4 – so giving the same overall focal ratio. At the expense of a longer tube than would otherwise be required, this design gives a flatter field of view than the standard design and shows less coma and off-axis astigmatism. It is sometimes said to be the SCT for those who do not like SCTs!

Mirror Shift

In the majority of Schmidt-Cassegrains, the image is focused by moving the primary mirror away or towards the secondary rather than having a focuser at the rear of the telescope. The mirror thus has to slide up and down a track and, as it slides, has a tendency to shift very slightly in its orientation with respect to the optical axis of the telescope. This causes the image to jump within the field of view. When one is visually observing relatively wide fields this is no real problem, but when one is attempting to focus an image of a planet onto the small sensor of a webcam for planetary imaging, this can be a problem. This 'mirror flop' could also ruin a long-exposure image, as the mirror might move as the orientation of the telescope tube changes over time. To overcome the first of these problems, a focuser can be added to the rear of the tube assembly so that, once approximate focus is made using movement of the primary mirror, final focus is achieved using the focuser. The SCTs recently made by both companies have far less mirror shift than earlier models, but it is still present to some extent, and in some of the very newest designs the mirror position can be locked in position once focus is found.

Cool-Down Times

As one can see from looking at the optical design, the light traverses the optical tube assembly three times. It is thus vital that before any serious observations begin the air inside the closed tube has had time to cool down to the outside temperature. If not, the image quality will be seriously degraded with bloated star images. (The way to check on progress towards this goal is to observe the out-of-focus image of a bright star: if the telescope is still cooling down, it will appear to be 'bleeding', with streamers extending from the edge of the Airy pattern.)

Two obvious aids to achieving thermal stability are to keep the telescope in a cool location, such as a garage, which will be closer to the outside temperature, and to bring the telescope out into the open air an hour or so before observations are intended to begin. To speed up the process, it is possible to purchase a blower which fits in place of a 2-inch eyepiece and which injects filtered air down a tube towards the secondary mirror.

Dewing of the Corrector Plate

One problem that observers will find when using a SCT is that after an hour or so of observing (sometimes less in very humid conditions) the image quality will often deteriorate and they will find that dew has formed on the corrector plate. This is because, as the telescope faces the sky, the corrector plate radiates heat away and can reach the dew point temperature, allowing the water vapour in the atmosphere to form a thin covering film. Under very cold conditions one might even (as I have) find frost, not dew, covering the corrector plate!

The simplest way of staving off dewing up of the corrector is to extend the telescope tube with what is called a 'dew shield'. Its length extending beyond the corrector plate should be at least one and a half times its diameter. A corrector plate without a dew shield is 'seeing' much more of the sky and so cools more quickly than when a dew tube restricts the area of sky visible to it. Many commercial dew shields lie flat for easy storage and are simply wrapped round the tube assembly and fixed with Velcro fastenings.

If the air is not too humid, a simple dew tube may be all that is required but often, even when one is used, the corrector plate will eventually dew up. What can you do then, short of deciding that it was time to end your observing session? Never attempt to wipe it away: you could easily remove the coatings on its surface, which are so important! One possibility is to warm the dewed corrector plate with a portable hair dryer running off a 12-volt supply. (Do not consider using a mains dryer under humid conditions!) This takes only a few seconds and will not significantly alter the corrector's shape.

A better solution is to continuously apply a little heat to the periphery of the corrector plate. This can be done by using a dew shield which incorporates an integral heating element (such as those made by Astrozap for Meade and Celestron SCTs) or by wrapping a heating element around the telescope tube before adding the dew tube. These do require a significant current, and a separate high-capacity battery should be used. Kendrick, who pioneered this procedure, is amongst the several manufacturers that provide the dew heating strips and the controllers which can be used to control the amount of heat emitted by them. Do-it-yourself designs for both the heating strip

Figure 5.4 The author's 9.25-inch Celestron Schmidt-Cassegrain equipped with a Celestron dew shield, TeleVue Starbeam finder, Starlight Instruments Crayford focuser and (*right*) a Dew It Yourself dew heater surrounding the corrector plate.

and controller are available on the Web, and a UK company, Dew It Yourself, provides kits for easily making heating strips, given a suitable soldering iron. The company has introduced a new design for catadioptric telescopes where the heating element (in the form of a tube rather than a flat strip) is placed inside the telescope tube just in front of the corrector plate. It is said that that this design should require less heat (and thus less current) to prevent dewing. Figure 5.4 shows this design mounted adjacent to the corrector plate of the author's Celestron 9.25-inch SCT. The current is controlled using a HitecAstro single channel controller. This has two outputs, the second of which could be used to prevent dewing of a finder scope.

It should be pointed out that a dew tube should have a matt-black- or black-felt-lined interior to help prevent light scatter. Even under conditions where dewing up is not expected, it is very sensible to add a dew tube. It will act to prevent unwanted stray light from falling onto the corrector plate and the interior of the optical tube which may be scattered into the eyepiece, so reducing the overall image contrast. It may also prevent unwanted finger marks on the corrector plate!

There can also be a problem when a telescope tube is brought back into the house after an observing session, because it can also dew up at that time. This will clear as the corrector plate warms up, but it is best to bring the telescope into a cool room and not place a cover (which will trap the water vapour) onto the front of it before the dew has cleared.

Overall Contrast and Micro-Contrast

In the past, SCTs were rightly said to give images of lower overall contrast than other designs. However, with the introduction of multi-coated corrector plates and mirror surfaces – UHTC in the case of Meade and XLT in the case of Celestron – this is now far less of a problem. Because the required secondary obstruction is larger than that of other designs, typically 34%, the micro-contrast will be reduced as well, so SCTs may not be quite so good for planetary observing, where both overall contrast and micro-contrast are important. But this problem is overstated. At very worst, with visual use the effective resolution is halved compared with that of an apochromat refractor of the same aperture, so an SCT of 200-mm aperture will be equivalent to a 100- or 127-mm refractor – and no one says that they are no good for planetary observing! In addition, there will be no diffraction spikes, and their larger aperture means that one has more light in the image to help prevent image degradation at high magnifications. (Part of this undeserved reputation is that many SCTs are not well collimated – good collimation is more critical than in other designs.)

The situation with webcam imaging is even better. As described in Chapter 1, the central peak of the Airy disk is actually reduced in diameter as the central obstruction is increased and so, during the brief moments of excellent seeing captured by the webcam images, the effective resolution may even be enhanced – as evidenced by the wonderful images produced by planetary imagers such as the UK's Damian Peach.

Wide-Field Imaging with the Celestron Fastar® Capability

Many of Celestron's latest optical tube assemblies and some earlier ones allow for the secondary reflector to be removed so that, in its place, a corrective lens assembly can be used to correct for spherical aberration, coma, off-axis astigmatism and field curvature. A CCD camera can then be mounted on the assembly to provide a very fast f2 system for wide-field imaging. Starizona produces a range of Fastar® compatible correction lenses to which can be attached a variety of CCD cameras.

Improving the Schmidt-Cassegrain

When the SCTs were first designed, 1.25-inch eyepieces with their limited field of view were being used, so the off-axis image quality, which is limited by coma and (for imaging) flatness of field, was not as important as it is now. To make their telescopes better suited for wide-field visual observing and imaging, both Meade and Celestron have updated their designs with the aim of reducing off-axis coma to improve the quality of star images towards the edge of the field of view and also, particularly in the case of the Celestron design, to make the field flatter. The former actually makes more stars visible, as their images are 'tighter', and the latter has become important with the large imaging sensors now being employed in DSLRs and CCD cameras. The two companies have, however, tackled the problem in different ways.

In the Meade ACF (Advanced Coma Free) series of optical tubes, the secondary mirror, rather than having a spherical surface, as in the standard design, is given a hyperbolic surface as used in the Richéy-Cretien (RC) design. The use of a different corrector plate than in the Schmidt-Cassegrain design coupled with the spherical primary mirror gives the same effect as a hyperbolic primary – also used in an RC design. So the resultant optical configuration gives a similar result to an RC telescope, which, as well as essentially eliminating coma over the field of view, helps to flatten the image plane, so allowing larger CCD sensors to be used for imaging. To eliminate the problem of mirror shift when focusing, Meade has made available a 'zero image-shift microfocuser' that fits between the base of the mirror assembly and the eyepiece. Once one has neared focus by adjusting the primary mirror position and locking it in position – so there will be no further mirror movement when observing – the fine focus can then be achieved with the microfocuser whose position is controlled by the telescope control handset. This is provided as part of the 16-inch system and is an optional extra for the smaller tube assemblies.

A different approach has been taken by Celestron, which has left the primary and secondary mirrors spherical and utilises a standard SCT corrector plate. To achieve the desired objectives of eliminating the off-axis coma and significantly flattening the image plane, Celestron has instead incorporated a dedicated optical assembly built into the baffle tube in what it calls an 'aplanic Schmidt-Cassegrain'. The two objectives have been well achieved, and the tube assemblies incorporate a number of new features as well: flexible tension clutches hold the mirror in place and prevent image shift as the telescope orientation changes, so keeping the image centred in the eyepiece (or sensor) whilst micro-meshed filtered cooling vents located on the rear cell allow hot air to be released from behind the primary mirror. Dedicated reducer lenses for each size of telescope are available to reduce the effective focal ratio to f7, thus giving shorter imaging exposures and wider fields of view. The EdgeHD optical tubes remain Fastar® compatible for ultra-fast, f/2, wide-field imaging, so each tube has been fitted with a removable secondary mirror to allow the Fastar® optics to be installed.

5.5   The Maksutov-Cassegrain

As the aspheric surfaces of the Schmidt-Cassegrain made the telescope difficult to manufacture (until a neat trick was invented by Celestron), opticians tried to design a corrector plate which would correct for the spherical aberration caused by using a spherical primary mirror but which would incorporate only spherical surfaces. The solution, which was first published by the Russian Dmitri Maksutov, was to use a 'meniscus corrector', which is a strongly curved negative lens having a very low power. Such a lens introduces spherical aberration of opposite power to that of the primary mirror and so largely eliminates it. The resultant telescope design has become known as the Maksutov-Cassegrain or, more commonly, the Maksutov. Maksutovs are very widely used in telescopes of 90- to 127-mm aperture, but the corrector plate becomes very thick and heavy for larger apertures, with few being produced with apertures greater that about 175 mm.

In 1958 John Gregory published a Maksutov design where the secondary mirror was simply an aluminised spot at the centre of the inner surface of the corrector plate, which made it very cost effective. His original design had a very high focal ratio of f23, but a year later he produced an improved design with improved colour correction and a less extreme focal ratio of f15. This design is known as a Gregory-Maksutov. An alternative to Gregory's design is to replace the aluminised spot with a separate convex mirror. This more complex design was introduced by Harrie Rutten and so is

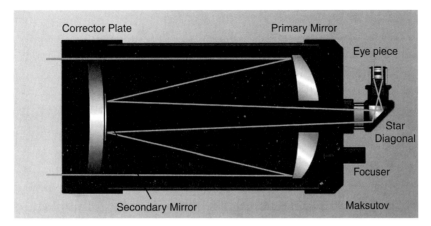

Figure 5.5 The optical configuration of a Maksutov telescope, in this case a Rutten-Maksutov. (Image: Starizona)

often called a Ru-Mak (Figure 5.5). It gives the optical designer an additional degree of freedom, as this mirror can have a different radius of curvature than the internal surface of the meniscus corrector. This extra degree of freedom enables the Maksutov to provide better off-axis images and gives a flatter focal plane, though the optical tube is longer than that in the Gregory-Maksutov. In some designs the effective focal ratio is reduced to f10, so allowing for a wider field of view.

The Gregory-Maksutovs have been made in very large numbers by Meade; they are now available with apertures of 90 and 125 mm and used in Meade's very popular ETX range of telescopes. Celestron also produces 90- and 127-mm Maksutovs as part of their SLT range.

The Russian company Intes-Micro produces several models ranging in aperture from 5 to 10 inches under the Alter brand name. They have superb optics with wavefront errors better that $1/6\ \lambda$ and hence produce excellent images, but their cost is substantially more than that of SCTs with equivalent apertures. The Alter range has internally baffled tube assemblies and air outlets surrounding the corrector plate to allow the interior to cool down more quickly. I own an Alter 500 (5-inch aperture) Ru-Mak and it is one of my favourite 'get up and go' telescopes.

5.6 Telescopes with Sub-Aperture Correctors

Though the vast majority of catadioptric telescopes use full-aperture correcting plates, it is perfectly feasible to either place a correcting element between the primary and secondary mirrors – the light passing through the element twice – or to include a lens assembly within the baffle tube. There are a number of such designs, two of which are manufactured by Vixen. In the VMC (Vixen Maksutov Cassegrain) series of telescopes with apertures of 95, 110, 200 and 260 mm, a meniscus corrector plate is mounted in the light path in front of the secondary mirror, so this design could also be called a 'sub-aperture' Maksutov. The dewing problems associated with full-aperture

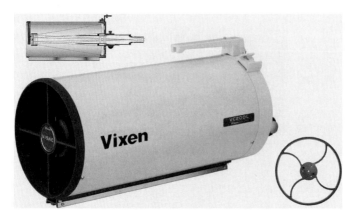

Figure 5.6 The 8-inch, f9 Vixen VC200L showing the three-element corrective optics mounted within the baffle tube. (Images: Vixen Optics) *Lower right*: A curved spider to reduce diffraction spikes.

corrector plates are avoided and, having an open tube assembly, the telescope will cool down to the ambient air temperature more quickly. (I would still advise the use of a dew tube to minimise the stray light that could enter the telescope tube.) However, bright stars will show diffraction spikes as the secondary assembly is supported by a spider. Primary mirrors of ~f2.5 are used with secondary magnifications of ~4, giving overall focal ratios of 10–11. This is typically less than that of Maksutovs of the same apertures, so giving well-corrected, somewhat wider fields of view. In the case of the VMC200L, the primary mirror is fixed – so cannot shift during long CCD exposures – and focusing is carried out with the use of a rack-and-pinion focuser behind the primary mirror, so that there will be no image shift during focusing.

Vixen also manufactures the VC200L reflector, which is an 8-inch, f9.0, highly corrected telescope ideal for astro-imaging (Figure 5.6). The design features a high-precision sixth-order aspherical primary mirror, a convex secondary mirror and a triplet corrector lens mounted within the baffle tube. This complex design provides point-like star images (smaller than 15 microns) out to the edge of a full frame (36 mm × 24 mm) CCD camera and boasts a 42-mm fully illuminated field. A dedicated f6.4 focal reducer is available to reduce imaging exposure times. Like the VMC200L it has a fixed primary mirror and rack-and-pinion focuser with their attendant advantage. The central obstruction of 40% makes it somewhat less suitable for planetary imaging, but it is designed primarily for wide-field imaging, where this is of no consequence. The spider which supports the secondary is rather thick and gives rise to prominent diffraction spikes. Since having recently acquired one of these interesting telescopes, I have obtained a replacement spider which incorporates curved vanes to improve the stellar images.

The Russian Novosibirsk Instrument-Making Plant produces three TAL designated telescopes of 150-, 200- and 250-mm aperture which are termed Klevzov-Cassegrains. They have spherical mirrors allied to the use of a sub-aperture meniscus correcting lens, combined with a Mangin mirror-lens secondary (a meniscus that has its back side aluminised to function as a mirror). These are supported on curved spider vanes to make the spider diffraction effects less noticeable. The primary mirror is fixed, and focusing is carried out using a rack-and-pinion focuser, as in the VMC200L. Their optical performance is said to be better than that of the standard Schmidt-Cassegrain (though not perhaps compared with the new ACF and High Edge designs), providing 14-micron star images out to a distance of 10 mm from the optical axis, and they are very cost competitive.

6

Telescope Maintenance, Collimation and Star Testing

Collimation is the process that is used to align the optics of a telescope in order to provide images of the highest possible quality, whilst star testing is a way of testing the optics of a telescope. These two topics are combined in this chapter because star testing can show that a telescope needs collimation and also provides the way of making the final collimation adjustments. Star testing can also highlight other optical problems and in some cases allow them to be corrected but, be warned, it is exceedingly sensitive and very few telescopes, though providing excellent images, will give perfect star tests. In more complex optical designs, star testing might well give somewhat odd results and will not be a good indicator of the telescope's optical performance.

Refracting telescopes rarely, if ever, need any maintenance but, after a year or two, the primary mirror of a Newtonian or the corrector plate of a Schmidt-Cassegrain may need cleaning. This is not too frightening a prospect, provided that it is done carefully, so let me first describe how best to do it. If the mirror surface is in really bad condition, with the reflecting surface beginning to erode, this might be the time to have it re-coated. If so, it is worth having one of the new multi-coated, high-reflectivity coatings applied, such as the Hi-Lux coatings provided by Orion Optics in the UK. Not only should such coatings last for many years, the overall contrast of the telescope will be significantly improved.

It is obviously necessary that one keep the front surface of a lens or corrector plate free of dew when observing, but what is not quite so obvious is that dew can form on either of these when the telescope is brought into a warm house. To prevent this and slow down the warming process, I wrap the telescope in a blanket before bringing it inside. This averts a rapid shock to the glass. If the temperature outside is near or below zero, I also shroud the telescope when first taking it outside to slow the initial cooling.

6.1 Mirror and Corrector Plate Cleaning

The cleaning part is the same for both. Here are a series of steps that should achieve a good result if the mirror is no more than 350 mm in diameter.

1. Prepare the area around a kitchen sink making sure that the sink and surroundings are scrupulously clean – abrasive cleaners can easily scratch a mirror! It is good to use a double sink if possible. The sink bottoms should be padded with a well-worn towel or rubber mats for the mirror/ plate to rest on or, during rinsing, to provide a soft landing should it be dropped. Fill one of the sinks with about 4 inches of lukewarm water and add a squeeze of washing liquid with, in the case of a really dirty mirror, some isopropanol.

2. Remove the primary cell or corrector plate from the telescope. In the case of a mirror, the mirror cell may well be difficult to remove from the telescope tube. To guard against it suddenly coming loose, place a pillow under the telescope to provide a soft landing! Note that Dobsonian telescopes will become unbalanced when the mirror cell is removed, so the front of the tube should be supported. In the case of a corrector plate it is vital that it be replaced with the same orientation; make a mark on its carrier and an adjacent part of the telescope tube so it can be replaced appropriately.

3. Remove the mirror from its cell and (though it should not really be necessary) apply some marks so, again, it can be replaced in its original orientation. It is a good idea to wear rubber gloves to provide a better grip and prevent any fingerprints from being placed on the surface, handling the mirror or plate as far as possible by their edges. Allow them to reach room temperature to prevent any thermal shock!

4. Use a mixer tap if possible to produce a stream of lukewarm water (or otherwise prepare a bucket of lukewarm water) so that with the mirror/plate held at an angle within the other half of the sink the can be rinsed thoroughly to remove any surface particles – this should not be hurried.

5. Then lower the mirror/plate into the prepared soapy water and move it around for a couple of minutes before leaving it to soak for a further 10 minutes. Lifting it out of the water, rinse it with lukewarm water to remove any trace of the detergent. Inspect its cleanliness and repeat if necessary.

6. Whilst holding the mirror/plate at a slight angle under water, use pads of surgical cotton to remove any remaining film or spots by gently sweeping the pads across the surface. New pads should be used for each sweep across the mirror/plate. Then rinse with a stream of lukewarm water from the tap.

7. A final rinse should now be done with about a gallon of distilled or de-ionised water with, preferably, a drop or two of Kodak Photo-Flow or, failing that, a clear washing-up liquid, the mirror being held near vertically. Then hold the mirror vertical to allow the water to drain off, and 'spot' of any remaining droplets with the corner of a paper towel, trying not to touch the surface.

8. Finally, place the mirror/plate face up on a soft surface (perhaps a second well-worn towel) to dry thoroughly and place sheets of soft tissue over it to prevent dust from falling onto it. When it is completely dry, replace it in the telescope.

The telescope will then need to be re-collimated, as described later.

6.2 Collimation

One good thing about refractors is that they rarely, if ever, need collimating and the majority do not even provide the capability of doing so. This is one reason that refractors are good telescopes for beginners. Newtonian and Schmidt-Cassegrain telescopes are most likely to need collimation, and one reason that Schmidt-Cassegrains have a somewhat undeserved reputation for providing less than optimum images is that they are out of collimation and their owners are somewhat afraid to collimate them in case things get worse! Collimation is a somewhat unnerving process the first time it is carried out but becomes far easier with practice and can make a real difference to the image quality. Some of the world's top astro-photographers may collimate their telescopes several times per night to maintain the highest-quality images. I hope that the following notes will help you overcome any fear of collimating your telescope so that you can get the very best out of it.

Collimating a Newtonian Telescope

How can you tell if your telescope needs collimation? During an observing session simply defocus the image of a bright star (this is star testing). As the image is taken further out of focus, a shadow of the secondary mirror and its spider will appear in the centre of the image, and if the telescope is perfectly collimated this should lie in the exact centre of the star image, as shown in Figure 6.1a. If the shadow is only slightly offset, the collimation will not be too bad and the final collimating adjustments can be made, as described under the section 'Final Collimation Steps'. If the offset is very bad (Figure 6.1b), then the preliminary stages of collimation should be carried out in daytime. Given a dark night with no visible stars, it is possible to use an artificial star such as that made by Geoptic. It produces a 'star' that is 50 μm in diameter, which must be placed 30 metres (100 ft) or more away from the telescope.

A very quick daytime test to see if the collimation is well out is to look through the focuser with your eye central in the field of view. Ideally, the whole – somewhat complex – image should appear symmetrical and your eye should be visible in the dead centre (Figure 6.1c).

I will outline the procedure using my 8-inch, f6 Newtonian telescope as an example. At this point it should be pointed out that the lower the focal ratio, the more critical the collimation becomes – an f4 telescope needs very precise collimation to give acceptable images. The collimation process can be carried out visually or, for Newtonian telescopes, aided by a laser collimation device.

The first thing to check in either case is whether the secondary mirror is dead centre in the optical tube. (This will not often be out.) One can simply measure the distance between the edge of the secondary support along each vane to the interior of the tube or, better, make a circular card whose diameter is that of the interior of the telescope tube and drill a hole in its exact centre. This should lie exactly over the central screw of the secondary holder. It may be necessary to remove a cover so that this central screw (which is used to set the secondary mirror on the axis of the focuser) will be

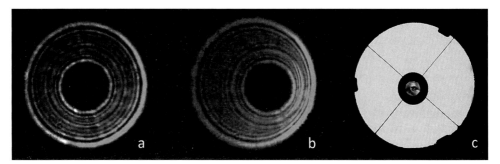

Figure 6.1 Tests to determine whether collimation is necessary.

visible. In the case of Newtonians of short focal ratio, the secondary mirror may be slightly offset in the direction away from the focuser. The telescope manual should tell you how much this offset should be. If the centre of the secondary support is not in the correct location, the screws at the end of each vane can be suitably adjusted.

Visual Collimation

There is one accessory that is required to start the collimation process, and this can be made from a 35-mm film canister (somewhat rare these days!) with a hole about 5 mm in diameter made in the centre of its base. It is called a 'collimation cap'. An interesting variant of the film canister device, the Easytester II, is made by Jack Schmidling Productions, Inc., which incorporates a mirror on the inner side of the eye aperture (the effect of this will be described later). A third device used to carry out collimation is a 'Cheshire eyepiece' but it is probably not best to use one for the initial stage, as its field of view is often not quite large enough to observe the whole of the primary mirror. All three are shown in Figure 6.2. The objective of all three devices is to ensure that the eye observes the interior of the telescope tube from the exact centre of the field of view.

The telescope should be pointed towards a bright, but not glaring wall, and it is sensible to have the telescope horizontal so that if an Allen key or screwdriver being used to adjust the secondary mirror is dropped it cannot fall into the tube and damage the mirror!

Aligning the Secondary Mirror

This may well not be necessary, as the majority of collimation adjustments need to be made to the primary mirror but, if necessary, this must be done first. The first requirement is that the secondary mirror appear dead central and circular as seen from the centre of the field of view. A simple film canister is best for this process. As seen through the collimation cap the secondary mirror outline should be seen dead centre within the outline of the focuser tube. If not, the central secondary adjustment screw may be needed to move the secondary mirror up or down the telescope tube. It may be necessary to loosen the three screw or Allen bolt tilt controls before this can be

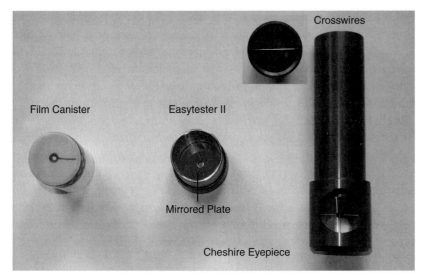

Figure 6.2 Devices to aid collimation. The Cheshire eyepiece has cross wires at its inner end to aid alignment.

done. This adjustment will almost certainly rotate the secondary mirror, and it must be brought back so that it faces the collimating cap and its outline is circular. It may well be necessary to adjust the three tilt adjusters to achieve this. This is probably the most fiddly part of collimating a telescope but, happily, is rarely required. Very fine alterations to the tilt adjusters may also be required in the next stage of aligning the secondary with the primary mirror.

Aligning the Secondary Mirror with the Primary Mirror

Now attention should be turned to the primary mirror as it is viewed through the secondary. It should be possible to see the full extent of the primary mirror because, if not, the telescope has been badly designed and the effective aperture of the telescope will be reduced. (It has happened!) The objective now is to make very fine adjustments of the secondary so that the view seen through the collimation cap is symmetrical. Usually fine adjustment of the three tilt screws will achieve this, but it may also be necessary to make a small rotational adjustment. This is probably the most time-consuming part of the collimation process. Once achieved, it is probably worth checking that the secondary still appears central within the focuser tube. If not, a second iteration of these first two steps will be required. The tilt adjusters should be tightened but not excessively so.

Aligning the Primary Mirror

This is the final and relatively easy part of the collimation process, though with telescopes of long focal length it would be nice to have far longer arms; as this is not the case, however, a helpful friend can be a great asset. It is also very useful to have a central spot or, even better, a ring for use when a laser collimator is used, in the exact centre of the mirror. Many telescopes are now so equipped. At the rear of the telescope, possibly under a metal plate which will need to be removed, will be found three 'primary tilt adjusters', each with a locking screw. To allow adjustment of the primary mirror to align it along the optical axis of the telescope, these locking screws must obviously be loosened. This is where a Cheshire eyepiece can be useful. When it is placed in the focuser so that the 45-degree flat is well illuminated, one can observe its reflection in the primary mirror. By means of corrections to the primary's adjustment screws, the reflection is moved until it appears centred on the primary mirror's centre spot. This is where an assistant can be very useful, as you can observe the movement of the reflection through the Cheshire eyepiece whilst he or she makes the adjustments. Usually only two of the adjusters will need to be altered unless one runs out of their adjustment, when the third will need to be used.

Many mirrors have a spot at the precise centre of the image. If so, when an Easytester II is used, one will see a reflection of the hole in the centre of the small perforated mirror mounted in the eyepiece along with the spot at the centre of the mirror. The primary mirror is adjusted to make these two spots coincident when, as if by magic, the view through the eyepiece goes black. Even if it is not used for making the adjustments when there is no spot on the mirror, it does provide a quick check to see if the collimation is accurate at the end of the process, and it can also be used as a first check to see if a telescope needs to be collimated.

Using a Laser Collimator

Following the initial procedure (if necessary) to centre the secondary mirror in the focuser aperture, a laser collimator can be used to adjust both the secondary and primary mirrors. It does require there to be a circular ring at the centre of the mirror. With the laser collimator mounted within the focuser, adjust the tilt screws on the secondary mirror until the laser spot falls centrally within the ring at the centre of the primary mirror. But beware! With many focusers, there can be a real problem with the alignment of the laser collimator within the focuser. A test for this is to rotate the laser collimator and see if the spot moves in a circle. The spot really does have to remain fixed at the centre of the ring. It may be possible to align it accurately and then gently tighten the eyepiece retaining screws or ring. At somewhat greater cost than standard laser collimators, the HoTech (US and UK) or Altair-Astro (UK) laser collimator incorporates a mechanism to produce a tight fit within the focuser and so produces a well-aligned laser beam to overcome this problem (Figure 6.3).

The laser beam having been centred within the circular ring, the primary mirror is now adjusted so that the laser beam returns to the centre of the collimator. Unless the primary is well off alignment, the laser spot will be seen falling on the 45-degree viewing surface. (If it is not, remove the laser and use a collimation cap to bring it close to alignment.) Final adjustments are made to bring the laser spot to the central hole of the viewing surface, when it will disappear. The HoTech collimator gives the laser spot an extended cross to help make this final adjustment easier.

Final Collimation Adjustment Tests

When these collimation procedures have been completed, a star test is carried out, which will also allow some fine corrections to be made.

1. Select a star that is at least third magnitude or brighter and 60 degrees or more above the horizon to minimise the atmospheric seeing effects.
2. Centre the star in the field of view of a medium-power eyepiece as precisely as possible – preferably using a reticule eyepiece.
3. Defocus the image of the star until it fills about half the diameter of the field of view. A dark shadow (that of the secondary) surrounded by a series of diffraction rings will be seen at, or close to, the centre of the defocused star image.
4. Use the primary mirror adjustment procedure previously described to make the fine adjustments to bring the secondary shadow towards the centre of the ring pattern. Bring the defocused star image back to the centre of the field of view after each adjustment. When the pattern is totally symmetrical the telescope is well collimated.

Collimating a Schmidt-Cassegrain or Other Catadioptric Telescope

As previously indicated, very minor collimation errors can make a real difference to planetary images, so precise collimation is a real necessity. One good point is that, as the primary mirror cannot usually be adjusted, the only corrections that might be needed are made to the secondary mirror. However, there is a problem that is not found with a Newtonian: the secondary mirror is convex and has the effect of multiplying the focal length of the primary by the magnification factor, often 5. The effect is that any adjustments made to it are magnified five times over those that might be made to the secondary mirror of a Newtonian. So, when making adjustments to the secondary, *never* adjust the three adjusting screws or Allen bolts by more than one-eighth of a turn at one time. These are not the easiest adjustments to make and can be made easier by replacing the screws or bolts with a set of three adjusting screws with knurled, large-diameter tops such as those sold as 'Bob's Knobs' and which are available for all popular Schmidt-Cassegrains.

The collimation of a Schmidt-Cassegrain requires the observation of a bright star that is high in the sky when the seeing is good. This latter requirement is a problem, and it can be worth investing in an artificial star such as the Geoptic Fiber Optic Star that I use. It provides a 50-micron aperture whose brightness can be adjusted. It has to be located far enough away from the telescope that the light source will appear star-like. In the case of an 8-inch Schmidt-Cassegrain a suitable distance is about 30 m. Ideally this should be used outside over grass so that local thermal air currents do not cause the image to be distorted, and the telescope should have been allowed to cool down to ambient temperature.

The real or artificial star should first be centred in the field of view, observed at ~100 magnification and defocused. If the star image appears to be 'bleeding' then tube currents are still present and the telescope must be given more time to cool. If the seeing is good enough one should see a dark central spot (the shadow of the secondary mirror) surrounded by a number of concentric circles. It is likely that the circles will not be concentric.

Now is the time to make very small adjustments to the three adjustment screws of the secondary mirror. It is probably best not to use a star diagonal when collimating a Schmidt-Cassegrain, but then you are likely to be viewing a high-elevation star from a rather awkward position. If you have a webcam, DSLR, integrating video camera or CCD camera with a reasonably short download time, you can observe the defocused star image on a laptop computer whilst making the adjustments. This can be a great help, but I found the use of a webcam with a small sensor somewhat of a problem, as with each adjustment of the collimation screws the star image tends to move out of the field of view. The process became much easier when I used a Canon DSLR in live mode and controlled by the Canon capture software, as the field of view is far greater. As the collimation reaches its final stages, with only very small adjustments being made, the central part of the field can be magnified to allow a more detailed view of the stellar image.

I have found one problem when using an artificial star. One needs to use the slew controls of the mount to bring the star image back to the centre of the field of view after each adjustment is made. With my Losmandy GM8 it is not possible to disable the tracking, so the image drifts out of the field of view, requiring the drive to be switched on to make a correction and then immediately off again.

If a screwdriver is held in front of the corrector plate its shadow will also be seen. It should be pointed radially towards each of the three screws in turn. The screw that one should first adjust is that where the shadow of the screwdriver is shortest. Make no more than one-eighth of a turn in one direction. The pattern may get more symmetrical, implying that this is the correct way to turn the screw. If not, take the screw back to its original position and make one-eighth of a turn in the opposite direction. Each time a screw is adjusted, the star image will move and it must be brought back to the centre of the field of view. By trial and error – where the error gets less with practice – it should be possible to centre the shadow. Collimation is now pretty decent, but you can probably do better.

This requires the observation of a fainter star, magnitude 2–3, which also needs to be high above the horizon to minimise atmospheric turbulence. When an artificial star is used, the brightness can be reduced. The magnification should be increased to ~400 in the case of an 8-inch telescope or about three to four times the aperture of the telescope in millimetres. If the focuser is racked back and forth through focus, a bright central spot surrounded by a complex system of rings will be seen in the out-of-focus patterns. Again use a screwdriver to point to each adjusting screw and first adjust the screw where its shadow is shortest. This time the corrections should be even less.

Figure 6.3 HoTech 1.25-inch laser collimator (*left*), Geoptic artificial star (*upper right*) and HoTech SCT collimation system (*lower right*). (Images: HoTech & IM)

Finally bring the star to focus, when the central Airy disk will be seen surrounded by a first bright ring and then fainter ones. When the telescope is perfectly collimated, this pattern must be totally symmetrical and adjustments of no more than one-twentieth of a turn should be made to correct it. Unless the seeing is very good this final step can probably not be made, but if it is necessary high-resolution planetary observations where precise collimation is most important will not be possible anyway!

Laser Collimation

Because the secondary mirror is not usually exactly aligned on the optical axis and this cannot be adjusted, laser collimators should not be used to initially collimate a Schmidt-Cassegrain or other catadioptric telescope. Even when the telescope is perfectly collimated, the reflected beam will not lie on axis and will be typically 0.125–0.25 inch off-centre. There is one, somewhat complex laser system that will collimate a Schmidt-Cassegrain. Manufactured by HoTech, it is called the HoTech Advanced CT Laser Collimator and uses a four-laser system mounted within a target screen and a mirror insert placed within the focuser. Light from the lasers passes through the optical system and is reflected back through it by the mirror. It is thus a 'double-pass' system to double the effective accuracy. There is a fairly steep learning curve to its use. The initial procedure is to use the lasers to set the target screen at the appropriate distance from the corrector plate and to be on axis and precisely at right angles to the telescope's optical axis. This done, the final process is to adjust the secondary mirror using the reflected beams from three of the lasers to determine its movement. Though not trivial to use, the system does allow one to collimate the telescope in daylight and gives very good results.

6.3 Using Star Testing to Uncover – and Hopefully Correct for – Optical Faults

You first need a night when the seeing is very good and a perfectly collimated telescope. Track a bright star or observe Polaris and rack the focuser from a little inside the focus through the focus to a similar point outside. If the image of concentric rings remains perfectly circular and appears nearly identical on both sides of the focus, you have a superb scope! The contrast of the rings tells you how smooth the mirror is – nicely delineated rings are what you hope to see.

A Word of Warning

Following the arrival of a 127-mm, f7 apochromat refractor (whose Strehl ratio was greater than .96 and so near perfect), the night was clear and, though the seeing was not that good, I was able to make the webcam image of Jupiter that will be described in Chapter 11. At the end of the observing session I decided to do a star test on Betelgeuse well up in the south. The intra-focal Airy disk appeared fine, but the extra-focal image was mushy and I could barely make out any of the rings. After a brief moment of panic, as I had spent a considerable sum of money on this telescope, I remembered something that I had read several years previously. When one is observing the extra-focal image, the telescope is focused on a point between the star (which is effectively at infinity) and the telescope. Depending somewhat on the telescope parameters the telescope can actually be focused on the turbulent cells in the atmosphere, so amplifying the effects of 'seeing' and destroying the image.

6.4 Using Star Testing to Improve the Optics of a Telescope

There are a few optical defects which are fairly obvious and which, in some cases, can even be corrected.

1. If the slightly out-of-focus disk looks elliptical rather than circular and its long axis moves through 90 degrees as you move through the focus, the objective (usually a mirror) is suffering from astigmatism. This is often caused by the mirror clamps being too tight (so distorting the mirror) and you may well be able to eliminate the problem by easing them off.
2. If the inside-focus image has a bright outer ring, whilst the outside-focus image has a more diffuse look, the objective is suffering from spherical aberration and is under-corrected. If the inverse is seen, the objective is over-corrected. Virtually all scopes show some over- or under-correction, so do not be too alarmed. The test is exceedingly sensitive.
3. If the outer ring appears to have little spikes radiating from it, rather like whiskers, this is an indication that a mirror has a turned-down edge. If placing a circular mask to block off the outer few millimetres of the mirror removes this effect and

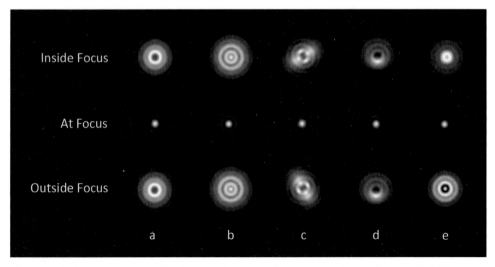

Figure 6.4 Star testing. (a) Perfect unobstructed aperture, (b) perfect with 34% obstruction (note that the first ring is brighter), (c) the effect of astigmatism, possibly due to a pinched mirror, (d) coma caused by an uncollimated telescope and (e) an over-corrected objective. (Images: Aberrator)

has a significant effect on the pattern that you see, then such a mask could permanently improve the quality of your telescope's images.

There are many Web resources – just Google 'star testing'. The freeware program 'Aberrator' will show you what you might observe and was used to derive the images of Figure 6.4. In addition, Harold Suiter's book *Star Testing Astronomical Telescopes* will tell you everything you could possibly want to know about star testing, and the Web page describing the book, www.willbell.com/tm/tm5.htm, includes some very useful star testing images.

6.5 Ronchi Testing

Another way of star testing makes use of what is called a Ronchi grating such as the 133-line grating mounted in a 1.25-inch barrel sold by Jack Schmidling Productions, Inc. The grating is placed at the focal plane of the telescope in place of an eyepiece, and a bright star, preferably near the zenith, is observed. The star will appear as a large bright disk with a pattern of parallel lines across it. As the position of the Ronchi screen is adjusted using the focuser, the line spacing changes as the grating is moved away or towards the focal point. The sensitivity of the Ronchi test is greatest when the grating is very near the focus point and the lines are farthest apart, with perhaps four to six lines crossing the stellar disk.

A perfect lens or parabolic mirror will give rise to a set of perfectly straight lines with no difference on either side of focus. The sensitivity can be increased with the use of a Barlow lens in front of the grating. If, when a Barlow lens is used, no 'bowing'

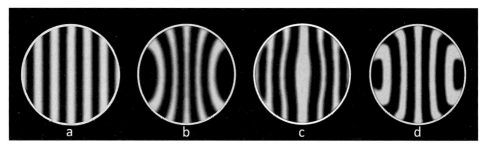

Figure 6.5 Ronchi test results as described in the text.

of the lines can be seen, it can be concluded that a fast scope, say, f4.5, will be within ¼ wave of full correction, whilst a telescope of f6 or greater will be ⅛ wave or better. One nice feature of the Ronchi test is that it can be used when the seeing is not good enough to do a star test. If the bowing is significant then a re-figuring of the lens or mirror is indicated.

Two other types of error can be determined from the test. If the lines are straight across the majority of the image but veer off close to the edge of the image, this is an indication that the edge is turned up or down. This is normally seen only with a reflecting telescope, where a turned-down mirror is relatively common. The way to improve such a mirror is to mask off its outer portion. If the lines are basically straight all the way across the image but include a symmetrical kink, the implication is that the overall figure is correct but that there is a 'zonal' error affecting part of the mirror or lens. There is no way of correcting such an error save having it re-figured. Finally, if the lines are straight but not well formed, this indicates a mirror which, though its overall figure might be correct, may have a rather rough surface.

The four images in Figure 6.5 indicate the types of results obtained with the Ronchi test. Figure 6.5a is perfect. Figure 6.5b, if seen inside focus, indicates the spherical aberration of a smoothly under-corrected optic or, if seen outside focus, an over-corrected optic. Figure 6.5c indicates a 'zonal' error, in this case towards the centre of the optic, whilst Figure 6.5d indicates a turned edge. If this pattern is seen inside focus the edge is turned up (rare), and if it is seen outside focus the edge is turned down – a common fault with amateur-made mirrors.

6.6 The WinRoddier Star Testing Analysis Software

An interesting program (but in the French language) has been developed to take a pair of intra- and extra-focus images and carry out an analysis to provide information about the quality of the telescope's optics. It is called WinRoddier (use Google to search for this *and* translate it). The procedure is a little tricky but well worth attempting if you have a webcam.

A night of good seeing is necessary when one takes two AVI sequences of the star when on either side of focus. The distance from focus must be noted and for most telescopes a distance of 4–5 mm is suitable, with 4.4 mm recommended by some. The

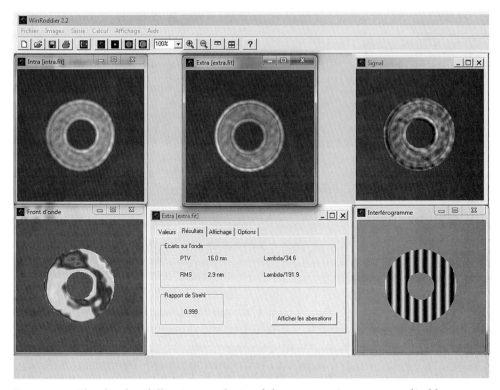

Figure 6.6 The display following analysis of the two test images supplied by WinRoddier. The intra- and extra-focal images and their combination ('Signal') are at the top, with the essentially perfect results shown below. (Image: WinRoddier Software)

exact distance is not critical, but it is important that the intra- and extra-focal distances be the same. There is a neat way to achieve this. Prior to the test, make a thin rectangle of cardboard 4.4 × 20 mm in size. To take the two images, first focus the webcam on the star and then use the cardboard finger to move the barrel of the webcam out by 4.4 mm and take the extra-focal image. Without adjusting its barrel position, bring the star back into focus. Then push the webcam barrel back home and take the intra-focal image. The way that this is done and how the files are processed in Registax are described in Chapter 11. The result will be the production of two TIFF images. These are then cropped in Photoshop so that the two images are approximately central in a suitable-sized image, perhaps 250 pixels on a side. They then have to be imported into the free imaging program called 'IRIS' and exported as FITS files – a process which gives the exact FITS file format expected by Roddier.

The program is downloaded from the WinRoddier Web site and opened. The 'Fichier' drop-down menu is opened and a new analysis sequence initiated by clicking on 'Nouveau test'. Then, in turn, the two intra- and extra-focal images are loaded into the software: 'Ouvrir l'image intrafocale, … extrafocale.' The 'Images' drop-down menu is then opened to allow, first, the images beyond the disk pattern to be

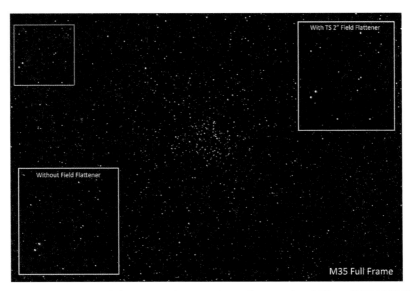

Plate 1.5 An image of the open cluster M35 taken with a Nikon D7000 DSLR with APSC sensor, with and without Telescop-Service 2-inch field flattener.

Plate 11.3 Single image of the Moon taken with a Nikon D7000 camera mounted on a 127-mm, f7 apochromat refractor.

Plate 11.8 Images of Jupiter and Saturn.

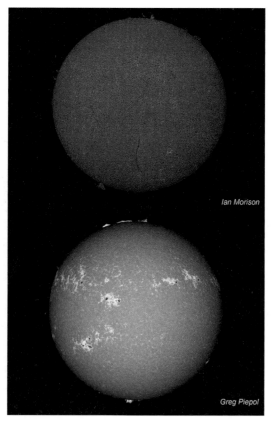

Plate 12.7 The solar disk imaged in the light of H-alpha (Ian Morison) and calcium K (Greg Piepol).

Plate 14.1 An 80-second total exposure image of the Plough (*top*); an enhanced image using a technique described in Chapter 18 (*bottom*).

Plate 14.12 An image of the Orion Nebula region combining an image taken with a Nikon D7000 with the red channel of an image taken with a modified Canon EOS 1100D added to enhance the H-alpha emission in the region.

Plate 15.7 The smoothed colour image (*top*) and the result of adding colour to the monochrome image of Figure 15.6 (*bottom*).

Plate 15.8 An image by Peter Shah of M31 taken with an 8-inch astrograph.

Plate 15.9 The colour image of M33.

Plate 15.11 Colour images of M51 (*top*) and NGC 4565 (*bottom*) imaged with the Rigel 14.5-inch Cassegrain.

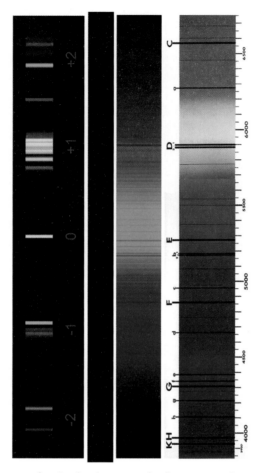

Plate 17.6 Spectrum of a single element of a fluorescent lamp (*top*). The solar spectrum taken using a needle to give the effect of a very narrow slit, along with a comparison spectrum (*bottom*).

masked off (and turn blue) with the command 'Extraire les plages', and then centred with 'Cocentrer les plages'.

At this point, the 'Saisie' drop-down menu is opened to enter 'Paramètres du test' and, in the box that is opened up, details of the telescope, its aperture and focal length, the distance of the two images from the focal point and the size of the CCD pixels in microns (typically 5.4) can be entered. In the 'Calcul' drop-down menu, click on 'Signal' to bring up a third image, which is a combination of the first two. Then click on 'Front d'onde' and, after a short time, a box opens which gives details of the optics such as the peak-to-valley (PTV) difference in nanometres and wavelength and the RMS accuracy (which will be less). The most important value is the Strehl ratio ('Rapport de Strehl'), as described in Chapter 1, Section 1.4.

If the button with vertical lines on a green background is clicked, a theoretical interferogram is produced. Figure 6.6 shows the WinRoddier screen display for the two test images downloaded from their Web page as FITS files.

7
Telescope Accessories
Finders, Eyepieces and Binoviewers

This chapter covers the two major accessories that are used with virtually all telescopes – finders and eyepieces – along with some possible upgrades, such as an improved focuser. A finder and, usually, a pair of eyepieces are supplied with all telescope/mount packages, but if an optical tube assembly is being bought by itself these may not be provided, so allowing you to choose those that will fit your needs best. When coupled with a 2-inch wide-field eyepiece, the short-focal-length refractors that are now in common use have sufficiently large fields of view that a finder may well not be needed. With the increasing use of computerised 'go-to' mounts a finder will be used only during the initial alignment on bright stars. This has made it possible for finders without 'magnitude gain' (i.e., not employing a small telescope) to be commonly used.

Over the years, telescope eyepieces have become increasingly sophisticated – but also more expensive. It is now very easy to spend more on a high-quality eyepiece than the initial cost of a telescope, particularly if it is bought second-hand, and this rather goes against the grain. It was only after I had spent a large sum buying my Takahashi fluorite refractor that I felt justified in spending a significant amount on eyepieces. They did not then seem quite so expensive! One general point with respect to eyepieces: the simple eyepieces that are often provided with beginner's telescopes, perhaps with just three elements, are not necessarily of low quality. The fact that fewer glass elements are used can be a good thing, in that there are fewer internal interfaces between the optical elements themselves and the air surrounding them, so minimising scatter and light absorption and thus giving higher-contrast images. 'Simple' eyepieces are much prized by planetary observers for whom contrast is so important. Their only real loss as compared with premium eyepieces is the fact that the available field of view will be smaller.

7.1 Telescope Finders

These are used to carry out a first, approximate alignment on a celestial object so that when one is viewing through the telescope with a low-power eyepiece the chosen object will appear in the field of view. Note that the title of this section is not 'Finder

Figure 7.1 Red dot finders.

Scopes', as the majority are no longer scopes! Let us have a look at the types of finders that are now available.

Red Dot Finders

These are now supplied with the majority of small telescopes, and some are shown in Figure 7.1. A red LED is observed reflected by a curved lens which projects a red dot floating above the sky. The brightness of the dot can be adjusted to suit the viewing conditions. The brightness control usually acts as the on/off switch as well, and I have often left the finder on so that the next time I use the scope the battery is dead. One manufacturer sells a version that automatically switches off after a set time and thus could save money in the long term. Some computerised scopes actually prompt you to switch the finder off after the initial alignment. The key aspect of these is that the position of the dot as seen on the sky does not change as the eye is moved. Those that are supplied with most scopes have a rather small circular aperture to look through. In some red dot finders the lens is semi-silvered, which makes the visible red dot brighter but can make the sky rather too faint to make out the stars. There are also more sophisticated red dot finders that have a far greater viewing window and these are somewhat easier to use.

The Rolls Royce of red dot finders is the beautifully engineered TeleVue Starbeam. Having a viewing window that is 39 mm in diameter and well recessed within a shield to minimise dewing, the Starbeam 'projects' a 10 arc minute diameter red dot onto the sky. It has one unique advantage in that a 'flip mirror' is incorporated so that the star field and red dot can be viewed from the side rather than on axis with the telescope – very useful for objects at high altitude. It is, however, quite large and heavy and suitable for use only with larger telescopes. Mounting brackets can be supplied for use with Newtonians, refractors and Schmidt-Cassegrain telescopes. Figure 5.4 shows one mounted on my Celestron 9.25-inch SCT.

Red 'Circle' Finders

Two variants of the red dot finder are very well regarded. These are the Telrad and Rigel finders (Figure 7.2). In contrast to the red dot types, these 'project' concentric

Figure 7.2 Rigel (*left*) and Telrad (*right*) 'red circle' finders.

circles onto the sky and are, perhaps, easier to use. The more substantial Telrad lies along the telescope tube, whilst the Rigel stands proud, making it somewhat easier to place one's head behind. I have one of the latter mounted on my 300-mm telescope and it performs very well.

All these finders have adjustment screws to align them along the axis of the telescope. This process can often be best carried out in twilight using a low-power eyepiece with the telescope to centre the top of a distant tree, pole or church spire in the field of view. The red dot or circle can then be aligned on it. However, at night it is easy to find the Moon or, by sighting along the telescope tube, a bright star or planet to use for adjusting the alignment.

Finder Scopes

These now tend to be provided only with larger and more expensive scopes and are essentially small refractors. They might be 6 × 30 or, more commonly, 8 or 9 × 50, and so have a field of view of a few degrees comparable to that of a pair of binoculars. They usually give an inverted image, which takes a little getting used to. In the image plane is mounted a glass reticule with etched cross wires which are used to align the scope. In our light-polluted skies these can normally be seen but will disappear under dark skies. One can usually centre the object in the field of view, but some have illuminated reticules which use a red LED to provide just enough light to make the cross wires visible. Until recently most such scopes have come mounted within two rings, each with three adjustment screws to make the initial alignment (Figure 7.3, right), but many now come in a mount which has one spring-loaded point of contact with the finder tube and two adjustment screws (Figure 7.3, left). Some have a right-angle

Figure 7.3 A straight-through finder (*left*) and right-angle, erect-image finder (*right*).

prism so that one observes at right angles to the telescope tube rather than along it. Often this is more comfortable to use, but those aligned along the tube make it easier to make an initial, approximate telescope pointing unless the object is high in the sky. At somewhat greater cost, one can even buy a finder scope giving an erect image and with an illuminated reticule!

Dewing Up

Just as the corrector plate of a telescope can dew up, so can the objective and even eyepiece of a finder scope. Heating bands for both can be purchased for use with a dew controller, and a suitable heating band is also available for use with a Telrad finder.

Laser Pointer Finders

The most recent finder type uses a green laser pointer mounted in a pair of rings (Figure 7.4). Such finders are very nice to use, as it is not necessary to bend down and one can instantly see where the telescope is pointing. They must be used with care and are forbidden at star parties, where the beam may be scattered into an astro-imager's telescope – astro-imagers do not like green lines appearing in their images!

There is one problem that I have found when using a laser pointer as a finder: they do not like the cold. There are only two solutions that I can offer. One is to not use a fixed mounting bracket as shown in the figure, but to make a trough, aligned along the telescope axis, that is mounted on the telescope in which the laser can sit in snugly when in use but that is otherwise held in a warm pocket. The second can be applied if the telescope has a dew heater system. A heating strip designed for a finder scope objective could be used instead to lie along the side of the laser.

Some General Comments

With the use of computerised scopes, finders are (hopefully) needed only for the initial alignment which will use bright stars or, in the case of some Celestron scopes,

Figure 7.4 A laser pointer finder and mount.

even the Moon and planets. A simple red dot finder is then really all that is necessary. When a non-computerised mount and 'star hopping' are used, a 9 × 50 finder scope can be very useful and, when using my 4-inch refractor on an equatorial mount, I mount a red dot finder in parallel with the finder scope. The red dot finder gives a rough alignment and then, when I am observing a faint object, the finder scope can be used to give a more precise alignment.

7.2 Focusers

Though not an accessory as such, with one always part of a supplied telescope, a focuser can be a useful addition to a Schmidt-Cassegrain telescope and is often bought as an upgrade to the original focuser. Three types of focusers are shown in Figure 7.5.

Helical Focusers

These are not as common now as they once were but can give a very smooth and backlash-free movement that can work very well to accomplish the fine focusing of a CCD camera. Many have a scale around their edge which acts as a vernier to allow the very precise re-positioning of a CCD camera's focus. There are two types, both being focused by the rotation of a ring as with a camera lens. With the simplest type the eyepiece rotates as the focus is changed, whilst in the more complex type the orientation of eyepiece and camera will not change. One excellent application is to add one to the rear of a Schmidt-Cassegrain telescope. Approximate focus is obtained by movement of the primary mirror, and then the precise focus is reached with the use of the helical focuser. The important point is that this is then achieved without any annoying image shift, which can be a real problem when one is using a small CCD sensor such as found in a webcam. Helical focusers are available from Hutech/Borg and Teleskop-Service, amongst others.

Figure 7.5 A Borg helical focuser (*left*), a Starlight Instruments Crayford focuser (*centre*) and a Long Perng Q2G-A helical rack-and-pinion focuser (*right*).

Rack-and-Pinion Focusers

These used to be very common and are still often found on Newtonian telescopes. The standard rack-and-pinion focuser is a simple design and works well for visual applications. However, inexpensive versions can suffer from stiffness, excessive play and backlash. Stiffness can result from the gear and teeth of the rack-and-pinion system being engaged too tightly or from poor-quality grease having been applied to lubricate the system. The tightness can often be adjusted and the grease replaced. A poor fit between the drawtube and outer walls of the focuser can result in excessive play, which can cause the image in the eyepiece to shift as the focus is racked in and out, making precise focus very difficult. Some backlash is inherent in the design, as otherwise the gears would bind, so even good-quality focusers can suffer somewhat as a result.

However, some of the world's best telescopes use rack-and-pinion focusers made to very high standards such as those found with Takahashi refractors. Starlight Instruments offers a very high quality focuser using a very fine pitch helical rack and pinion that has the teeth of the rack and pinion cut at an angle. As several teeth are in simultaneous contact, extremely fine tolerances are required and so these are expensive to produce. The reward is that these focusers provide a very smooth movement with significantly reduced backlash. I was delighted when I found that my CFF telescope's 127-mm refractor was equipped with the company's 3-inch version, easily capable of supporting 6 pounds of imaging equipment. It is shown in Chapter 2, Figure 2.7. Revelation also provides excellent 2-, 2.7- and 3-inch helical rack-and-pinion focusers, which are beginning to replace Crayford focusers on a number of good-quality refractors.

Crayford Focusers

These are now perhaps the most common type of focuser, being found on virtually all new refractors. The focuser was invented by John Wall, a member of the Crayford Manor House Astronomical Society, who named it after his society and decided not to patent the idea. The design is similar in appearance to a rack-and-pinion focuser, but employs no teeth. Instead, a round axle is pressed against a flat plate on the side of the focuser drawtube, relying solely on friction to move the drawtube as the axle is turned. The opposite side of the drawtube is pressed against a set of ball or roller bearings against which it can move smoothly with very low friction. The pressure applied by the axle can be adjusted for smoothest operation and, once focus is found, the drawtube may be locked in position to support heavy eyepieces or cameras. Crayford focusers can be used, as with helical focusers, to minimise the image shift found with Schmidt-Cassegrain telescopes. Very high quality versions are produced by Moonlight Telescope Accessories and Starlight Instruments.

7.3 Eyepiece Basics

A telescope produces an image of the sky or an object like the Moon in what is called its 'focal plane'. This lies at a distance from the objective lens or mirror given by its focal length, F. The purpose of the eyepiece is simply to allow one to observe this image in detail, rather like using a magnifying glass to see fine detail in an object. (Indeed, a magnifying glass could be used as an eyepiece, though it would not be a good one!) An eyepiece too has a focal length, say, f_e, which can vary from ~2.5 mm up to 55 mm. Eyepieces with shorter focal lengths allow one to observe the image in greater detail but also reduce the area of the image that can be observed at one time. The ratio F/f_e gives what is called the magnification. For example, a 10-mm eyepiece used with a telescope of focal length 1,200 mm gives a magnification of 120.

The Field Stop and the Apparent Field of View

Dependent on the design of the eyepiece, the image quality will become poor at some angular distance from the optical axis. The designer will thus introduce a field stop into the design to limit the field of view to that which gives reasonable star images at its edge. (However, the image quality will still tend be poorer at the edge than at the centre.) The field stop gives a sharp edge to the field of view and, when the eye is suitably placed to observe through the eyepiece, this should be visible. In some eyepiece designs, the field stop is in front of the glass elements that make up the eyepiece and so can be seen, whilst in other designs it lies within the interior. The design of the eyepiece and the diameter of the field stop will determine what is called the 'apparent field of view' of the eyepiece, which can range from about 30 degrees up to slightly more than 100 degrees.

Normal eyepieces such as the Plössl have apparent fields of view of around 50 degrees. Wide-field eyepieces like the TeleVue Panoptic and Delos have apparent fields of view close to 70 degrees. Ultra-wide-field eyepieces such as the TeleVue Nagler have fields of view of slightly more than 80 degrees and, with the Ethos eyepieces, this extends to more than 100 degrees! One can hold up different eyepiece designs to the sky and, if one eyepiece is in front of one eye and another eyepiece is in front of the other eye, the difference in their apparent fields of view can be readily seen.

Eyepiece Barrel Size

For an eyepiece of given focal length, a design that can provide a wide apparent field of view will naturally have a larger-diameter field stop. However, it is pretty obvious that the diameter of the field stop cannot exceed that of the interior diameter of the eyepiece barrel. In the case of the standard 1.25-inch barrel eyepiece, this is ~27 mm. Until low-power, wide-field eyepiece designs became available, this limitation was not really a problem but, as such eyepieces became available, a larger, 2-inch barrel size was introduced to accommodate them. This allows the field stop to have a diameter of up to about 45 mm.

It is worth pointing out that, unless the field stop needs a diameter greater than ~27 mm, 2-inch barrels are not needed – 2-inch eyepieces are not 'better'. However, some of the latest ultra-wide eyepieces are quite large and heavy, so that even though they have a field stop of less than 27 mm they are mounted in dual diameter 'skirted' barrels so that they can be used with a 1.25-inch focuser but would be better supported in a 2-inch focuser.

Actual Field of View

When an eyepiece is used with a telescope, the apparent field of view will translate into an actual field of view (F_{ov}) on the sky. This is given approximately by the apparent field of view divided by the magnification, so if the 10-mm eyepiece mentioned earlier had an apparent field of view of 52 degrees then the actual field when used with the telescope of 1,200-mm focal length (giving a magnification of 120) would be of the order of 52/120 = 0.43 degree. If the diameter d of the field stop in the eyepiece can be measured, then a more accurate value can be obtained by dividing d by the focal length of the telescope (F) and multiplying by 57.3 to give the result in degrees:

$$F_{ov} = d \times 57.3 / F = \text{degrees}$$

If the field stop is internal to the eyepiece and so cannot be measured, one can observe a star close to the celestial equator (Mintaka, the top right star of Orion's belt, for example) and time how long it takes to cross a diameter of the field of view with the telescope stationary. As the sky on the celestial equator moves at 15 degrees an hour, the field of view (in degrees) is obtained by dividing the number of seconds (t) required to cross the field by 3,600 and multiplying by 15:

$$F_{ov} = t \times 15 / 3,600 = \text{degrees}$$

Eyepieces with wider apparent fields of view will enable the observer to see more sky at one time. This is a nice feature but one which can also actually help one to see more when observing globular clusters or compact open clusters. Why should this be? If I were observing the globular cluster M13 in Hercules, I would naturally select an eyepiece so that the cluster nicely filled the field of view. Let us suppose that I was using my refractor of 820-mm focal length. I would find an eyepiece to suit by gradually increasing the magnification. M13 has an apparent diameter of 40 arc minutes, so I suspect that I would choose an eyepiece with a field of view of about 1 degree. If I were using a Plössl eyepiece with an apparent field of view of ~50 degrees, I would find that it had a focal length of ~16 mm, as this would give a magnification of 51 and so an actual field of view of ~1 degree. Now suppose that I were using a TeleVue Nagler or equivalent eyepiece with an apparent field of view of 82 degrees. To give the 1-degree field, I would use a magnification of ~80, so the focal length would be ~10 mm. The key point is the increase in magnification. As the stars in a cluster are point sources of light, their brightness does not change as the magnification is increased. (The brightness depends only on the aperture of the objective.) However, the apparent brightness of the sky background caused by light pollution or sky glow *does* decrease as the magnification is increased, so enabling the stars to stand out against a darker background. This will not hold true when one is observing extended nebulae or galaxies, but the 'porthole on the universe' effect given by the ultra-wide-field eyepieces can still make their significant cost worthwhile.

Eye Relief

The second key parameter is what is called the 'eye relief' and is basically how close your eye must be to the eye lens of the eyepiece to see the whole of the field of view. (The edge of the field should be sharply defined.) With simple eyepieces the eye relief gets less the shorter the focal length of the eyepiece and will often not allow the observer to wear glasses when observing through them. (This is not a real problem unless one's eye is badly astigmatic.) Eyepieces with a long eye relief (16–20 mm) also tend to be easier for those unfamiliar with telescopes, so are very useful at star parties.

Curvature of Field

In the same way that telescope optics can suffer from curvature of field, so can eyepieces and this means that, if the stars in the centre of the field of view are in focus, those near the edge will not be and vice versa. Some eyepiece designs assume that a small amount of field curvature will be present from the objective and introduce it in an opposite sense to compensate.

Distortion

This is quite common, particularly in wide-field designs, and is more obvious when eyepieces are used with telescopes of short focal ratio. It is often noticed when one is sweeping across a star field when the rate at which the stars appear to move varies as they cross the field. This would also somewhat distort the image of the Moon if it nearly fills the field of view.

Chromatic Aberration

The centre of the field of view of virtually all eyepieces is free of any false colour, but the simple three-element designs can show some lateral colour around bright stars towards the edge of the field of view.

Lens Coatings

The lenses in all eyepieces should be coated to reduce reflected light from their surfaces, which could be scattered within the eyepiece and hence reduce the overall image contrast. A single coating will reduce the reflected light from ~4% per surface down to ~1.5%, but the use of dielectric multi-coatings on all the surfaces, as will be found in premium eyepieces, will reduce this to ~0.5% per surface. This will result in more light passing through the eyepiece, so giving somewhat brighter images, and, more importantly, will reduce the amount of light that is scattered within the eyepiece so that the contrast will be higher. To minimise the effects of scattered light, the interior of the eyepiece and the edges of the lenses should be blackened. It is vital that the many elements employed in the ultra-wide-field eyepieces be given multi-coatings.

Parfocal Eyepiece Ranges

The designer of a series of eyepieces covering a range of focal lengths will often go to the trouble of making them parfocal. This means that, once one eyepiece in the series has been focused, it should not be necessary to adjust the focus (at least not by much) when swapping from one to another – a nice, if not absolutely necessary, feature.

More Can Be Less

As we will see, many eyepiece designs now use seven or even more elements to provide stunning wide-field views of the heavens. There is no doubt that they are very good indeed, but one should not overlook simple four- or even three-element designs thinking that they cannot be so good. An optically good and well-manufactured four-element design could, and indeed should, give better colour fidelity and contrast than an expensive multi-element design. An eight-element system must have sixteen precision-polished surfaces, and each element must be perfectly centred and cemented. This is obviously easier to achieve with a simple design and the light will have passed through far less glass – that will both scatter and absorb light. A simple design should give images which are brighter and show more contrast. So when, such as for planetary observing, wide fields of view are not required, do not overlook the simple and often far less expensive designs such as Plössls or Orthoscopics – they will almost certainly provide you with the best views!

Simple Eyepieces

For many years the standard eyepieces were either Ramsden or Huygenian, which used just two glass elements to form an image. Their apparent fields of view were very small and so they have fallen out of favour. They are, however, useful for projecting images of the Sun, as there are no cemented glass elements which could be damaged by the Sun's heat.

Low-cost telescopes tend to come equipped with three-element eyepieces such as the Kellner (K), Modified Achromat (MA) or Rank Modified Kellner (RKE). (The eyepiece type is engraved on the eyepiece barrel.) At the centre of their somewhat modest (~40 degrees) field of view, they can give excellent images but tend to suffer from what is called 'lateral colour' towards the edge of the field of view. As only three elements are used they can, if these are multi-coated, give very high contrast images and so can be very good for planetary observations where only the centre of the field of view is of importance. The RKE designs are much prized by planetary observers.

Monocentric Eyepieces

There is an eyepiece design, called a monocentric, which was introduced by Adolf Steinheil in the late 1800s; it uses three elements that are cemented together and so have only two air/glass interfaces in order to give the highest possible contrast and light transmission. Monocentrics have an apparent field of view of just 32 degrees and suffer from off-axis astigmatism, but neither is important for their intended use – that of planetary observing – where their performance is acknowledged to be as good as, if not slightly better than, that of any other eyepiece design. Zeiss-manufactured monocentrics were once regarded as the ultimate planetary eyepiece. In 2003 Thomas M. Back of TMB Optical produced a range of these eyepieces using the latest Schott high-refractive-index glasses, which, though expensive, carry on the Zeiss tradition.

The Plössl Four-Element Design

Higher-priced telescopes may well come with two Plössl eyepieces made from two achromatic doublets. These can be excellent performers and usually have an apparent field of view of ~50 degrees. Though they were developed in 1860 by a Viennese optician, Georg Simon Plössl, the modern eyepieces are based on a design by Albert König of Carl Zeiss. Ets Clavé acquired the König Plössl design and tooling, and commenced manufacture under its own name. For some time, these were some of the very best eyepieces available – and had a price to match! The Plössls of shorter focal length have little eye relief and are, perhaps, best avoided. So instead of purchasing, say, a 6-mm Plössl, buy a ~12-mm Plössl and a ×2 Barlow lens – to be described later in this chapter.

In 1984 Al Nagler set up the TeleVue company to market his Plössl eyepieces. These are probably the best now available and all but the 40-mm have apparent fields of view of 50 degrees. That of the 40-mm eyepiece is only 43 degrees, and the actual field will be the same as that of the 32-mm, which is a far better buy as the contrast will be higher due to the greater magnification. Their cost is not unreasonable, and they are available in focal lengths from 8 to 55 mm. Only the 55-mm eyepiece requires a 2-inch focuser. I make very good use of my 32-mm TeleVue Plössl. It has a field stop diameter of 27 mm – as large as it is possible to fit into a 1.25-inch barrel – and thus gives the widest true field of view available with a 1.25-inch focuser. It is thus the best possible 1.25-inch eyepiece to use for first locating an object or for observing open clusters and other objects of large angular diameter.

Super-Plössls

A number of companies have produced derivatives of the Plössl design using one or more additional elements to give better eye relief. Two such series were the original Celestron Ultima range (I have a set) and the Meade 5000 Plössls. The Antares Elite Plössls use five or seven elements to provide a 52-degree apparent field of view and are said to give bright, high contrast images. The Parks Optical five-element Gold series are also excellent but at a somewhat higher cost.

Orthoscopic Eyepieces

Eyepieces of this four-element design – shown with other eyepiece types in Figure 7.6 – use one triplet element group and one single element. They have apparent fields of view between 40 and 45 degrees and do not show any curvature of field or spherical or chromatic aberration. Apart from having a limited apparent field of view, they are near perfect eyepieces and their cost is very reasonable. There are two types, both imported from Japan.

The traditional varieties have 'cone' or 'volcano' tops to fit snugly into the eye socket and do not come with eye cups. Imported from Japan by University Optics, amongst others, they remain a superb choice for lunar and planetary observers. Another mechanical design, though with an identical optical design, has flat tops and may incorporate a rubber eye cup, which can be removed if necessary. Baader Planetarium is one importer of this type. The use of multi-coatings on the lens surfaces is said to increase their contrast very slightly over the volcano type. Pentax produces a modified design called XO orthoscopics which uses five or six elements and is available in 5-mm and 2.5-mm focal lengths. Their image quality is exceptional, but so is their price!

Eyepieces with Long Eye Relief

For observers who, as they suffer from astigmatism, need to wear glasses when observing, several manufacturers produce long-eye-relief eyepieces, which have a typical eye relief of 20 mm. One very well regarded line is that made by Vixen in its LV series. All have 20 mm of eye relief, even those with focal lengths down to 2.5 mm. The 2.5- to 7-mm eyepieces have apparent fields of view of 45 degrees, whilst those with longer focal lengths have 50-degree fields, except the 40-mm eyepiece at 42 degrees. Celestron markets a range of X-Cel LX eyepieces having a 60-degree apparent field of view and an eye relief of 16 mm. These are quite reasonably priced. At greater cost, Takahashi produces a range of LE (for long eye relief) five-element eyepieces having typical eye relief of 9 mm. They are superb eyepieces but I would not really regard them as long-eye-relief eyepieces.

Figure 7.6 'Normal' field of view eyepieces: University Optics 'volcano' 4-mm orthoscopic, Takahashi LE 7.5 mm, Celestron Ultima 13.5-mm super-Plössl and TeleVue 20-mm Plössl.

Wide-Field Eyepieces

There is now a lot of interest in what are called 'wide-field eyepieces' (Figure 7.7). These have wider than normal apparent fields of view, ranging from ~60 degrees up to ~100 degrees, so enabling one to observe a wider field of view for a given magnification. Eyepieces with an apparent field of view of ~80 or more degrees give a rather nice 'spacewalk' feel which can be very satisfying. To achieve such wide apparent fields of view, up to eight glass elements are required so such eyepieces are not cheap!

The less expensive of such eyepieces tend to have apparent fields of view in the range 60–68 degrees (compared with ~52 degrees for Plössls). An apparent field of view of around 70 degrees is said by some to be the largest that can be used without causing eye strain. Vixen provides a set of Lanthanum LVW eyepieces which have apparent fields of 65 degrees. Celestron has an Axiom range which have apparent fields of 70 degrees, as do the Meade Series 4000 QX and 5000 SWA eyepieces. TeleVue Radian eyepieces have a field of view of 60 degrees and their Panoptic eyepieces, 68 degrees. The 24-mm Panoptic is a very useful eyepiece, as its field stop of 27 mm is as large as it is possible to include within a 1.25-inch barrel. As with the 32-mm Plössl, it thus gives the widest possible actual field of view when a 1.25-inch focuser is used but with a greater magnification.

Thomas Beck of TMB, a superb optical designer whose refractor objective lenses are very highly regarded, designed the Paragon range of three eyepieces with 30-, 35- and 40-mm focal lengths each and a 68-degree field of view. Sadly, following his untimely death, these are no longer available under the TMB Optical brand, but

Figure 7.7 Wide-field eyepieces: TeleVue Nagler 7 mm, Hyperion 13 mm, TeleVue Panoptic 22 mm and TMB Optical 40-mm Paragon.

Teleskop-Service in Germany now imports identical eyepieces under the TS-Optics Paragon brand. These are superb performers and their price is very reasonable. The 40-mm eyepiece is mentioned later.

Beyond 70-degree fields of view, eyepieces get very expensive. Some of the best are the TeleVue Naglers, the Meade series 5000 UWA and the William Optics UWAN range, all with 82-degree apparent fields of view. For a given focal length, wide-field eyepieces will have larger field stop diameters and with focal lengths of 24 mm or more, these may well be greater than the ~27-mm maximum allowed with a 1.25-inch barrel. The wide-field eyepieces of longer focal length will thus use 2-inch barrels.

Extreme Wide-Angle Eyepieces

It was some time before eyepieces with an apparent field of view (AFV) greater than 82 degrees became available, but then TeleVue brought out the Ethos range with focal lengths of from 3.7 to 21 mm and an AFW of 100 degrees. More recently, the company has produced the Ethos SX eyepieces of 3.7- and 4.7-mm focal length with an AFV of 110 degrees! Since then, Meade has introduced its XWA range with 9-, 14- and 21-mm focal lengths, all having an AFV of 100 degrees; Orion offers GiantView eyepieces, which have a focal length of 16 mm and AFV of 100 degrees; and Explore Scientific has introduced a range of eyepieces of 100-degree AFV, ranging from 9- to 25-mm focal length at very competitive prices.

Wide-Field Eyepieces with Long Eye Relief

Baader Planetarium produces a range of excellent-value Hyperion eyepieces having a field of view of 68 degrees. Ranging from 3.5 to 24 mm, all have an eye relief of 20 mm. These are termed 'modular eyepieces' in that they have the interesting property that, by inserting a spacer in between the front concave doublet and the remaining six elements, it is possible to alter their focal lengths. At somewhat greater cost, TeleVue has recently introduced a new eyepiece range called Delos having focal lengths from 3.5 to 14 mm. They have an apparent field of view of 72 degrees and an eye relief of 20 mm. These eyepieces are very comfortable to use and absolutely perfect for use at star parties where inexperienced observers may find it difficult to see the image through an eyepiece with little eye relief.

The Maximum Possible Field of View

As previously mentioned, given a 2-inch eyepiece barrel, the maximum field stop diameter would be ~45 mm, so an eyepiece with a field stop of this diameter would give the observer the maximum possible field of view with a given telescope and 2-inch focuser. One such is the TeleVue 31-mm Nagler, which has a field stop of 42 mm. This eyepiece is massive and weighs more than 2 pounds but does provide spectacular views. It is also very expensive. At lower cost, but also highly regarded, is the William Optics UWAN 28-mm eyepiece, which has an 82-degree field of view and a field stop diameter of 43.5 mm. Another eyepiece which has a 42-mm field stop is the TS-Optics 40-mm Paragon. This has an apparent field of view of 68 degrees rather than the 82 degrees of the Nagler and so the magnification is lower, but it gives superb images out to the edge of the field. The 40-mm Paragon is perhaps the eyepiece I use the most, giving an actual field of view of more than 4 degrees when used with my 80-mm refractor, so giving a wonderful view of the Pleiades Cluster and, even when used with my 300-mm Maksutov, it gives a field of view of slightly less than 1 degree. This is just enough to encompass the Perseus Double Cluster, and my first view using this combination under dark skies was one of the most beautiful that I have ever seen, with pinpoint stars sprinkled out to the edge of the field.

With eyepieces of long focal length, it is important to ensure that the exit pupil is not significantly larger than your dilated pupil. Suppose that an eyepiece of 55-mm focal length were used with a short-focal-length 80-mm refractor which might have a focal length of as little as 400 mm. The magnification would then be slightly more than 7, and the exit pupil of the telescope would be 11 mm. If your dark-adapted pupil were only 5 mm, more than three-quarters of the light would be lost! Unless the focal length of your telescope is greater than 800 mm, an eyepiece of this focal length is not a worthwhile purchase.

The maximum field of view that can be observed with a given telescope obviously depends on its focal length – the shorter the focal length the wider the field – but it also depends on whether the focuser has a barrel size of 1.25 or 2 inches. If your telescope focuser has a 1.25-inch barrel size, then a 32-mm Plössl or a 24-mm TeleVue

Panoptic will give the widest possible field of view as their field stops are 27 mm across – about the maximum you can fit into a 1.25-inch barrel. A 2-inch barrel can accommodate a field stop of ~46 mm, so you can see significantly more. The TeleVue 31-mm Nagler is perhaps the ultimate 2-inch wide-field eyepiece.

Zoom Eyepieces

In the past, most observers would not countenance the use of a zoom eyepiece – its optical performance was likely to have been very poor. But things change, and TeleVue provides two highly regarded planetary zoom eyepieces with a focal length range of 4–2 and 6–3 mm. They both have a constant apparent field of view of 50 degrees and 10 mm of eye relief. However, the wide-range zoom eyepieces do *not* have a constant apparent field of view and, at the larger focal lengths, this may be as little as 40 degrees. They do have one very good use. When observing a small object such as a planet or globular cluster, one tends to start with a low-power eyepiece and then move up in magnification until the image quality falls off due to the limitations of the 'seeing' at the time and the reducing brightness in the image. This takes time, and it is quite useful to carry out this step with a zoom eyepiece, increasing the power until the optimum magnification is found. One then simply reads off the focal length on the barrel and selects the appropriate eyepiece of fixed focal length.

Barlow Lenses

A Barlow lens is probably the first accessory that one should buy following the purchase of a telescope and its usual two eyepieces (see Figure 7.8). It is a concave lens that is placed in the light path of the telescope just before the eyepiece and causes the light cone arriving from the mirror or objective lens to diverge, thereby effectively increasing the focal length of the telescope. As the magnification of the eyepiece is directly proportional to the focal length, this has the effect of increasing the magnification provided by the eyepiece. Barlow lenses are normally specified by their magnification factor, which can range from ~1.5 to 5. The most common have a magnification of 2, and so are called ×2 Barlows.

For good image quality, Barlow lenses are usually made from a cemented achromatic doublet. Some, using more exotic glass or three elements rather than two, are termed 'apochromatic'. As well as producing some excellent Barlow lenses, TeleVue also produces Powermates, which include a second doublet lens to give somewhat enhanced performance.

The magnification produced by the convex lens depends on how far it is from the eyepiece, and so Barlow lenses of the same magnification could use a higher-power concave lens in a short barrel or a weaker concave lens in a longer barrel. The latter is preferable as the image quality, particularly towards the edge of the field, is likely to be better than with a short Barlow lens. The shorter design may also introduce some vignetting (darkening) towards the outer field of view when used with eyepieces of longer focal length.

Figure 7.8 A ×2 'short Barlow', a TeleVue ×2.5 Powermate and a William Optics binoviewer.

One might think that, rather than buying a ×2 Barlow lens, one could simply buy an eyepiece of half the focal length, but a Barlow does give some real advantages. Perhaps the most compelling is that the purchase of a Barlow effectively doubles the number of eyepieces that you have available. A telescope will often come supplied with 26- and 10-mm eyepieces. Using a ×2 Barlow with each gives the effect of adding 13- and 5-mm eyepieces. A good Barlow will cost no more than a single eyepiece, so this is a very cost effective way of enhancing your eyepiece collection.

Telescopes of short focal length need eyepieces of very short focal length to give the high magnifications needed to observe the planets. For example, with my 80-mm, f6.8 refractor, I would need an eyepiece with a 3-mm focal length to give a magnification of ~200. As already mentioned, inexpensive eyepieces of this focal length have very little eye relief. However, the use of a ×2 Barlow enables a 6-mm eyepiece to give the same magnification and is far easier to use. (In fact, the long-eye-relief eyepieces of shorter focal length nearly always incorporate a Barlow lens in their construction.) The rapidly converging light cone from a fast (f4 or f5) telescope is difficult for mid-range eyepieces to handle, giving rise to obvious aberrations towards the edge of the field. Expensive eyepieces are better able to handle this but, if the effective focal length – and hence focal ratio – is doubled using a ×2 Barlow, inexpensive eyepieces will perform well.

The eye relief of many eyepiece designs is some fraction of their focal length. For example, the eye relief of a standard Plössl is typically 65% of its focal length and, for a 10-mm Plössl, may be only 5.5 mm. Many observers find eyepieces with eye reliefs shorter than ~12 mm difficult to use, so instead a 20-mm Plössl could be used in conjunction with a ×2 Barlow. This will give the same magnification, but now the eye relief will be ~13 mm. The use of a Barlow with eyepieces of long focal length can actually make the eye relief too great, so is not to be recommended. With Newtonian telescopes, it is often not possible to bring the sensor of a DSLR or CCD camera to the focal plane. Often, the addition of a Barlow will allow focus to be reached, albeit at the cost of reducing the field of view.

Narrowband interference filters, such as the OIII filter often employed for observing planetary nebulae, work best when a parallel beam of light passes through them. Their performance is reduced when used with a telescope of short focal length with its rapidly reducing light cone. Their performance can be greatly improved if a Barlow is used prior to the filter, which then 'sees' a less rapidly reducing light cone.

Binoviewers

A binoviewer is a beam-splitting device that enables the light from a telescope to be passed into two eyepieces, so that the effect is like observing though a pair of binoculars (a binoviewer is shown in Figure 7.8). When the brain processes images from two eyes rather than one, subtle details, as on the surface of a planet, become more distinct. It is also more relaxing to observe through two eyes than to ignore or close one eye. The brain may also be fooled into giving images some three dimensionality that can add to the observing experience. When one is observing at high magnifications, floaters in the eye can become a problem. As these will be different in the two eyes, the brain can partially eliminate their effects.

There are some downsides. Due to the additional path length (4–5 inches) within binoviewers, their use with refractors and Newtonians can be a problem, as the focuser cannot always be racked in enough to achieve focus. The use of a Barlow lens may (not necessarily) overcome this problem, and a suitable concave lens that can be screwed into the front of the binoviewer barrel may be provided along with the binoviewer. Splitting the light into two and passing it through the required prisms (hopefully fully multi-coated) will mean that the light passing into each eye will be slightly less than half that when a single eyepiece is used. However, the brain has more information to process, and this counteracts the problem so that the effective sensitivity remains about the same, with some observers actually gaining a slight increase in sensitivity. Of course, two eyepieces will be required, so obtaining a range of magnifications can be expensive, with the use of a Barlow lens becoming an even more cost effective way of doubling up the effective number of focal lengths. The size of the prisms in the less expensive binoviewers can also limit the maximum possible field of view, as the light path diameter may be restricted to 22 mm or even less. This precludes the use of low-power, wide-field eyepieces, which may have field stops greater than this, as in one of my favourite eyepieces, the TeleVue 32-mm Plössl, which has a field stop of 25 mm. If a pair of matched eyepieces are not provided with the binoviewer (the William Optics binoviewer comes with two 20-mm wide-field eyepieces), a pair of 25- or 20-mm Plössls will work very well.

The more expensive binoviewers, such as those made by TeleVue, Baader Planetarium and Denkmeier Optical have prisms 27 mm in diameter, so allowing the use of all 1.25-inch barrel eyepieces without vignetting. The latest Denkmeier, the Binotron-27 Super System, uses a 2-inch (rather than 1.25-inch) focuser barrel to minimise vignetting and has prisms polished to 1/10 wave. It incorporates a sliding stage to bring one of three lens elements into the optical train. The effects are somewhat complex and depend on the type of telescope. One nice feature of this binoviewer is that it provides a straightforward way of collimating the two eyepieces so that the brain can easily merge the two images. Perhaps the ultimate binoviewer is the Seibert Optics Black Night Elite 45. This indicates that the prisms are 45 mm across and so can be used with wide-field, 2-inch eyepieces. At the time of writing there is a one year waiting list!

Schmidt-Cassegrains are best suited for use with a binoviewer, as they normally have sufficient focus travel to accommodate a binoviewer without the use of a Barlow lens. But, of course, even with relatively long focal lengths, wide field views are not really possible. A small Schmidt-Cassegrain, such as the Celestron C6 with a 1,500-mm focal length coupled with a William Optics binoviewer (which is very similar to those sold by Celestron, Orion and Sky-Watcher), is a combination that works really well and provides stunning views of the Moon. If one stops the tracking on the mount, one can almost believe that one is flying over the lunar surface!

8
Telescope Mounts
Alt/Az and Equatorial with Their Computerised Variants

No telescope can be used sensibly unless it is supported by a mount which is sufficiently sturdy to hold it steady. This is an obvious statement, but many mounts supplied with less expensive telescopes are not really up to the task. Often the tripod supports are so light that, when one attempts to move the telescope, the tripod moves! (The solution to this problem is to fill the tripod legs with sand.) As telescope apertures increase, it may well be that the cost of a suitable mount will exceed that of the telescope tube assembly, but this is a price worth paying as a poor mount will cause one endless frustration when observing. A really solid mount is of prime importance for astro-imaging, and some authors state that one should halve the nominal load capacity of a mount for this use. Many telescopes at the lower end of the price range are sold only as a package with an included mount, but for more expensive refractors or reflectors, the tube assembly can usually be bought separately and the user can chose a suitable mount.

Mounts come in two basic types: altitude/azimuth (Alt/Az) or equatorial. Before the advent of computer-controlled drive systems, most mounts were equatorial. The reason was simple. Once an equatorial mount has been aligned on a star, drive – at a fixed sidereal rate – need be applied to only one axis to track it across the sky. Thus a simple electronic controller, based on a crystal oscillator to give an accurate time base, could be used. In contrast, the Alt/Az mounts have to be driven in two axes at variable rates in order to track. For example, when a star is rising in the east, the altitude drive rate will be quite high whilst the azimuth drive rate will be fairly low, but as the star crosses the meridian, due south, the azimuth rate will be high but the altitude rate will be zero and will change from a positive rate whilst the star was rising to a negative one whilst it is setting.

For large professional telescopes the Alt/Az mount is far preferable to an equatorial, as the telescope tube does not need to be counterbalanced and the weight of the mirror and the observing equipment is kept as low as possible so, as soon as digital computers became available, they became universally used. This technology, at little cost, is now used for amateur Alt/Az mounts, and it is easy to extend their abilities to provide a 'go-to' facility so that, once aligned, the telescope can find planets, stars and nebulae with little user intervention. So now the majority of smaller telescopes are provided with Alt/Az mounts.

Alt/Az mounts suffer from two problems. The first is that it is almost impossible to track an object as it goes close to the zenith. Consider an Alt/Az-mounted telescope following a star that rises in the east and passes directly overhead. Up to the zenith the altitude will be increasing but, unless the mount can move the telescope into the opposite arc (which is not usually possible), the mount must then slew at its highest possible speed through 180 degrees before beginning to reduce the altitude of the telescope. The second problem is that Alt/Az mounts suffer from 'field rotation', which means that, over time, the orientation of an object will rotate within the field of view. Observing the constellation Orion as it rises in the east, one will see it on its side with the hunter's feet to the right. However, when seen due south, it will appear vertical, and as Orion sets in the west its feet will be to the left. Orion will have thus appeared to rotate in the sky. For visual observing with an Alt/Az mount this is no problem but, with astro-imaging, the allowed exposure times will be limited. Ironically, the greatest rate of field rotation is when the object passed through the meridian – when it is highest in the sky and so best placed for imaging! Astro-imagers will thus tend to use equatorial mounts, although a program called 'Deep Sky Stacker' can be used to stack a number of short-exposure images having first corrected for the rotation between the individual images. Professional telescopes are equipped with a computer-controlled 'field rotator', which will rotate the CCD camera anticlockwise during the observation and so eliminate the problem. These are now available to amateurs. An example, the Optec Pyxis LE derotator, has a 2-inch barrel to insert into the telescope focuser and a standard male T-thread to allow a camera to be mounted onto it using the appropriate bayonet adaptor.

8.1 Altitude/Azimuth Mounts

Besides using the Dobsonian mount described in Chapter 4, there are three ways to achieve an Alt/Az mount. Two of these are fork-mounted configurations: smaller telescopes will often be mounted on a single-fork mount, whilst larger, but compact telescopes such as Schmidt-Cassegrains will use a dual-fork mount (see Figure 8.1). The great advantage of a fork mount is that there is no need to counterbalance the telescope, so keeping the total weight of the moving parts down and thus requiring a lighter mount and tripod. In the case of single-fork telescopes it may well be impossible to observe towards the zenith in order to avoid the telescope tube fouling the mount. With dual-fork mounts, visual observing may be possible near the zenith, but there may be insufficient clearance to allow imaging cameras of any size to be used.

A third design is used for heavier and longer tube assemblies such as medium-sized refractors and Newtonians. These employ a counterbalanced system where the telescope lies on one side of the mount whilst a balancing counterweight is located on the opposite side. The mount has to thus support a greater weight than the telescope tube alone. An interesting point is that the counterweight can be a second telescope, so one could employ, say, an 80-mm refractor balanced by a 127-mm Maksutov to

Figure 8.1 Single- and dual-fork mounts – Celestron NexStar SLT, Meade Lightbridge, Meade ETX 90 and Celestron CPC 1100. (Images: Celestron and Meade)

provide both wide fields of view for clusters and higher-magnification views of the planets. For solar observing, one could mount an 80-mm refractor with a Baader Solar filter counterbalanced by a 60-mm H-alpha solar telescope.

Single-Fork Mounts

Examples of these are the two Celestron NexStar computerised Alt/Az mounts that are supplied with their smaller telescopes but can also be bought separately. The smaller, the NexStar SLT, would make a very good mount for an 80-mm refractor, whilst the more substantial, the NexStar 6/8 SE, can even support an 8-inch Schmidt-Cassegrain. Both use the 'Sky Align' method of alignment described later. A single-fork mount is also used in the self-aligning SkyProdigy range. Sky-Watcher produces a range of single-fork mounts that include simple manual designs, some (SupaTrak) that will track an object once manually found and also computerised (SynScan) 'go-to' models. Single-fork mounts are also supplied with many of Meade Instrument's smaller telescopes.

Dual-Fork Mounts

These are used by Meade for a very wide range of telescopes, from its small-aperture ETX range of telescopes up to its LX90, LX200 and LX600 series of Schmidt-Cassegrains. Celestron supports its larger Schmidt-Cassegrain telescopes from 8-inch aperture upwards in its CPC range of 8- to 14-inch telescopes. They are not sold without a telescope. It is possible to purchase equatorial wedges which fit between the tripod and mount so that they can be converted into equatorial fork mounts.

Counterbalanced Alt/Az Mounts

These can be either manually or computer controlled. Examples of the former are the SkyTee-2 Alt-Az mount, the Sky-Watcher AZ4 and HDAZ Alt-Az mounts, the

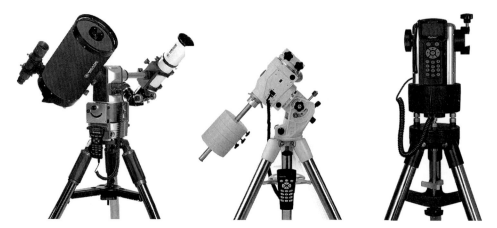

Figure 8.2 Convertible Alt/Az to equatorial mounts: Meade LX80 in Alt/Az mode, Sky-Watcher AZEQ6-GT in equatorial mode and iOptron Minitower Pro in Alt/Az mode. (Images: Meade, Sky-Watcher and iOptron)

Williams Optics Easy Touch mount and the Vixen Porter mounts. They are ideal for 'grab-and-go' astronomy with short tube refractors and 127- or 150-mm Schmidt-Cassegrains or Maksutovs. Some have two Vixen-style dovetail saddle plates on either side of the mount, so allowing two telescopes to be mounted at once. Computerised mounts of this type are made by Meade and iOptron.

The Meade LX80 Alt/Az or Equatorial Mount

Meade calls this a multi-mount, as it can be used either in Alt/Az mode with the head horizontal or as an equatorial mount by tilting the Alt/Az head so that it is polar aligned (see Figure 8.2). The mount weighs 35 pounds and can be operated with a single optical tube assembly (OTA) with a total weight of up to 40 pounds in Alt/Az or equatorial mode. When used in dual OTA mode (two telescopes counterbalancing each other) it can support a combined weight of 75 pounds. The tripod is robust and stable. A nice feature is that the tripod levelling adjusters are at the *top* of each leg, which makes it far easier to set the mount horizontal, as one can easily observe the built-in bubble level whilst carrying out the adjustment. The tripod legs have rubberised bases, so acting like vibration pads. The mount is equipped with the AudioStar controller and includes audio commentaries on 500 of the objects within its 30,000-object database. It has a pointing accuracy of 10 arc minutes, which is improved to 5 arc minutes when the high-precision mode is activated. The mount can be powered by eight AA batteries or an optional 12-volt DC power supply.

One very nice feature of an Alt/Az computerised mount is that, when it moves to the first alignment star and perhaps does not find it immediately, the only error should be in azimuth, so making it easy to bring the star into the field of view using the azimuth motion controls. (This does, of course, require that you have accurately levelled the mount.) Users report that the tracking is very good, with, for example,

Saturn observed through a 90-mm refractor staying within the field of a 7-mm eye-piece plus ×2 Barlow (a ~20 arc minute field of view) for 30 minutes. The photographs on the Web do not give a feel of its actual size – which is significantly greater than I was expecting. This is a very substantial mount and tripod!

Sky-Watcher AZEQ6-GT Alt/Az or Equatorial Mount

This type of dual mount seems to be the 'in' thing at the present time and, perhaps not surprisingly, Sky-Watcher has also produced an excellent, though more expensive, version called the AZEQ6-GT, which can support 18 kg for imaging use or 25 kg for visual use (Figure 8.2). In Alt/Az mode the counterweights can be replaced with a second telescope, so greatly increasing the telescope load capacity and ideal, say, for observing the Sun in white light with one telescope and in H-alpha with a second.

iOptron Minitowers

iOptron produces two versions of the Minitower: the Minitower V2.0 and the Minitower Pro (Figure 8.2). They look similar, but the Minitower Pro employs a sturdier tripod and enhanced bearings to support a payload of 33 pounds rather than the 25 pounds of the Minitower V2.0. It is possible to add an equatorial wedge so that the Minitower Pro can be used in equatorial mode. They can be used in a single OTA mode with the use of a counterweight to balance the OTA if it is more than a few pounds in weight. An alternative is to use the supplied Vixen-style dovetail saddle plate to support a second OTA, as with the Meade LX80. The mount has an integral GPS module to set the accurate location and time.

The Minitower provides three options for its power supply. Perhaps the neatest is to use eight AA batteries that can fit into the tower unit. Rechargeable Ni-MH batteries can also be used, thus reducing the potential running costs. Alternatively, a mains adapter is provided but, preferable for safety reasons, a 12-volt plug and cable are provided for use with rechargeable battery packs. I was able to review and then purchase the original Minitower, and the following will give a feel of its use.

Using an iOptron Minitower

The mount has first to be set into its 'home' position with the Alt/Az head horizontal and aligned to the south with the telescope vertical. The tripod is first set up with its top approximately level (I use a vintage mahogany/brass level for this). The head is then located on it, held up by three variable-length coupling rods which can be finely adjusted by the user whilst being able to easily observe and centre the bubble in its inverted cup, so quickly giving an accurately levelled mount. There is an azimuth clamp under the head that must be loosened half a turn whilst this is done, and this also allows the head to be rotated so that an arrow (beside the letter *S*) points south before it is locked. As will be indicated later, this does not have to be too accurate.

First light was with an 80-mm ED refractor at the Society for Popular Astronomy's annual astronomy weekend under dark skies in Shropshire. Polaris was hidden behind a tall building and so our south alignment was very inaccurate. The mount was powered up, and after a minute or so the hand controller display showed 'GPS OK'. Jupiter was low in the west and was used to do a 'one-object align'. The 'Select and slew' mode was chosen in the top menu and Jupiter chosen in the 'Sun, Moon and Planets' section. The telescope immediately slewed round towards Jupiter but ended up a little to its west – not surprising, as the alignment to the south was only approximate. But, as the head was accurately horizontal, the elevation should be correct and the only significant error would be in azimuth. A low-speed azimuth slew quickly brought Jupiter into the centre of the low-power field of view obtained with a TMB 40-mm Paragon eyepiece. Moving to a higher power and some very minor tweaks in position accurately aligned the mount on Jupiter. At this point, the second line in the menu, 'Sync. to target', was activated and the computer then internally compensated for our inaccurate alignment south. This took less time that it takes to read this, and we had effectively carried out a 'one-star align' but with the object of our choice. If one was going to observe this object anyway, then the alignment had essentially taken no time. The seeing was surprisingly good, and the belts and zones showed up well. The mount was then slewed to Andromeda high in the west – an impressive sight with the Paragon eyepiece, nicely centred in the field of view. Becoming more ambitious, I slewed the mount to M33 in Triangulum. Heavy showers earlier in the day had cleared the atmosphere, and M33 was easily seen with averted vision, but then clouds ended our observing. Selecting 'Park telescope' in the top menu brought the telescope back to vertical with the head aligned due south – so now its direction was known accurately.

The following night the clouds cleared again and a new observing site was found with good views from south-east round to north. This time the first target was the Pleiades Cluster, which was used to align the mount as before. Observations then took in the Orion Nebula, clusters in Auriga and Gemini and finally, as the clouds encroached from the west, the galaxy pair M81 and M82. In all cases, when a TeleVue 20-mm Plössl was used, the objects were very close to the centre of the field of view. The 'go-to' precision displayed by the mount on those first two nights had been very impressive, but what impressed our group the most was how fast the mount could be set up and aligned.

Unlike many computerised mounts, the Minitower *does* allow one to move to and then synchronise on the Sun – after a suitable warning, of course. This time, on the following morning, an 8-inch Celestron Schmidt-Cassegrain (with a 1-inch aperture mask in front of the corrector plate) was mounted and the supplied counterweight along with a second of similar weight was used to balance it. The mount had no problems slewing or tracking smoothly, indicating that the Minitower can support a scope of this size, though I suspect that the Minitower Pro would be more suitable. For safety, the finder objective was covered and the telescope eyepiece pointed down to project an image onto card using a simple two-element eyepiece. The Sun's image was easily centralised and the mount aligned using the 'Sync. to Target' command.

Then we were able to do something that I, personally, have never been able (or dared) to do before – find Venus in daylight just 9 degrees from the Sun. A slew having been executed to Venus, it was immediately seen as a tiny gibbous disk virtually in the centre of the field of view!

The mount uses a GOTONOVA 8401 hand controller, which has a large eight-line display giving, along with the date and time, the target's demanded and actual right ascension (RA) and declination (Dec), as well as the altitude and elevation of the mount. It can track at lunar and solar rates, as well as the normal celestial rate. Pressing the 'Menu' button brings up the main menu, which has eleven options. As described earlier, the most frequently used will be 'Select and slew' and 'Sync. to target'. The former opens up the object menu, which has eight sub-menus to reach the ~130,000 objects in its database. These include the planets, Sun and Moon, the Messier, Caldwell, Abell and Herschel catalogues and the NGC, UGC and MGC catalogues, 190 comets, 4,096 asteroids, 88 constellations, up to 256 user objects and the ability to enter the RA, Dec of an object. Should you select an object that is below the horizon at the time, the display tells you when it will rise and, should it be an object which never rises at your latitude, it will simply not accept it.

The mount uses 12-volt DC servo motors which are very quiet (great if you have close neighbours!), and the positioning is monitored and controlled using optical encoders having a resolution of 1 arc second. With an accurately levelled and aligned mount, the go-to accuracy was excellent, with the selected object always close to the centre of the field of view of a medium-power eyepiece.

The next important requirement is that, once an object has been centred in the field of view, it remain there as the mount tracks the object across the sky. For this test, a Celestron 6-inch Schmidt-Cassegrain was set on the mount and a TeleVue 7-mm Nagler used to view the planet Jupiter, low in the south. The mount having been carefully set, Jupiter was centred in the 20 arc minute field diameter and the scope left. Pleasingly, Jupiter was still centred in the field of view well over an hour later as it was about to set behind some trees.

Meade 'Level North' and GPS Technology

Some of the Meade series, such as the dual-fork LX and single-fork LT ranges, incorporate three elements to make the initial alignment process easier: a GPS module is used to determine the precise location and time and so saves having to manually enter these before use, an electronic compass is used to determine north and a system is used to measure and then correct for any tilt of the mount away from horizontal. Once this initial mount alignment is completed, the pointing accuracy is sufficient to 'go to' a first alignment star, which should appear within the finder scope. Once manual alignment corrections are made on this star, the process is repeated for a second star to achieve what is called a 'two-star align'. Once this is complete, the mount can then 'go to' any of the planets, stars or nebulae that are contained with its catalogues.

The Celestron NexStar SLT Range

In 2005 Celestron introduced an innovative method called 'SkyAlign' for implementing the initial alignment of a scope. It first became available with the company's low-cost NexStar range, which uses a single-fork Alt/Az mount. It is now available with many of Celestron's mounts. I was able to review, and then bought, the 130 SLT, which is equipped with a 130-mm (5.1-inch) Newtonian tube assembly of 650 mm focal length. I can honestly say that I was surprised and impressed by the f5 Newtonian tube assembly, which came with a spider made of *thin steel* tensioned by screws around the optical tube – so giving less obvious diffraction spikes. The focuser came with adapters for both 1.25-inch eyepieces *and* 2-inch eyepieces. This means that for wide-field views a 2-inch eyepiece of 30- to 40-mm focal length could be used.

I was also very impressed with its optical quality. Star testing showed that the surface was both smooth and better than ¼ wave – the nominal requirement for good performance – excellent for a budget (or any) scope. The telescope tube is equipped with a red dot finder (good) and comes with two eyepieces. The 25-mm eyepiece gives a magnification of 26 with a field of view of 2 degrees. The 9-mm eyepiece gives a magnification of 72 with a field of view of ~40 arc minutes. For planetary use a ×2 Barlow will be needed to give sufficient magnification.

The cover on the base of the mount opens to allow the insertion of eight AA batteries. What pleased me greatly was that the mount works fine with a set of rechargeable batteries. The hand controller/computer is Celestron's latest, which allows for the software to be updated. The mount came already mated to its tubular stainless steel tripod. This is certainly better than the aluminium tripods that have been used on this class of telescope before, but it is still rather light. A bubble level is incorporated into the base so the mount can be levelled but, if SkyAlign is used, levelling the mount is not too critical.

The first task is to set the time and location (the nearest city in a provided list will do fine). One then uses four keypads on the controller to drive the telescope so that the red dot finder points, in sequence, to three bright objects in the sky. (I use the word 'object', as bright planets and even the Moon may be used as well as bright stars, provided that the date, time and location have been entered into the handset.) When the red dot appears over each object one presses 'Enter'. One is then asked to centre it in the field of view of the telescope. The computer automatically reduces the slewing speed at this point. When the object is centred, the 'Align' key is pressed. One should try to get at least 90 degrees between the objects, and it is also best not to use objects in a straight line. Once the magical 'Align successful' prompt appears, that's it, the sky is yours! The simple idea is this: any pair of objects will have a defined angle between them. With three objects there will be three measured angles and there will be only one possible set of three bright objects which will give these three angles, so the mount can determine which they are and even tell you their names. The first time I aligned the mount, the sky was cloudy and I was not sure what was the third object aligned on, as it was seen in a gap in the clouds – it was Saturn!

The numeric keys have letters or words on them relating to the various catalogues held in the computer's memory. These include the Messier, Caldwell and NGC catalogues of deep-sky objects as well as, of course, named and catalogued stars and the planets. The telescope will also take you on a tour of the best objects visible in the sky that night.

The scope brought all the objects I observed into the field of view of the 25-mm eyepiece. Perhaps the most critical test was Saturn. The 9-mm eyepiece (at ×72) could not really give enough magnification to see Saturn at its best, and so I added a TeleVue ×2.5 Powermate into the focuser to give ×180. During moments of very good seeing I could see the Cassini division and possibly a hint of a band around the surface. This confirmed the star testing which showed that the optics of the scope were clearly up to the mark! The telescope and mount performed better than I ever expected. The SkyAlign alignment procedure never failed to give a successful alignment first time, and always thereafter it brought the chosen objects into the field of view of the 26-mm eyepiece. The optics are of such good quality that it will virtually always be the atmosphere that limits the image quality. This is an excellent low-cost telescope.

8.2 Self-Aligning Alt/Az Mounts

A really interesting development has been the introduction by Meade and Celestron of mounts that carry out the initial alignment process without any user intervention.

Meade LightSwitch Mounts

These single-fork mounts are equipped with an impressive array of technology to enable the mount to align itself. The mount and fork are very sturdy and can support up to 8-inch Schmidt-Cassegrains (as shown in Figure 8.1). Worm drive gears, nearly 5 inches in diameter, are used to find and then track objects across the sky. It is a development of the 'Level North' technology described earlier. In addition to the system to measure the tilt of the mount so that it can be compensated for (the 'Level' part) and a magnetic compass to find north, it has two further elements. The first is a GPS module so that it can find the precise location and time. Given the location, time and direction of true north, it is able to approximately align the telescope on a suitable bright star, which should be visible given a clear sky. The second is a 640 × 480 colour Eclipse camera which is mounted, on axis, below the telescope.

Once the mount has approximately aligned the scope on the first star, the Eclipse camera images the star field and the computer measures the offset from the nominal position of the star. Pointing corrections are applied until the centre of the camera image (and hence telescope) is aligned on the star. The telescope then slews to a second bright star and the process is repeated. The mount has thus completed a two-star align and the telescope can then be used to 'go to' any other objects within its (100,000 objects) catalogues and above the horizon. The whole process from turning power on takes between 8 and 10 minutes, the mount first rotating to find north and tilt and then moving to the two stars and aligning on them. A nice feature is that the images taken by the Eclipse camera can be either viewed on an optional monitor or saved onto an SD card for later viewing. An 'astronomer inside' feature of the mount will give audio (and visual if an optional monitor is installed) information about the objects being observed!

The Celestron SkyProdigy Mounts

These were introduced in 2011 and have become available with two refractors of 70- and 102-mm aperture, a 90-mm Gregory-Maksutov, a 130-mm Newtonian and a 150-mm Schmidt-Cassegrain.

Mounted on the opposite side to the telescope on the Alt/Az mount is a small scope and CCD camera aligned on the telescope axis. Having powered up the mount (a battery pack for eight D-cells is provided, or one can use an external battery) one simply presses the 'Align' button on the hand controller, and the mount, in 'StarSense' mode, moves automatically to three areas of the sky, images them and locates their position on the celestial sphere by matching the stars it sees with its database. The whole process takes just 3 minutes before the display says 'Align successful'. One can then immediately start seeking out celestial objects to observe. It needs a reasonable arc of clear sky (clockwise from its initial position) to achieve this by itself but, if you are observing from a site with a restricted view of the sky, then a second alignment mode will allow you to manually point the scope to three areas of sky that *are* visible to you.

Once alignment is achieved, the user can immediately choose objects to observe from the usual range of catalogues containing more than 4,000 objects. Observing the Moon or planets takes just a little more effort. To tell where and when these will be visible in the sky, the computer requires a knowledge of the location, date and time, so when you first select 'Solar System' in an observing session, the computer prompts you to check the 'Lat/Long' of your observing site and the date and time, editing these if necessary. The computer does retain these when powered down so, once set, they will rarely need adjusting. You then simply select the solar system object you wish to observe – but only from those that will be above your horizon.

There is one further alignment method, linked to the solar system, which is very useful when, for example, Jupiter or Venus can be seen in a twilight sky but it is not dark enough for the StarSense mode to be used. One uses the red dot finder to manually slew to the planet, centre it in the eyepiece, and align on it. This then enables the computer to track it across the sky. I have been able to evaluate the 130-mm Newtonian SkyProdigy, and the following description may give a feel of what it is like.

Using a SkyProdigy

First light for the SkyProdigy was at the Kelling Heath Equinox star party in the autumn of 2010 (see Figure 8.3). Under dark skies with the Milky Way arching overhead the scope was set up and the 'Align' button pressed. Three minutes later, the scope having pirouetted around the sky, 'Align successful' came up on the display

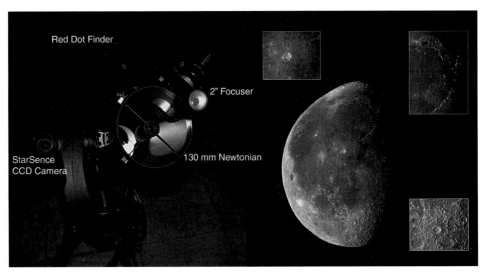

Figure 8.3 The Celestron SkyProdigy 130 and a composite lunar image taken using it.

and observing began. The Dumbbell Nebula, M27, came to mind and the scope slewed round to it and came very gently to a halt. It was immediately visible as a slightly elongated smudge just off the centre of the field of view of the 25-mm eyepiece. Once it was centred by means of a slow slew rate, the 9-mm eyepiece was used and the higher magnification helped it to stand out better. The galaxies M81 and M82 and the Double Cluster in Perseus were found equally easily, but the real highlight of the night was a glimpse, using averted vision, of the galaxy M101. I am absolutely certain that a speck of light was visible within it – the supernova first seen a few weeks earlier!

A thorough night of visual observing followed a few days later from my home location within the confines of a town. I concentrated largely on the objects that I know and have observed through many telescopes. High overhead, in Lyra and Vulpecula, I observed the Ring and Dumbbell Nebulae and then found the Coathanger in Vulpecula and the globular cluster M13 in Hercules. One significant point: the height of the well-positioned focuser made the scope easy for me to use whilst sitting on a chair for comfortable observing. Moving round towards the south and east, I spotted the globular cluster M15 in Pegasus and then M31, the Andromeda Galaxy. Finally, expecting to end my observing session with a real highlight, I slewed to the Double Cluster in Perseus. It is useful to know that the two clusters, NGC 884 and NGC 869 (whose NGC numbers I cannot remember), can be easily found as C14 (which I can remember) within the Caldwell Catalogue menu. All appeared exactly as I expected them to do when viewed through a 130-mm scope. But, as I was about to put the scope away for a few hours, I spotted Jupiter through a gap between the houses and had to move the scope to a far corner of the garden to observe it. This time I used the red dot finder for the first time and did a 'Solar System Align' on Jupiter. The image through the 9-mm eyepiece was very pleasing, even though it was still relatively low in the sky, but I did allow myself the use of a ×2 Barlow to increase the magnification. The bands were easily seen with a hint of structure within them when the seeing allowed.

Following a few hours' sleep, I rose well before dawn and first observed Jupiter now higher in the sky, giving a very pleasing view. The Pleiades Cluster was next. The 25-mm eyepiece field could not encompass the whole, but showed the central region well. Then, using the 9-mm eyepiece, I homed in on my favourite part of the Pleiades, that which lies between the stars Alcyone and Maia. Halfway between the two there is a close pair of stars, with that nearest to Maia obviously red in colour, whilst, just beside Alcyone, there is a lovely triplet of stars making a perfect equilateral triangle.

Orion was now well up in the south–south-west and so M42, the Orion Nebula, was my final target. The 25-mm eyepiece showed the overall nebula, somewhat lost in the urban sky, but the 9-mm showed the central region very well, with the dark region of the 'fish's head' intruding into the bright region surrounding the central Trapezium. The region was nicely seen with the 9-mm, but was better when a ×2 Barlow was used, showing the four brighter Trapezium stars cleanly and clearly resolved. As dawn broke, it was time for an early-morning cup of tea!

Later, in early evening, I had a brief chance to observe the first-quarter Moon low in the southern sky. As it was still quite light, I first slewed to the Moon's position and then used the 'Solar System Align' mode. The 9-mm eyepiece just encompassed the Moon, which showed commendable contrast given the elevation and rather hazy sky. Using a ×2 Barlow to give a magnification of 144 made it possible to observe the craters along the limb in detail.

My final observations with the telescope came when I took it to the Peak star party. The Saturday daytime had sunny spells, so the Sun was an obvious target as, nearing solar maximum, it was getting increasingly interesting to observe. Given *very great care* this is possible with the SkyProdigy, but there are two critical points necessary to protect the telescope. Firstly, please do not even think about using eyepiece projection as, even if the brightness is reduced by use of the small aperture in the tube cover, the heat from the Sun can be enough to break down the cement bonding the doublet in the eyepiece; secondly, should the cover of the StarSence CCD camera not be present, the CCD will be destroyed – I taped the cover in position so it could not possibly fall out! The only really safe way to observe the Sun is to use a full-aperture solar filter as described in Chapter 12. They can be bought ready-made or produced at home using Baader Solar Film, which is widely available from telescope suppliers. With a Baader filter in place, a very nice grouping of sunspots could be seen. That evening, the sky was partly cloud covered and somewhat hazy. I had not expected the scope to be able to self-align and was about to use the 'Solar System Align' mode on Jupiter but decided to try it anyway. Under the amazed gaze of some onlookers it *was* able to align itself and I was then able to 'go to' Jupiter to wow them even more. The three alignment methods covered every eventuality and could not be faulted. The 'go-to' mount found all the objects that I sought and, once they were found, tracked them with very high precision across the sky.

The Celestron StarSence Automatic Alignment System

As just described, the Celestron SkyProdigy telescopes use a small camera to image the sky in three directions and matches the observed star fields with its internal database of 40,000 stars. Having typically matched the positions of 100 stars in each field, the control computer then knows the precise direction in which the telescope was pointing as each star field was imaged and is thus able to carry out the initial alignment of the telescope. This (pretty amazing) technology has now become available as an accessory to add to a wide range of Celestron telescope mounts both new and, in many cases, old.

8.3 The Equatorial Mount

The full name of the type of equatorial mount most frequently used by amateur astronomers is 'German equatorial mount', or GEM for short. Mounted on a tripod is an equatorial head supporting a short tube called a 'polar axis', which has to point towards the North Celestial Pole (NCP). (The terminology used in this section is defined in Chapter 1.) This is driven by a motor and gear system so that it rotates once each sidereal day. The telescope tube is mounted on this rotation axis so that the telescope's position can be adjusted in declination. Once set so that the telescope is following a star, this should not need more than minor adjustments.

Let us first consider how an equatorial mount works. Imagine that you are at the North Pole. The Sun, planets and stars will appear to rotate round at a constant elevation, so that if you had a vertical column (which would point to the NCP above your head) and mounted a telescope on the side so that you could change its altitude, then, by rotating the pole and adjusting the altitude, you could move to any object in the sky. Once the object had been centred in the field of view, it would remain within the field if you simply rotated the pole once per sidereal day. (Note that the Moon moves quite rapidly across the sky and thus needs to be followed at a somewhat different rate. Many equatorial mounts have a 'lunar' drive rate as well as a sidereal rate.)

Provided that you have a column pointing at the NCP, you do not have to be at the North Pole for this to work. In many mounts, a small telescope, the polar scope, is mounted within and aligned along the polar axis to enable precise alignment on the NCP. Good features of an equatorial mount are that the orientation of a star field remains fixed, so allowing long-exposure photography, and also that only one motor has to be running at a constant speed to track an object across the sky.

Periodic Error Correction

The rack which is used to rotate the telescope at sidereal rate is usually driven by a motor-driven worm meshing with a worm gear. For each 360-degree rotation of the worm, the worm gear rotates by one tooth. Depending on how well the worm has been machined, the rotation rate of the worm gear will vary very slightly as the worm rotates, giving rise to a small tracking error. With my Losmandy GM8 this has been found to cause a tracking error of about ±5 arc seconds. However, the effect repeats very closely for each turn of the worm, and so this is called a 'periodic error'. Some mounts allow this to be corrected in what is termed, not surprisingly, 'periodic error correction', or PEC. In the case of my Losmandy, the worm rotates once each 11 minutes. Before the start of an observing run where high precision is required, a recording mode is initiated when, whilst a bright star is being observed, very minor guiding corrections are made using a very slow slew rate – perhaps twice that of the drive rate – to keep the star centred in the field of view. After 11 minutes the 'Recording' mode is stopped and thereafter these fine drive corrections will be applied automatically as the worm rotates, thus correcting the periodic error.

8.4 How to Align an Equatorial Mount

Before an equatorial mount can be used, the polar axis must be aligned so that it points towards the NCP. This can be done in largely daylight, so maximising your valuable observing time. There are two steps to this process. The first step is to set the correct altitude for the polar axis. This depends on your latitude. This can be found from a map or can be found on Google Maps by bringing up a map including your location (perhaps by entering the ZIP or post code). Then, when you right-click on your precise location within the map and drop down and click on the 'What's here?' line of the menu that appears, the longitude and latitude appear in the 'Location' box above the map in place of the ZIP or post code.

On the side of the mount is a scale and pointer with adjusting screws to set the polar axis to the correct altitude by setting the pointer to the latitude. Of course, in use, the altitude of the polar axis will not be right unless the equatorial head is level, so the tripod legs should be adjusted whilst a spirit level (in two directions at right angles) is used to make sure that the base of the head is horizontal. Usefully, some equatorial heads incorporate either a single bubble level or two levels at right angles.

The second step is to point the polar axis towards true north. An obvious way is to use a compass, but this involves knowing the magnetic variation at your location. There is a superb website, www.ngdc.noaa.gov/geomag-web/#declination, which will give the deviation (called 'magnetic declination') from true north at your location. The latitude and longitude can be entered directly or found from your country and nearest city. The date is automatically entered but can be changed to a future date if required and then, when you click on the 'Calculate' box, the current magnetic declination and its rate are given to an accuracy of half a degree. For example, at the time of writing, the magnetic declination at my home location was 2 degrees 20 arc minutes west, so that I should align the polar axis this amount to the right of north as given by the compass. The page also gives the rate of change, which, for my location, is currently 9.5 arc minutes per year.

If you usually use the mount in a fixed location, perhaps your garden, then a very accurate north–south line can be found, assuming that you have a chance to observe the Sun around noon. If the Sun is used to cast a shadow when it is exactly due south, the shadow will lie precisely north–south. The exact time when the Sun is due south depends on your longitude and the 'equation of time' and involves a tricky calculation where it is very easy to get a sign wrong!

There is a far easier way. The time that the Sun is due south can be found with a program like 'Stellarium' (free to download). Set it to your location in latitude and longitude, and then to about noon local time on a day when the Sun is shining. Remove the atmosphere and put on the meridian line to show the position of due south. Observing due south, you will see the Sun and adjust the time until the Sun is slightly to the east of the meridian. The image scale is increased so that the Sun's disk is quite large, and the exact time when it crosses the meridian is noted. Due to the equation of time and the longitude difference from the local meridian (in the case of the UK, the Greenwich Meridian) the time could be significantly different from noon even when daylight saving time (BST in the UK) is not in force. A watch should be set to an accurate time standard – perhaps using the US time standard (time.gov/widget.html) – which determines and corrects for the time delay in the transfer of time to your computer. It gives Universal Time so ignore the hour part unless you are in the UK in winter! Place a bamboo pole to cast a shadow over the observing platform and wait for the time as determined from Stellarium. The shadow then provides an accurate north–south line along which to align the equatorial axis. When I tried this at home, having done the alignments in daylight, it was very pleasing to look through the polar scope (as described later) of my mount that night and immediately see Polaris in the field of view!

What if you bring out your scope just in the evening? Set the head horizontal and the polar axis to the appropriate altitude, as already described, and adjust the azimuth to point the polar axis up towards Polaris. I have made an adapter so that I can mount a green laser pointer in the polar axis tube (Figure 8.4) and simply move the mount so that it points up at Polaris – but not quite. The NCP is a little way – 42 arc minutes, or 0.7 degree – from Polaris towards the star Kochab, the second-brightest star after Polaris in Ursa Minor – so the laser has to be pointed somewhat to the Kochab side of Polaris. If you have done this, then your alignment is pretty good and fine for visual observing, when small adjustments to the pointing can be made as you track an object. If, however, you are going to do long-exposure astro-photography, you will need to be more accurate and I would then use the polar telescope to offset Polaris the right distance and direction away from the NCP. This scope must be rotated so that Polaris is in the right position. Put the point (often a small circle) where Polaris is to be placed so that it is directly opposite Kochab. Some polar telescopes show the direction of the Plough and Cassiopeia and, in this case, the polar scope is rotated so that they lie in the appropriate directions. If not, a planisphere or the Stellarium program can be used to find the relative orientation of Kochab to Polaris if you cannot see it at the time of observation. Then, when you adjust the mount slightly so that Polaris is within the offset circle, the equatorial axis will be the correct distance away from Polaris towards Kochab, just as required. A detailed description of how to use the star Kochab to align an equatorial mount has been described by Clay Sherrod at www.weasner.com/etx/ref_guides/polar_align.html.

A neat alignment trick can be used with a computerised mount. First ensure that the equatorial head is horizontal and align the mount so that the polar axis points reasonably close to the NCP. Once you have put the scope in the 'home' position, the

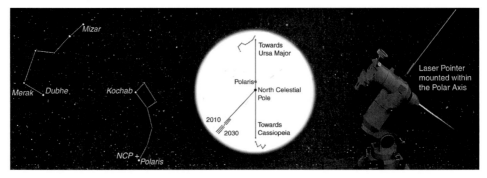

Figure 8.4 The position of the NCP towards Kochab, the view through a typical polar telescope and the author's use of a green laser pointer to make a close alignment to the NCP.

computer will drive the scope to the first alignment star – which may well not even be in the field of view. Instead of using the slow motion drives to centre it in the field of view as you are told to do, adjust the azimuth and elevation of the polar axis to get it approximately right. Then switch off the mount, move back to the home position and start the alignment process over again. It should be a lot better. After, perhaps, a third try you should be pretty well aligned and can begin to enjoy observing.

A similar technique has been proposed by Michael A. Covington for use when a scope can be one-star-aligned and then synchronized on. Having approximately aligned the mount, set the telescope in the home position and do a one-star align to a bright star well away from Polaris. Use the slow motion drives to centre it in the field of view and 'Synchronise to target'. Then slew, under computer control, to Polaris, but this time centre it using the mount adjustments. Switch off the mount, set the telescope in the home position and carry out same process a second time. This time, when you return to Polaris it should be closer and it should be in the field of view of a wide-field eyepiece. Adjust the mount to centre it in the field and repeat the whole process until both Polaris and the star lie at the field centres as you slew between them.

If you want to carry out long-exposure imaging without the use of a guide scope, then it may be necessary to refine the alignment using drift scans in a process that is described in Chapter 16.

8.5 Locating Objects with an Equatorial Mount

In principle, using dials on the mount axes one can arrange the telescope to point directly towards an object given its right ascension and declination and knowing the sidereal time (which is used calculate the 'hour angle' of the object). I have never done this and do not regard it as a particularly rewarding process. As the telescope can be moved in both declination and right ascension (i.e., around the polar axis), in many cases it is possible to 'star-hop' to the target. But supposing the objects to be observed are somewhat dim, such as the pair of galaxies M95 and M96 in Leo? If they are found on a star chart, it will be seen that they lie on almost exactly the same declination (+12 degrees) as the bright star Regulus, Alpha Leonis. Regulus can be centred in the field of view and the declination axis locked. From the star chart it can be seen that the pair of galaxies are 37 minutes of time in right ascension to the east of Regulus. Using the hour angle scale one could, in principle, rotate the telescope in right ascension by this amount to bring them into the field of view of a wide-field eyepiece. Again, I do not tend to do this, but simply sweep the telescope slowly eastwards whilst observing though a wide-field eyepiece until they appear.

Star Hopping with a Computerised Mount

On the face of it, this seems nonsensical: star hopping is carried out with non-computerised mounts! But, in fact, it is not quite so stupid as it might appear. If a computerised mount has to slew a long way to find a target object, its precision may only be sufficiently good that an object would only be visible in a wide-field, low-power eyepiece. Usually this will be fine and the object can be centred before one uses a higher-power eyepiece if so desired. But if the object has a very small angular diameter and so will appear as a star in a wide-field eyepiece, there can be a real problem. One then really needs the object to be within the much smaller field of view of a high-power eyepiece so its true identity can be seen.

An excellent example is the Eskimo (or Clown) Nebula in Gemini. This planetary nebula has an overall size of 48 × 48 arc seconds, but its bright central core is only half this size and cannot be distinguished from a star in a low-power eyepiece. So how can it be easily found? The answer is to star-hop. It is not that far from the 1.1-magnitude star Pollux, Beta Geminorum, which will usually be found in the bright-star catalogue of the controller. First slew to Pollux, centre it in a reasonably high power eyepiece and synchronise to it. This will greatly improve the pointing accuracy in this part of the sky. The Eskimo Nebula is only 8 degrees away, so the accuracy of most mounts should be such that it can be placed near the centre of a medium-power eyepiece. Its true identity can then be seen and so, having centred it in the field of view, one can use a high-power eyepiece. If there is still a problem one can move nearer still by slewing to NGC 2420, a faint open cluster slightly more than 2 degrees away from the Eskimo Nebula. An alternative is to slew to Delta Geminorum, also slightly more than 2 degrees away. It may well not be in the controller's star catalogue, but it is usually possible to enter the coordinates of a user-defined object and slew to that. (Delta Geminorum lies at RA: 07:20:07, Dec: +21:58:56.)

When you are observing, it is sensible to move round the objects from west to east – to first catch those that will be setting in the west and waiting for those towards the east to rise higher in the sky. Then if the mount is synchronized as each object is found, its pointing accuracy will be maintained throughout the observing session.

8.6 Digital Setting Circles

In their simplest form, these consist of positional encoders mounted on the two axes of the telescope which are used to drive a pair of digital displays. However, in virtually all cases, they are incorporated into a computerised system which allows the mount to be used in a 'Push-to' mode. They are thus ideal for use with Dobsonian mounts – one example being the Orion Intelliscope. The alignment procedure is very similar to that with driven computerised scopes. First, one is asked to point the telescope vertical and then press 'Enter'. A list of bright stars is shown on the display, one is chosen (best not to choose Polaris) and the telescope is centred on the star and 'Enter' is pressed again. The process is repeated with a second star. The display then shows a 'warp factor', which ideally should be less than ±0.4, indicating that a good alignment has been found (±0.6 is not too bad). One can then choose from a number of menus, such as the Messier Catalogue (M), the New General Catalogue (NGC), stars, nebulae, clusters and galaxies. It also allows one to store 99 user-defined objects. When a celestial object has been selected, the display gives two arrows to show you in which direction to move the mount and the current error in both azimuth and altitude. When these are both zero, the object should lie within the field of view of a medium eyepiece (~25-mm focal length). As with all computer-controlled telescopes that are not equipped with a GPS system, you will need to input the location and time the controller before finding the planets. (I doubt that you would need help in finding the Moon, Venus and Jupiter!) One interesting feature of this system is that, simply by pressing the ID button, you will find the names of objects that you might happen to come across as the telescope is used to scan the heavens. There are quite a number of similar systems, such as those sold by JMI in the United States.

Equatorial mounts can be equipped with them too. Many years ago I purchased a non-computerised Losmandy GM8. Thinking recently that I would like help in finding some of the more difficult to locate objects such as faint galaxies, I decided to computerise it. Losmandy sells a Gemini computer control system but, as the motors on the non-comprised mount are not sufficiently powerful to slew the telescope, new motors as well as the computer control system are required to computerise the mount, which makes the upgrade somewhat expensive. I thus decided to install a digital encoder system using Losmandy encoders and the Wildcard Innovations Argo Navis control handset, which includes positions for 29,000 objects and allows for 1,100 user-defined objects. The computer indicates the direction in which to move the mount in right ascension and declination and gives the reducing error in position as one does so. The controller uses a single rotary dial in an innovative way along with just two control buttons, 'Enter' and 'Exit'. This can be purchased directly from Wildcard Innovations in Australia or from JMI in the United States (under the NGC-SuperMAX name) and Intercon Spacetec in Germany, all of which will supply suitable encoders for a range of mounts.

A Useful Survey of Equatorial Mounts

An astronomy blog that is well worth reading is written by the author Rod Mollise – search for 'Uncle Rod's Astro Blog'. The August 2012 contribution contains a very useful survey of many of the equatorial mounts that are currently available. Other monthly contributions provide useful comments about the amateur astronomy scene and mini-reviews of many interesting products.

Figure 8.5 Vixen Sphinx, Sky-Watcher EQ6, Losmandy GM8 and iOptron ZEQ25GT equatorial mounts. (Images: Vixen, Sky-Watcher, Losmandy and iOptron)

A Chinese Equatorial Mount

iOptron has brought out an interesting variant of the equatorial mount, the ZEG25GT, which uses a 'Z' configuration of telescope and counterweight shaft. It has been called a 'Chinese equatorial mount' and provides a more stable configuration than a standard German equatorial mount and, as a result, allows a mount weighing only 4.7 kg to support a telescope weighing up to 12.3 kg. A polar telescope (which is never obstructed) is included along with a 32-channel GPS system to aid setting up. iOptron can also provide a PowerWeight™ 8-amp-hour battery that takes the place of the counterweight to provide a very neat system.

Premium Equatorial Mounts

For the really keen amateur, and particularly the astro-imager, a number of companies produce very high quality equatorial mounts (Figure 8.5). Losmandy, Astro-Physics and Takahashi, all of which have been established for many years, provide a wide range of mounts capable of carrying heavy telescope loads. Losmandy manufactures the GM8, G-11 and Titan mounts – with the latter having an instrument capacity of 100 pounds. Those made by Astro-Physics Inc. range from the lightweight Mach1GTO, which is designed for portability but is extremely rigid and has excellent tracking ability, through the 900GTO and 1600GTO mounts up to the 'El Capitan' 3600GTO mount, which is conservatively (!) rated at up to 300 pounds. Takahashi's mounts range from the lightweight EM11, through the EM200 and EM400, up to the EM500 rated at 100 pounds.

There are some new companies producing very attractive mounts; for example, 10 Micron Astro-technology manufactures mounts ranging from the GM 1000 HPS up to the GM 4000 HPS. They incorporate very high precision encoders to provide excellent 'go-to' ability and achieve typical tracking errors of the order of ±1 arc second. This allows unguided exposures to be made exceeding 600 seconds.

Avalon Instruments produces an innovative single-fork equatorial mount, the M-Uno. This has abandoned the traditional worm and gear drive in favour of one that uses toothed belts. These provide a very smooth motion and the complete elimination of backlash. The single-fork design does not require a long counterweight arm, and there is no 'meridian flip issue' as is the case with German equatorial mounts. This means that the mount will track a target from east through the meridian into the west uninterrupted. It is equipped with the Sky-Watcher Synscan 'go-to' controller.

Perhaps the ultimate amateur mount for observatory use is the Software Bisque Paramount ME II equatorial mount capable of handling an instrument weight of up to 240 pounds. It uses belt-driven drives to eliminate backlash. The company has recently brought out the Paramount MX equatorial mount, which can be mounted either on a pillar for observatory use or on a dedicated tripod for dark-sky excursions. The uncorrected periodic error is less than 7 arc seconds before correction (1 arc second after correction) and the use of the 'T Point' alignment system (which uses stars across the sky to build up a precise map of any pointing errors) will provide a 'go-to' precision of 30 arc seconds or better.

9
The Art of Visual Observing

The objective of this chapter is to show you how to get the very best out of your visual observing. I think that it is somewhat of a pity that many amateurs have almost given up using their eyes to view the heavens to concentrate on astro-imaging. It's nice to think that the photons being detected by your retina from M31, the Andromeda Galaxy, left the galaxy some 2.5 million years ago – you are literally looking back in time! There are many images that remain in my memory: the scars on the face of Jupiter as fragments of Showmaker-Levy 9 impacted its surface, the globular cluster M13 appearing almost three dimensional when I first viewed it with my newly constructed 10-inch Dobsonian telescope, the Double Cluster in Perseus when observed with my 12-inch Maksutov and the iridescent green of the Dumbbell Nebula when seen through a 16-inch Schmidt-Cassegrain. But telescopes do not have to be so large to observe some memorable sights: when testing a 5-inch Newtonian from a dark site I was delighted to observe a supernova that had recently occurred in the galaxy M101.

9.1   The Eye

We should not forget that the eye is an important part of the imaging chain (see Figure 9.1). Understanding its strengths and weaknesses can help make the best use of its (impressive) abilities. The aspheric lens has a focal length of around 24 mm but, by use of the ciliary muscles, its focal length can be changed to enable the eye to accommodate both near and far distances (at least in young people). The lens has an 'aperture stop', the pupil, to help the eye accommodate a range of brightness. Increasing in aperture from ~2 to 7 mm, this can increase the eye's sensitivity by about 12 times. But, sadly, as we grow older the pupil no longer increases in diameter so much, and 5–5.5 mm may be its maximum. As the collecting area is proportional to the area of the pupil's diameter, the difference between a 7-mm pupil and a 5-mm pupil is a factor of 2 in area; thus young people may be able to see stars with their unaided eyes which are about two-thirds of a magnitude fainter than those that can be seen by people in their 60s.

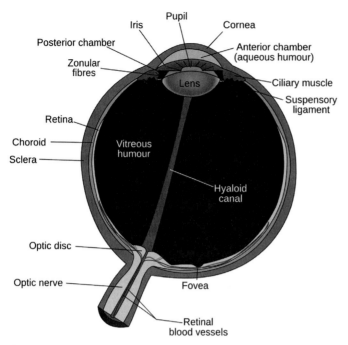

Figure 9.1 The human eye. (Image: Wikimedia Commons)

Like camera lenses, the human lens suffers from optical aberrations that become more apparent as the pupil opens up. These aberrations will reduce the ability of the eye to resolve detail in an image and, when the pupil is 6 mm or more in diameter, the fundamental resolution of the lens drops by about half from its optimum value when it is ~1.5–2 mm in diameter. Experiments have shown that when the pupil has a diameter of 2 mm, the eye can resolve targets spaced approximately 30 arc seconds apart but that this drops to ~60 arc seconds when the pupil fully dilates. No matter what the ability of the lens, the actual resolution of the eye will also depend on the spacing of the receptors in the retina. Their spacing in a region of the retina, called the foveola, where we observe our sharpest images, is 1.5 microns and, as one might expect, evolution has made this a very good match to the inherent resolution of the lens, so that effectively all the detail present in the image that falls onto the retina can be captured. The foveola lies within the fovea and is where we will image any object that we wish to observe in detail. Fifty per cent of the information transmitted to the brain from the eye comes from the foveal region. As the eyes provide ~90% of all the information from the senses, 45% of all the data that our brain processes comes from just two areas that are only 1 mm across!

The eye utilises two types of light sensor embedded in the retina – rods and cones. It is interesting to note that these are towards the back of the retina behind the network of nerves that bring their signals to the brain. As these are brought together and exit the eye, there is a region where there can be no light receptors, giving a 'blind spot' about 15 degrees off the optical axis of the eye towards the ears. The rods are monochromatic, whilst the three types of cones are sensitive to bands of colour in the red, green and blue parts of the visible spectrum. From their signals, the brain constructs a colour image of the scene. The foveola, which is about 0.35 mm across, contains only cones, but away from this rods appear and the density of cones begins to diminish. The spacing of the receptors is greater, so our peripheral vision will be less sharp. This probably matches the fact that, due to lens aberrations, the image quality will fall away from the optical axis so a denser matrix of receptors would not give any higher resolution. (One advantage that the eye has over a camera or telescope is that the retina is curved rather than flat, and it is interesting to note that the 48-inch Schmidt camera used curved photographic plates to image across a 6.4 degree × 6.4 degree field of view.)

There is a point one should note when using the eye in conjunction with a telescope. Assuming the eye is dark adapted and thus ~5–7 mm in diameter, one might at first sight think that its resolution will not be at its best. However, what matters is the effective aperture, and this is determined by the diameter of the exit pupil of the telescope/eyepiece combination. If the exit pupil is only 3 mm in diameter, the outer parts of the lens are not being used and so cannot degrade the image falling on the retina. As one increases the magnification, the exit pupil reduces in diameter, and one often finds that a particular magnification gives the clearest image. One reason for this is that the exit pupil is then matched to the optimum aperture of the eye's lens. Zoom eyepieces may not be the best optically, but they can be quite useful when, for example, one is about to observe Jupiter or Saturn. One can quickly change the magnification of the image and see which will be best on a given night and choose an appropriate eyepiece to give a similar magnification. Often this will be limited by the seeing, but on nights of very good seeing one will often find that the exit pupil is close to the eye's optimum value of about 1.5 mm.

As magnification is increased with a small-aperture telescope, the exit pupil may well get below 1 mm in diameter. When a 4-inch (102-mm) telescope is used, a magnification of just 100 will equal this, and with a magnification of, say, 200 it will drop to 0.5 mm in diameter. Two problems then arise. Firstly, diffraction will begin to limit the resolution, and so increasing the magnification (which reduces the diameter of the exit pupil further) will not allow one to see more detail. This is one reason it is often said that there is no point in using a magnification of more than 50 per inch of telescope aperture, as then the exit pupil will become less than 0.5 mm and so begin to degrade the image. I nearly always find that a magnification of about 25 per inch of aperture gives me the clearest and highest-contrast images when observing the planets, as this then gives a best match to the capabilities of my eye. The second reason is more applicable to older observers; as we age, the amount of debris, called floaters, within the eye increases and this can seriously affect the observing experience. Their effect becomes more obvious as the magnification increases and the exit pupil is reduced. There are two ways round this: one is to purchase a larger-aperture telescope as then, for a given magnification, the exit pupil will be greater, and the second is to use a binoviewer as the brain, given two images to process, is able to reduce their effect.

9.2   Dark Adaptation

The increase in diameter of the pupil is the eye's first, and rapid, approach to what is termed 'dark adaptation', increasing the eye's sensitivity by up to 12 times (but less when one is older). However, if the eye is kept in darkness for longer periods, a second, chemical process takes place within the eye when rhodopsin, also known as visual purple, is created in part from vitamin A, beta-carotene (a reason that visual observers should eat plenty of carrots!). This increases the sensitivity of the cones to some extent, but that of the rods to a far greater extent, and this is why night-time viewing is essentially monochrome. It takes about 30 minutes for this process to become efficient, and longer periods of darkness may well improve the eye's sensitivity further. White light of any significant level will instantly destroy the rhodopsin and it will then take a further 30 minutes to recover.

Dark-adapted night vision is called 'scotopic vision' and will be hindered by a deficiency of vitamin A, which gives rise to what is called 'night blindness'. I was recently asked why a gentleman had not been able to see the Aurora Borealis when all others had seen it. He had then had his eyes tested, with no obvious problems detected. I was not immediately able to think of any reason but later realised that a vitamin A deficiency may well have been the cause and was pleased to find that this is a recognised condition.

Rhodopsin absorbs very strongly in the blue/green part of the spectrum (and thus reflects reddish-purple light; hence its alternative name) but is far less sensitive to the effects of red light. This is why red lights are used at star parties and for lighting the interior of observatory domes. Telescope control programs often have a red 'night mode' capability, and laptop screens may be covered with a red film for use with other software. Even bright red lights can partially destroy one's dark adaptation, and I have two lights: a powerful lamp using 24 bright white LEDs covered by a red film and used to point at the ground when moving around and a small low-power torch having a single red LED for use in picking out eyepieces or looking at star charts. Head straps with a red LED lamp attached to the front are also widely used.

An aspect of dark adaptation that is not so widely known is that spending the previous day under bright skies, perhaps on a beach, will reduce the ability of the eye to dark-adapt. If you are exposed to bright skies, be sure to wear dark glasses. It is not unheard of for some keen observers to keep an eye patch over their observing eye prior to a night's observing, and some even use one to place over the eye when not at the eyepiece during an observing session. If there are some nearby lights at an observing site, by covering your head with black cloth you can eliminate extraneous light when at the eyepiece.

Some stars are bright enough to show colour, but the sensitivity of the red- and blue-sensitive cones does not increase as much as that of the green, so that the eye then has a very peaked response in the green, with a significant fall-off towards the red and blue. This perhaps explains why red giants tend to look orange rather than red. Planetary nebulae, like the Dumbbell Nebula mentioned earlier, emit a significant amount of light in the green OIII emission lines (discussed further in the section on narrowband colour imaging in Chapter 15). This, coupled with the eye's peak colour sensitivity in the green, is the reason these are some of the very few deep-sky objects in which we can perceive colour. When a telescope with an 8-inch aperture or above is used under transparent skies, the centre of the Orion Nebula can also be seen to have a greenish tinge.

When dark adapted, the rods become significantly more sensitive than the cones, corresponding to a difference of several magnitudes, so it is not surprising that faint celestial objects are seen only in black and white. Thus in order to see the faintest details we need to focus our eyes on the rods rather than the cones. The majority of the rods are situated outside the middle 10% of the retina, where the cones are concentrated, and so the eye is most sensitive away from the centre of its vision. Thus, to observe faint objects, such as a galaxy, an observer will intentionally look to one side of the object rather than straight at it. The peak concentration of the rods occurs at about 20 degrees off-axis. The most sensitive region is located on the side of the retina towards the eye with the least sensitive part towards the ear (remember too that the blind spot is on this side), so moving the eye around the object but concentrating on the object (quite hard to do) should enable you to find the optimum off-axis viewing angle and direction.

The sensitivity of the eye depends on its supply of oxygen. Some observers will hyperventilate, taking several deep breaths through the nose or puckered lip, before putting their eye to the eyepiece, but it is best to then continue breathing normally rather than to holding the breath as one can be tempted to do. This technique is even more important at high altitude. Here there are two opposing factors: the higher one observes from, the less atmosphere there will be above, thus enabling one to see fainter objects, but the atmosphere becomes thinner so less oxygen will reach the eyes, reducing their sensitivity. Six to nine thousand feet is probably the optimum height without the use of an oxygen bottle to take breaths from before each look through the eyepiece. I shall never forget observing from the visitor centre at a height of 9,300 ft (2,835 m) on the flanks of Mauna Kea. A three-day-old Moon was setting behind the mountains but, when the thin crescent had set, the 'old Moon' illuminated by light reflected back from the Earth (earthshine) appeared as bright to me as would a full Moon as seen from sea level!

Probably due to the useful ability of sensing missiles or attackers that might approach from the side, evolution has made the peripheral parts of the retina sensitive to motion. One can use this fact to confirm the presence of a faint feature (or even to first see one) by gently tapping the telescope tube to induce a small vibration. The side-to-side motion of the field of view can often make a faint object become visible. Or one can simply use the slew control of the mount to slowly move the object across the field of view. This technique is very useful for observing, for example, the dark dust lane that arches across one side of M31, the Andromeda Galaxy. In the same way, by moving M31 well away from the field of view and gradually moving in towards it, you may well make its full extent more apparent.

Another factor comes into play in the observation of low-contrast objects, for example, the spiral arms of a faint galaxy. An object whose surface brightness is such that the contrast between it and the sky is below the eye's detection threshold will be invisible. Probably as a result of the way the brain processes the data from the eyes, higher magnifications may facilitate the spotting of faint structure against the sky background. This is somewhat counterintuitive, as using a lower power will give a brighter image, but remember that the sky background will also be brighter and the contrast ratio between an extended object (such as a galaxy or nebula) and the sky background will not change. However, as the area on the retina covered by a low-contrast feature is increased (by upping the magnification), the ability of the brain to perceive it first increases and then decreases. (This is not the same as when higher magnification reduces the background light level but leaves a star's brightness constant and so makes it possible for fainter stars to be seen.) There is an optimum magnification which depends on the sky conditions, the surface brightness of the object, its size and the telescope aperture. A smaller telescope will tend to require a higher magnification to detect a low-contrast object, and the optimum magnification for a 150-mm telescope may be some 40% greater than that for a 200-mm telescope. For small objects such as galaxies with a low surface brightness (and hence low contrast against the sky background) a moderate magnification may well be best, ranging from ~100 for a 350-mm telescope, ~150 for a 200-mm scope to ~200 for a 150-mm scope. The obvious strategy is to use a number of eyepieces to gradually increase the magnification, so making the image fall on a larger area of the retina to find the appropriate eyepiece for the object.

For large extended objects such as M31, the Andromeda Galaxy, and M42, the Orion Nebula, a low magnification tends to be best – and will also allow a wide field of view to be observed. But remember, the lower the magnification, the larger the exit pupil and there is no point in having a telescope exit pupil greater than the dark-adapted diameter of the eye's pupil. The minimum magnification is simply the telescope aperture divided by the pupil diameter. But this can be a problem with telescopes of long focal length such as an f10, 203-mm Schmidt-Cassegrain having a focal length of 2,032 mm. Given a dark-adapted pupil of 7 mm this corresponds to a magnification of 29, which would require the use of an eyepiece with a 70-mm focal length. The longest focal length of eyepieces is 55 mm, so one would need to add an f6.3 focal reducer to give an effective focal length of 1,280 mm, as this would then require an eyepiece with a focal length of only 44 mm. With a dark-adapted pupil of 5 mm the problem is even worse.

An interesting question is whether the eye can detect a single photon. Experiments carried out in the 1940s showed that they can. Dark-adapted subjects were shown very brief (1-millisecond) flashes of light which subtended an angle of 10 minutes of arc within their peripheral vision, where there is the highest concentration of rods. The photons would thus fall on the order of 350 rods. The flashes became visible when the light pulse contained 90 photons, which is thus the number that must enter the eye to trigger a response. In fact only 10% of these reach the rods within the retina, so it is highly unlikely that more than one would fall on a single rod which must thus be able to detect a single photon. The fact that a number must be received very closely spaced in time means that the brain is not overloaded by random triggering which would give rise to a background noise level.

This suggests that the 'quantum efficiency' of the eye, that is, the number of detections compared with the number of incident photons, is of the order of 1–2% – far less than that of a cooled CCD camera, which can approach ~80%. When a very faint star is just below the nominal threshold of vision, the statistical nature of their arrival times means that, every so often, a bunch of photons will arrive sufficiently closely in time for an instantaneous detection. Thus if, when observing a star field, one appears to see brief, very faint 'flashes' of light at a particular spot occurring reasonably often, then a star will have been detected! It is generally stated that this technique will allow stars of one greater magnitude to be detected so that, if one could continuously observe a 13th-magnitude star in the field, stars of 14th magnitude might well become apparent over time.

Given a superb observing site and transparent sky, a star with a visual magnitude of zero will, over the visual bandwidth, provide about 2 million photons per square centimetre per second. A fully dilated eye has an area of ~0.5 cm^2, so about 1,000,000 photons will enter the eye per second. Assuming that the eye has an exposure time of 1/30 second, some 33,000 photons are available for detection. If the eye could detect a star when 100 photons entered it during this time interval, it should be able to detect a star that is 1/330th as bright. As 5 magnitudes is equivalent to a factor of 100 in brightness, this would correspond to a star of slightly more than magnitude 6 and so agrees with observation. When recently observing M101 using a telescope of 127-mm aperture (from a dark site where I could see fifth-magnitude stars) I should have been able to see stars about 500 times fainter than with my unaided eye. This is about 6.5 magnitudes, so giving me a limiting magnitude of about 11.5. At the time, the supernova in M101 had a visual magnitude of 11.1, so I have some reason to believe my detection was real. Some observers can detect fewer photons and the observation of stars of approximately seventh magnitude is not uncommon from a truly dark site.

There is one final, and highly important, factor in detecting faint details in an object – the amount of time one spends observing it. As one observes an object over a period of time – which could easily stretch for an hour or more – the amount of detail that can be perceived increases. This is presumably due to the brain 'learning' the object in some way. This knowledge appears to remain, as on later occasions one will tend to see more when first viewing it again than when it was first observed. The effect is often noticed when an experienced observer shows an object to a newcomer. It will often take quite a while for the newcomer to 'see' what was immediately apparent to the experienced observer. When the noted visual observer Stephen James O'Meara produced his wonderful companion to the Messier objects, he observed each object for an average of 6 hours or more spread over several nights!

As I was researching the contents of this chapter – having previously decided on its title – I came across this quotation by Sir William Hershel:

You must not expect to see at sight…. Seeing is in some respects an art which must be learned. Many a night have I been practicing to see, and it would be strange if one did not acquire a certain dexterity by such constant practice.

9.3   Choosing an Observing Site

There are two obvious points: try to find an observing site from which a large part of the sky can be seen and which is also as dark as possible. This may very well mean leaving your home and venturing into the countryside. Again it is pretty obvious that if you wish to observe to the south, where (from the Northern Hemisphere) the planets will be highest in the sky, you should find a location to the south of the town or city where you live to avoid observing over its light pollution. With the large Manchester conurbation to my north, there is no way that I can avoid light pollution when observing in this direction, but I have found locations to the south, west and east of my home that I go to depending on what I wish to observe. I thought I had found for myself a dark and almost hidden spot which has an excellent low horizon to the west, but when I went there to photograph a nice grouping of Jupiter, Venus and Mercury along with a thin crescent Moon I found that two other local astronomers had beaten me to it!

Keen astronomers may well travel far afield to carry out a night's observing and then it is good to know where the light pollution will be least. Phillips has produced a 'Dark Sky' map of the United Kingdom which shows the limiting magnitude of stars as will be typically seen with the unaided eye. The darkest locations agree with what one might well think: be as far away as possible from large conurbations in locations where the population density is low. In the United Kingdom, the national parks are good choices, with the Cotswolds, north Devon, Mid Wales and south-western Scotland being some of the darkest locations within reasonable distances of major conurbations. The national parks in the United States are also obvious choices. An overall view of the effects of light pollution across the globe (see Figure 9.2) is provided on the 'Night Sky in the World' Web site, which also provides high-resolution maps giving the limiting visual magnitudes that might be attained.

There are now national and international Web sites that plot the readings made by what are called 'SQM' meters made by the Unihedron company. These small battery-operated devices act as very sensitive light meters which, when pointed up to the zenith, integrate for a while to give a measure of the sky brightness in magnitudes per square arc second. (They also measure the temperature so that they can eliminate the effects of dark current in the sensor.) This is a logarithmic measurement and thus large changes in sky brightness correspond to relatively small numerical changes. Under a bright Moon or near streetlights, a value of ~16 might well be recorded, whilst under the darkest skies the value may increase to more than 21. A UK map with SQM readings can be found in the 'MyDarkSky' Web site, and a global map within the 'Globe at Night' Web site. Both include conversion tables to give limiting zenithal magnitudes. As an example, within rural Cheshire to the south of Manchester I have recorded a value of 18.6 corresponding to a limiting magnitude at the zenith of 4, whilst at my holiday home in Mid Wales, the values were 20.9 and 6.1. Figure 9.3 gives plots for both Ursa Minor and Crux which can give an indication of the limiting visual magnitude at their altitudes in the Northern and Southern Hemispheres respectively.

Figure 9.2 Light pollution. (Image: NASA/GSFC, NOAA, NGDC and DMSP)

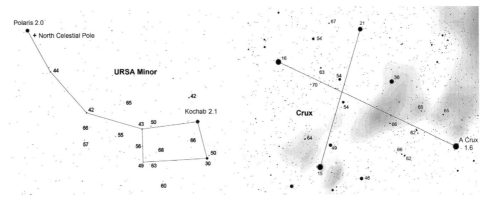

Figure 9.3 Limiting magnitude charts for the Northern and Southern Hemispheres. (For the fainter stars the decimal points are removed to avoid confusion with stars; e.g., 57 means 5.7.)

You might think that from a superb dark-sky location when your eyes have become fully dark adapted the sky would look black. Surprisingly, it can look quite grey! At the Mauna Kea Observatory at a height of 13,700 ft (4,175 m), where, perhaps with the aid of oxygen, it is possible to detect eighth-magnitude stars, new observers sometimes complain that the site is poor because the sky does not look dark. This is the result of natural sky glow, which is made up of three components. The first is called 'air glow', which is caused largely by oxygen atoms glowing in the upper atmosphere that have been excited by ultraviolet light during the previous day. Much of the emission is at a wavelength of 5,577 angstroms in the green and is the same colour as that seen in auroral displays. Its brightness varies throughout the solar cycle, being greatest around solar maximum, when it can be the dominant cause of sky glow. When Stephen James O'Meara produced his classic companion to the Messier objects using just a 4-inch telescope, the skies above Big Island Hawaii were at their darkest for 10 years – his observing period happily coinciding with solar minima. The second

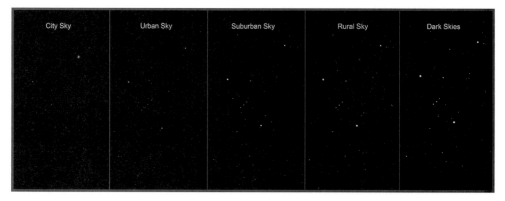

Figure 9.4 The effects of light pollution on observing the constellation Orion.

component is light scattering caused by dust particles in the plane of the Solar System. This can be observed as the zodiacal light most obviously seen as a broad arc rising above the western horizon after sunset (or seen in the east before dawn). A further effect of this dust is called the *Gegenschein* (German for 'counter-glow'), which is the result of light scattered back towards us from the region in the sky opposite the Sun. The third component is starlight (including that from the Milky Way) scattered by the atmosphere just as sunlight is scattered in daytime. As air molecules preferentially scatter blue light, the night sky has a very faint blue colouration, though it is below our threshold of colour vision. Sadly, for many of us such considerations are of little consequence as the sky glow is totally dominated by light pollution (see Figure 9.4).

9.4 Making the Most of Your Observing Site

Observe when the objects are highest in the sky. This is a pretty obvious statement! If a celestial object is observed when just 10 degrees above the horizon, it will be seen through 5.6 times more atmosphere – termed 'air mass' – when it would be just 1 air mass if observed directly overhead. At an elevation of 30 degrees this has reduced to just 2 air masses (see Figure 1.13). Not only will a reduced air mass, containing water vapour, dust and aerosols, absorb less light from the object being observed, there is less to scatter light pollution, so away from dark skies there is a double advantage!

The Use of Filters to Combat Light Pollution

It is not always practical to leave one's home for an observing trip in the country, but there are ways of improving one's observing from a light-polluted location – and even from a dark-sky site. I would like to make one point that I have never seen stated, as perhaps it does not help with the (highly important) dark-sky campaigns. The problem with light pollution is not so much the light emitted itself (unless it is very close to your observing site) as the fact that water vapour, aerosols and dust scatter that light back towards the Earth. If there were no atmosphere, light pollution would not be a problem. Even from an observing site within the borders of a town, there are some nights – rare though they might be – when the sky is so transparent one *can* see stars not normally visible. The best view that I have ever had of the Orion Nebula from my home location was when low clouds were scudding across the sky. They were trapping the light pollution, so the sky seen through breaks between them was wonderfully dark!

But it is possible to use filters to help mitigate its effects. Some broadband filters are designed to attenuate the light emitted by the common causes of light pollution (such as high-pressure sodium lamps) allowing that across other parts of the spectrum to pass through. Examples of these are the Lumicon DeepSky filter, the Orion SkyGlow, the Meade broadband and the Celestron LPR. However, it has to be said that with the increasing use of white lights for sports grounds and security lights, broadband filters are not always that effective. They can give some improvement in contrast on deep-sky objects but, to observe faint galaxies whose light covers the full spectrum, a dark-sky site is really the only option.

However, for some other deep-sky objects such as planetary nebulae, narrowband filters can come to one's aid and really make a difference. This is due to the fact that they emit significant light at specific emission line wavelengths, most notably the OIII lines and H-beta lines in the green. If a filter is designed to allow only these wavelengths and reject all others, they can be made to stand out against a virtually black sky even from a heavily light-polluted site. Excellent examples are the Lumicon ultra-high-contrast filter (UHC), the Meade narrowband and the Orion Ultrablock filters. These pass the two OIII lines at 4,960 and 5,010 angstroms and the H-beta line at 4,860 angstroms but will block virtually all light pollution and air glow. They are excellent for observing the Orion, Swan, Lagoon and other extended nebulae as well as planetary nebulae. One trick is to align (using a go-to scope) on a faint planetary nebula which will probably not be seen. Pass one of these filters side to side between the eyepiece and the eye – the sky will get much darker and the planetary nebula will blink in and out of view! If one were to buy only one filter, these or their equivalents would be the ones to get.

Two further filters can be used for visual observing that have even narrower bands. These are the OIII and H-beta filters. The OIII can work better than the UHC or Ultrablock filters on some emission nebula and is superb on planetary nebulae. Other objects which respond greatly to the use of an OIII filter are the Veil Nebula in Cygnus (a supernova remnant) and the Rosette Nebula in Monocerous. This should be the second filter that you purchase. Filters passing only the H-beta line are also available but have a more limited use. They will enhance a few emission nebulae – most notably IC434, which lies behind the dust cloud forming the Horsehead Nebula in Orion. Their use, given a really dark sky and a telescope of reasonably large aperture, is the only way that it is normally possible to see the silhouette of the horse's head.

Should One Wear Glasses?

Sadly, many of us need to wear glasses to correct our sight, but we may not necessarily need to wear them at the eyepiece, as the image can easily be brought to focus on our retina. Indeed, having one's eye close up to the eyepiece helps to reduce any extraneous light and, if the eye relief of the eyepiece is short, it may even be necessary in order to view the whole field of view. The contrast of the image may well be improved too, as two lens surfaces are eliminated. Incidentally, this does alter the effective focal length of the eyepiece somewhat and, as I am significantly short-sighted, with a given eyepiece I will gain a slightly higher magnification but see a slightly smaller field of view.

However, if your eyes suffer from severe astigmatism, stellar images will be distorted and then glasses are a must when you are using low-power eyepieces where the exit pupil (and hence the diameter of the eye lens) is large. With higher powers where the exit pupil is smaller, astigmatic effects within the eye are less noticeable and glasses may not be needed. But there is one situation in which it would be best to use glasses. This is when you are showing others the view through your telescope. The focus achieved with the use of your glasses should mean that others will find the image in focus as well – assuming, of course, that they use their glasses if needed.

9.5 Deep-Sky Objects

Deep-sky objects are astronomical objects other than individual stars and objects within the solar system such as the Sun, Moon, planets, comets and asteroids. The term is used by amateur astronomers to denote generally faint objects of which a few can be seen with the unaided eye but the vast majority of which require the use of a telescope.

Types of Deep-Sky Objects

Open Star Clusters. These are groupings of young stars which gradually disperse over time so become less obvious features. A classic example is the Pleiades Cluster in Taurus. There are more than 1,000 catalogued open clusters visible within our Milky Way galaxy.

Globular Clusters. These are compact spherical groupings of stars. They have the same origin and many are very old, dating from the time of the formation of the galaxy. They may contain anything from tens of thousands to millions of stars, and their diameters range from 50 to 300 light years. The best example in the Northern Hemisphere is the cluster M13 in Hercules. In the Southern Hemisphere, 47 Tucanae is one of the best – the bigger and brighter object Omega Centuri is now thought to be the nucleus of a small galaxy that has been stripped of its outer stars.

Diffuse Nebulae. These are dense regions of gas and dust which tend to be locations where stars are being born. There are two types:

1. Emission Nebulae. As the name implies these are bright, due largely to the emission of pinkish-red light at the H-alpha wavelength of 6,563 angstroms. This is emitted by hydrogen atoms that have first had their electron stripped off by an

ultraviolet photon emitted from a nearby blue (very high surface temperature) star. When the hydrogen atom re-captures an electron, it cascades down the allowable energy levels. The most prominent transition between these levels gives rise to the H-alpha emission. (See Chapter 17 for a discussion of the hydrogen spectrum.) As massive blue stars burn their hydrogen fuel very quickly, these stars will be young and will usually form part of what will become an open cluster. So emission nebulae are regions of star formation. The best example in the heavens is the Orion Nebula, which can be seen from both hemispheres.

2. Reflection Nebulae. These are bluish regions in which dust particles are scattering light from nearby young stars. They appear blue because the brightest stars emit most of their energy in the blue part of the spectrum and because the dust scatters blue light more efficiently. There are reflection nebulae around the brighter stars of the Pleiades Cluster (which is thought to be passing through a dust cloud) and in the Running Man Nebula just north of the Orion Nebula.

Dark Nebulae. These are regions which are seen in silhouette against the light of the Milky Way. Perhaps the best example is the Coal Sack Nebula in Crux, beside the Southern Cross. The most prominent in the Northern Hemisphere is the Cygnus rift – seen as a deep gash lying along the plane of the Milky Way.

Planetary Nebulae. These are so named because some appear to look a little like planets having a circular disk. They are, however, the remnants of stars similar to our own. Our Sun is currently converting hydrogen within its core into helium. At the end of what is called its 'main sequence' life, when the hydrogen has been fully converted, the core collapses somewhat, its temperature increases from ~15 million K up to ~100 million K, at which point the helium is converted, first, into carbon and, then, into oxygen. Depending on a star's mass, heavier elements may be produced as well. When all the allowable nuclear reactions have been completed, gravity causes the core to collapse, leaving what is called a 'white dwarf', a very dense and initially extremely hot object about the size of the Earth. The collapse is halted by what is called 'electron degeneracy pressure' – a quantum mechanical effect. At the same time the outer parts of the star are blown outwards, giving rise to a shell of excited gas (excited by ultraviolet light emitted by the white dwarf) surrounding the white dwarf. The Ring and Dumbbell Nebulae, in Lyra and Vulpecular respectively, are two of the best examples.

Supernovae. The explosions of giant stars, these are rarely seen in our own galaxy – the last in 1604 – so are found in external galaxies. The most prominent in recent years was supernova 1987A, observed in the Large Magellanic Cloud and easily visible to the unaided eye. Requiring telescopes to observe them more recently were those in the galaxy M51 (in 2005 and 2011) and the galaxy M101 (in 2011). The discovery of the 2005 M51 supernova was made by an amateur astronomer using a Meade LXD75 mount and Schmidt-Newtonian telescope!

Supernova Remnants. These are the remnants of massive stars in which electron degeneracy pressure cannot halt the collapse of the core, which reduces in size until it has a diameter of typically 20 km. The electrons and protons have fused to give neutrons that make up the vast bulk of the star, which is thus called a 'neutron star'. These cannot be seen visually, but many emit rotation beams of radio waves – rather like interstellar lighthouses – and are called 'pulsars'. The outer parts of the star are dispersed into space and have a far more chaotic structure than planetary nebulae. Few can be seen with amateur telescopes, but the Crab Nebula – the remnant of a star that exploded in 1054 – is one, lying close to the easternmost 'horn' of Taurus.

Galaxies. Galaxies are the fundamental building blocks of the universe. They contain from between hundreds of millions to hundreds of billions of stars in a wide variety of forms which were characterised by Edwin Hubble.

- Spiral Galaxies. These are some of the most beautiful in the universe. They have a spherical central bulge surrounded by a flat disk containing spiral arms. These are caused by a rotating density wave which compresses the gas as it passes, so triggering star formation. The very luminous young blue giant stars – whose light dominates the more numerous but far less luminous yellow and red dwarfs – and the emission nebulae in which they are formed make the spiral arms stand out. Prime examples are the Andromeda Galaxy, M31, M33 in Triangulum and M51 in Venes Canitici. We see them because of the stars within them, but they contain far more mass in the form of gas and dust between the stars and even more in the form of dark matter.
- Barred Spiral Galaxies. These have a central bar instead of a spherical bulge. It is believed that our own Milky Way is a barred spiral and that such galaxies make up two-thirds of all spirals. One of the easiest to observe is NGC 1300, having an integrated apparent magnitude of 11.4. It is thought that the Large Magellanic Cloud may be a barred spiral, although the arms are vestigial – its initial spiral structure largely destroyed by the gravitational pull of our own galaxy.
- Giant Elliptical Galaxies. These are the largest known galaxies and are probably the result of mergers between smaller galaxies. They have an elliptical outline but no obvious internal structure. They are often found in the core of galaxy clusters, an example being M87 in the Virgo cluster. The stars are generally old, and they contain far less gas and dust than spiral galaxies.
- Dwarf Ellipticals. These are elliptical galaxies that are far smaller than others. They are often seen in association with other larger galaxies such as M32, a satellite of the Andromeda Galaxy.
- Starburst Galaxies. These are galaxies in which a great burst of star formation is in progress. The prime example is M82 in Ursa Major. It is believed that they are caused by a collision or close encounter of two galaxies. (In the case of M82, M81 – a normal spiral – is the second galaxy.)
- Irregular Galaxies. These have no obvious form (as their name implies). The Small Magellanic Cloud is perhaps the best example visible, but some now regard it as a type of barred spiral whose structure was destroyed by the pull of our own galaxy. They are chaotic in appearance, with neither a central bulge nor any trace of spiral arms. Irregular galaxies make up a quarter of all galaxies but, as they are generally small, and hence faint, are not often observed by amateurs.

Clusters of Galaxies. Galaxies are found in groups of up to ~50 or in clusters ranging up to several thousand. Our own Milky Way is the second-largest (after the Andromeda Galaxy) in our 'local group', which contains slightly more than 40 galaxies. We cannot be quite sure, as some may be hidden behind the band of the milky way. (As mentioned in an earlier chapter, 'milky way' is lower case when the term refers to the band of light seen across the sky rather than our galaxy.) The nearest giant cluster is the Virgo cluster, which contains at least 1,300 galaxies and contains some giant elliptical galaxies such as M87. The Virgo cluster subtends an angle of 8 degrees across the sky to the lower left of Leo. This region, often called the 'realm of the galaxies', is a very rich one to observe with a large telescope.

The Virgo cluster lies at the heart of the Virgo Supercluster, one of millions of superclusters observed across the universe. It contains at least 100 groups and clusters within a volume 110 million light years across. Our local group lies towards its periphery.

9.6 Catalogues, Star Charts and Planetarium Programs

The Messier Catalogue

Messier, a French astronomer, who made his observations from the Hotel de Cluny in Paris, observed all types of astronomical phenomena such as occultations, transits and eclipses, but his great love was discovering and observing comets. His 13 comet discoveries brought him great fame, and he was made a fellow of the Royal Society and elected to the French Academy of Sciences.

Whilst scanning the heavens in August 1758 in search of new comets, Messier came across a faint nebulosity in the constellation of Taurus, the bull. Thinking first that this might be a comet, he observed it on following nights and soon realised that it was not a comet as it did not move across the sky. To prevent both himself and others from wasting observing time in the future, he decided to produce a catalogue of nebulous objects that might be first thought to be a comet. This object in Taurus thus became the first object in his catalogue, M1. It is the remnant of a supernova that was observed in 1054 and is now known as the Crab Nebula, as a great nineteenth-century astronomer, the third earl of Rosse, having observed it with his great 72-inch Newtonian, thought that it resembled a horseshoe crab. Messier did not begin compiling his catalogue in earnest until 1764, and within seven months he had added a further 38 entries, including the globular cluster M13, the Dumbbell Planetary Nebula, M27, and the Andromeda Galaxy, M31.

The first of Messier's catalogues, listing 45 objects, was published in 1774, and his final list of 103 objects was published in 1781. Since then the list has grown to 110 objects, as astronomers and historians have found evidence of another seven deep-sky objects that had been observed either by Messier or by his assistant Pierre Méchain not long after the final list had been published.

With his list, Messier bequeathed a wonderful resource on amateur astronomers, as it contains many of the most beautiful celestial objects that can be seen in a small telescope – covering every type of object from open and globular clusters, diffuse and planetary nebulae to many of the brightest galaxies that can be observed from mid-northern latitudes. Because Messier discovered them with relatively small refractors (albeit without the light pollution that many of us now suffer) these objects can be observed visually with virtually all amateur telescopes and are among the most attractive deep-sky objects in the heavens.

A book that has been a constant companion since the early 1980s, and is still available new or second-hand, is *The Messier Catalogue* by John H. Mallas, who observed and drew them using a 4-inch refractor, and Evered Kreimer, who photographed them with a 12½-inch Cave reflector. It is published by Sky Publishing Corporation and includes a facsimile of Messier's 1784 catalogue – well worth practising one's French for! Two other 'Messier' books, both published by Cambridge University Press and highly regarded, are *Messier's Nebulae and Star Clusters* by Kenneth Glyn Jones and, more recently, *The Messier Objects*, by Stephen James O'Meara.

The Messier Marathon

Each year, around the time of the new Moon in mid-March to early April, it is theoretically possible to observe all 110 Messier objects during a single night. Amateurs who attempt this feat are said to be undertaking a Messier Marathon – which requires locating and observing an object on average once every 5 minutes during around 10 hours of darkness. The ease of achieving this depends on the precise date of the new Moon, and the best opportunity occurs when the new Moon is in late March, ideally close to a weekend. If the marathon is attempted earlier than mid-March, Messier objects that rise close to dawn, such as M30, will be difficult to spot against the lightening sky. Conversely, after the end of March, objects such as M74 will be lost in the evening twilight. There is also a period around the autumnal equinox when many of his objects can be seen.

The location where the attempt is made is obviously pretty important and, as Messier observed from Paris, latitude 48.9 degrees north, observers from northern Europe and Canada may never be able to spot some of his southerly objects and, likewise, those observing from southerly latitudes will never be able to spot the more northerly objects such as M81 and M82 in Ursa Major. An observing site at latitudes close to 25 degrees north is optimum. A high location with very low horizons in the east and west is critical – essentially at 0 degrees elevation. The elevation of the southern horizon is nearly as important, particularly when one is observing from more northern latitudes – no more than 4 or 5 degrees of elevation. The northern horizon is not quite so critical except, of course, when one is observing from more southerly climbs. An obvious point is to go to a dark-sky location well away from towns and even further from cities.

Several books and Web sites give sequences and star charts for carrying out the marathon, and an excellent description of a particular event is given in the superb book *Astronomy Hacks* by Robert and Barbara Thompson, published by O'Reilly. (This may well put you off even trying!)

Due to Messier's latitude, his catalogue could not include many of the beautiful objects in the southern skies but, in recent years, a list and catalogue have addressed this.

The 'Astronomical A-List'

Available on the Web (just Google 'Astronomical A-list') or in the book *Pocket Guide to Stars and Galaxies* is a list with descriptions of (arguably) the best 50 objects in the heavens. Covering both hemispheres, it includes, for example, the Large Magellanic Cloud and the 30 Doradus star cluster and Tarantula Nebula within it. The active galaxy Centaurus A, the globular cluster 47 Tucanae and the double star Alpha Crucis are also included in the 10 southern objects within the list, which was compiled by the author. Many of those in the Northern Hemisphere are, of course, Messier objects.

In Chapter 3, I recommended the book *Touring the Universe through Binoculars*, which indicates what might be seen with binoculars up to 15×70 in size. Such an instrument is very similar in ability to a small refractor or Newtonian telescope so, if one is using one, the book could prove a very useful reference work of more than 1,100 celestial objects.

The Caldwell Catalogue

In 1995 Sir Patrick Moore produced a catalogue of 109 objects within both hemispheres that are not included in Messier's list, which was first published in *Sky and Telescope* magazine. He could hardly give them an 'M' designation, so he used the first part of his full surname, Caldwell-Moore; thus these objects have a 'C' designation. It is a truly excellent addition to Messier's list and includes many superb objects to observe. Examples are the Hyades Cluster and the Perseus Double Cluster, C41 and C14 respectively, and the Eskimo Planetary Nebula, C39, in Gemini. More than 80 objects from the southern skies are included so, together with Messier's catalogue, Caldwell's could keep an observer occupied for some considerable time. An excellent book covering all objects within the Caldwell Catalogue is published by Cambridge University Press and forms part of the Deep-Sky Companions series. Again written by Stephen James O'Meara, it is entitled *The Caldwell Objects*.

The NGC and IC Catalogues

In 1888 the astronomer John Louis Emil Dreyer published the New General Catalogue of Nebulae and Clusters of Stars (NGC). It was an updated version of John Herschel's Catalogue of Nebulae and Clusters of Stars and contains 7,840 objects. Many of these are galaxies, but it includes all types of deep-sky objects. Later, in 1895 and 1908, he added a further 5,386 objects in two additional Index Catalogues (IC). Many objects in the Southern Hemisphere are included, having been observed by John Herschel or James Dunlop, but the coverage is somewhat less thorough.

The NGC included many errors, and the NGC/IC Projects (Google 'NGC/IC Project') are aiming to eliminate these and provide up-to-date images and data on them. This is a wonderful resource for amateur astronomers. My favourite NGC object is NGC 2903 – a bright, ninth-magnitude spiral galaxy in Leo, and I am a little surprised that it was overlooked by both Messier and Moore. It is worth knowing that the two clusters forming the Perseus Double Cluster are NGC 884 and 869 if one wishes to use a computerised scope. Not all include the Caldwell Catalogue but, if so, one is simply able to 'go to' C14.

9.7 Star Charts and Planetarium Programs

One of the most treasured 'books' in my possession contains A2-sized charts covering the entire sky in what was originally called the *Skalnaté Pleso Atlas of the Heavens*. It is named after the Skalnaté Pleso Observatory in Slovakia, where it was produced. The charts were hand-drawn by Antonín Bečvář (1901–65) and show, against a background of stars down to 7.7 magnitude, virtually all the objects that were visible in an 8-inch telescope. Different types of deep-sky objects were colour-coded using red for galaxies, yellow for clusters and green for nebulosities. The Milky Way was shown in two shades of blue. Called *Atlas Coeli 1950*, it showed the stars as they were positioned in 1950 (Figure 9.5).

Long out of print, the atlas was updated by Wil Tirion to account for the precession from 1950 to 2000 to produce the *Sky Atlas 2000*, which includes 26 A2-sized star charts to cover both hemispheres. It very closely follows Bečvář's beautiful design. Tirion is, without doubt, the world's best celestial cartographer at the present time, and his star charts appear in many books. The same charts at a somewhat smaller and more manageable scale are published in his *Cambridge Star Atlas*, which is now in its fourth edition, published by Cambridge University Press. It is well worth owning. An excellent set of star charts, also by Wil Tirion, along with some 'photo-realistic' maps of the same area of sky are included in the *Phillip's Night Sky Atlas* written by Robin Scagell and available at very low cost.

The other 'classic' star atlas is that due to Arthur Norton and J. Gall Inglis, who produced *Norton's Star Atlas*. Its 16 charts include 8,000 stars down to magnitude 6, along with clusters and nebulae, as well as lists of 500 other interesting objects. The other great asset of this star atlas, now edited by Ian Ridpath, is that it provides a wonderful reference work of all things astronomical.

There are a number of superb planetarium programs such as 'Redshift 7 Premium', 'The SkyX Serious Astronomer Edition' and 'Starry Night Enthusiast' that, deservedly, have to be paid for, but there is one open-source freeware program that many amateur astronomers use. This is 'Stellarium', which is being regularly enhanced and has now reached the point where it is a worthy competitor of the others. I have found only one problem in that on some of my computers it runs so slowly as to make it totally impossible to use, whilst on others with seemingly similar power it runs perfectly. If this is found to be the case, then earlier versions which are still available to download may solve the problem. Be careful downloading it, as you can be 'directed' to download programs you don't want; the

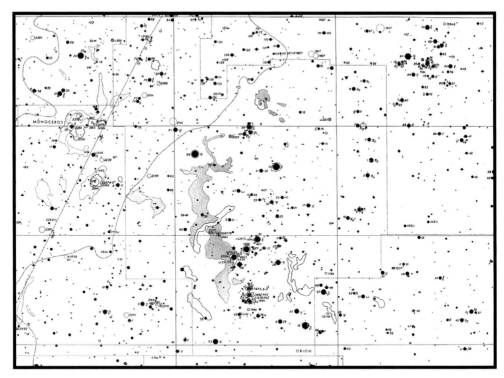

Figure 9.5 A section of Antonín Bečvář's *Atlas Coeli 1950* showing the Orion region. (Image: Antonín Bečvář, Skalnaté Pleso Observatory)

biggest, often green, download button is usually not for the program that you actually want!

All of these programs give 'photo-realistic' views of the heavens and, for example, show the positions of the Galilean satellites when zoomed into Jupiter and bring up images of many of the celestial objects when 'high magnification' is used. I use Stellarium extensively to provide the star charts that appear on my monthly 'Night Sky' page. This gives details of what may be seen in the heavens each month (just Google 'Night Sky Jodrell Bank'). An audio version also appears within the Jodrell Bank Podcast – termed the Jodcast!

An excellent sky atlas for tablet computers is called 'SkySafari', produced by Southern Stars. It is available in three levels: a 'Light' version, a 'Plus' version and the ultimate 'Pro' version; the last-named includes the full Hubble Guide Star Catalogue, which contains 15.3 million stars, 740,000 galaxies and more than 580,000 solar system objects! There is also a version of Stellarium provided for tablets.

9.8 Visual Observing Under Light-Polluted Skies

Sadly, this is something that applies to many of us. The major problem comes when one is trying to observe diffuse objects such as galaxies which are dim, low-contrast objects. The brightness of a galaxy is usually given as its integrated brightness expressed as an apparent magnitude; for example, a galaxy whose apparent magnitude is 13 would give the same total amount of light as a star of apparent magnitude 13. But the problem is that this light is spread over an area of sky so that the apparent brightness to our eyes is far less. What is far more important is the surface brightness, usually expressed in magnitudes per square arc second so, as galaxies typically extend over minutes of arc, this will be far lower than its apparent magnitude. The fundamental point is that if one attempts to observe a galaxy under light-polluted skies its surface brightness must exceed that of the sky glow; otherwise it will not be visible no matter how large a telescope one uses.

When we observe a point object such as a star or a star cluster, the sky glow is dimmed as the magnification is increased but that of the star is not, so increasing the magnification is a way of combating light pollution. Other objects such as the Moon and planets, though not point-like, do have a sufficiently high surface brightness to overcome all but the most severe light pollution, so these are the objects that an urban astronomer must pursue, leaving the search for faint galaxies for observing excursions into darker skies. It should be pointed out that even when one is observing the Moon and planets the sky glow does act like a veil hanging in front of the object, so that the image contrast will be reduced and so even planetary observing will still be best from dark-sky sites.

A second major factor is the source of the air above you. If, in the UK, it has come from the south it may well contain sand from the Sahara Desert. The sky transparency on such nights will be very poor, but there is often compensation. On such hazy nights of low transparency, the atmosphere is often extremely stable, making them excellent for lunar and planetary observing. In contrast to air reaching the UK from the south, we also have periods when the 'polar maritime' air comes southwards from the north. This is fairly dry and will not carry dust, so is very transparent and will not reflect nearly so much light pollution back to Earth. However, polar maritime air is usually very unstable, so the 'seeing' can be very poor, with the Moon, for example, appearing to be boiling! Sadly, nights of both high transparency and good seeing are very rare. (The image of the Moon shown in Plate 11.3 was taken on such a night.)

Planets are obviously rewarding targets, and all can be seen under urban skies. Uranus and Neptune show small coloured disks, and Mercury is visible after sunset or before dawn at times of the year but shows no features. Neither does Venus, whose thick atmosphere allows only its phase to be determined. In contrast, Mars (near closest approach), Jupiter and Saturn are highly rewarding objects and deserve extended study – spotting the more prominent markings on Mars, observing the Galilean moons as they weave their way about Jupiter and observing the passage of its red spot across the surface along with the changing aspects of the equatorial bands and zones. Saturn, too, shows faint bands on its surface, and one can search for details in Saturn's rings. On virtually all clear nights, lunar and planetary observers will have something worthwhile to observe.

Of course, any object with a high surface brightness can be seen from light-polluted skies. Some observers track down double stars such as the beautiful blue and gold pair making up the star Albireo and the 'double-double' Epsilon Lyrae, with its two tight doubles providing quite a test of seeing and telescope optics. The Astronomical League's 'Urban Observing Program' provides lists of objects that can be observed from urban skies, and one of these contains 13 of the best double and variable stars (such as Algol) to observe.

A number of the brighter galaxies have sufficiently high surface brightness to be seen when conditions are not too bad, and open and globular clusters make excellent targets, as they can be made to stand out against the sky glow by the use of higher magnification. Planetary nebulae can also have quite high surface brightness, though they are often hard to track down, as many (such as the Eskimo Nebula) appear star-like until high magnification is used. The use of an OIII narrowband filter can be of real use when one is observing these, as they emit quite a lot of light in the OIII line, which the filter passes whilst rejecting virtually all the sky glow. A telescope of 8 inches or more in aperture will be needed to use such a filter, but some help for those with smaller telescopes can be obtained by using one of the 'sky pollution' filters such as the Orion SkyGlow or Lumicon DeepSky filters. Their effect is, however, relatively small. A second 'urban observing program' list contains nearly 90 of the best such objects to observe (many are Messier objects) and, together, these two lists would give an urban observer enough objects to keep him or her occupied for many years.

Apart from the selection of suitable observing objects as indicated, there are some general tips to be made aware of:

- Stay up late if you can, as after midnight many lights will have been turned off; one good outcome of the economic downturn is that some towns are now turning off the streetlights after midnight, so making a significant difference! The best times are often the hours before dawn, as the air is often more stable then.
- Observe objects when they are highest in the sky. You will be observing through the least atmosphere, so the light pollution will be reduced.
- Encourage your neighbours to come and see a spectacular object like the Moon, Saturn or Jupiter through your telescope, so that they can appreciate the problem of light pollution and may be willing to turn off their exterior lights for you. In particular, if they have security lights, ask them to angle them down so that they do not light up the sky.
- Use a dark cloth to cover your head and eyepiece to shield your eyes from stray light.
- A dew shield, even if not needed (and even on a Newtonian), will help reduce the stray light scattered into the telescope.

10

Visual Observations of the Moon and Planets

This is a wonderful branch of the hobby and one that can be pursued even in heavily light-polluted skies. What does one need? Well, of course, any telescope will do but, as has been discussed earlier, telescopes which have an inherently high contrast are best. Many of the planetary surface features are of low contrast, and for viewing the Moon, which will be very bright both inside and, at higher magnifications, outside the field of view, the overall contrast of the telescope becomes quite important. Refractors are best for both overall contrast and micro-contrast but are very expensive in larger apertures. If a reflecting telescope is to be used for lunar observations, mirrors with high-reflectivity coatings will prove significantly better, and these will also improve planetary observations – but to a lesser extent, as there is far less light entering the telescope and hence less that can be scattered by the mirror surfaces. Maksutov-Newtonians are a very close second to refractors in terms of their micro-contrast and greater affordability in larger apertures. Newtonians with focal ratios of f6 and above and Maksutovs come next, and even Schmidt-Cassegrains, if well collimated, can give very worthwhile views. So this is not something to get hung up about. The best telescope for observing is the one that you have!

As the angular sizes of the planets are small, high magnifications are needed, with the use of up to 200 on a reasonable night when the seeing is good. Such magnifications would also be used for exploring lunar features. Higher magnifications can be used on nights of exceptional seeing. One useful accessory is a zoom eyepiece. This can help you find the optimum magnification to suit your telescope, eye and the seeing. Reduce the local length until the image quality begins to fall off and then choose the nearest focal length eyepiece in your collection. Simple eyepieces, such as Plössls and orthoscopics, contain less glass and fewer air/glass interfaces than wide-field types, so they can be very good for lunar and planetary observing. It is probably best not to use eyepieces with too short a focal length, as their eye relief can be very small; instead, use a greater-focal-length eyepiece with a ×2 Barlow lens. Using a binoviewer can also add something to the visual experience.

Coloured filters can enhance the contrast of certain regions on a planetary surface, and suitable filters will be discussed with each planet where they can be of use. A set of six high-transmission planetary filters is produced by Baader Planetarium with the following colours: dark blue, bright blue, green, yellow, orange and red. In addition, Wratten filters of various colours such as yellow and orange can be bought to supplement a set of RGB filters that you might have for imaging. With larger telescopes, the Moon's brightness can be somewhat overwhelming, and a useful accessory is a neutral-density Moon filter. Polarising filters, which feature a pair of linearly polarised filters whose relative orientation can be adjusted to control the light transmission, can also be useful for lunar observations. It can be somewhat of a nuisance to screw and unscrew filters into the eyepiece each time you wish to use one, so it can be worthwhile buying a manual filter wheel which could hold four coloured filters and a clear filter. It is then easy to quickly swap between them to find the one that makes specific details of the planetary surface more prominent.

10.1 The Moon

Many amateur astronomers, particularly those interested in deep-sky observing, do not like the Moon, as its light near full Moon obscures many of the fainter deep-sky objects. Remember, if the sky glow exceeds the surface brightness of an object, you will not be able to observe the object, even with the largest of telescopes. Rather than cease observing when it is 'in the way', why not enjoy it? There are many interesting regions of the lunar surface to observe, and it is by far the easiest object with which to make one's first attempts at astro-imaging.

There is one resource that all amateurs should have: the freeware program 'Virtual Lunar Atlas'. Be a little careful when downloading it, as it is very easy to click on obvious download buttons which will install quite different programs! It provides a 'zoomable' lunar map which shows the illuminated part at any time and gives details of the lunar features. These are best seen when near the terminator (the line between the light and dark areas of the Moon), as the long shadows there help to give them 'relief', so the program can be used to determine when best to observe a particular feature. I enjoy observing and imaging the Moon, and in the following subsections I describe some of the sights that I have observed most often on the surface of the Moon.

A wonderful asset for a lunar observer which, though long out of print, can still be obtained second-hand is the *Times Atlas of the Moon*. It is based on the US Air Force 1:1,000,000 lunar charts derived from the images sent back to Earth from the five unmanned *Lunar Orbiter* missions during 1966 and 1967 and is, I believe, the most detailed lunar atlas that has ever been made available to the amateur astronomer. Although now very difficult to obtain, the best descriptive work detailing the lunar surface is that written by Patrick Moore, which accompanied a large-scale chart drawn by H. P. Wilkins in the Faber and Faber book *The Moon* by H. P. Wilkins and Patrick Moore. It includes some beautiful drawings, and if you can track one down it is well worth buying. A very nice four-quadrant lunar map is included in the *Night Sky Atlas*, written by Robin Scagell and published by Phillips which is superb value.

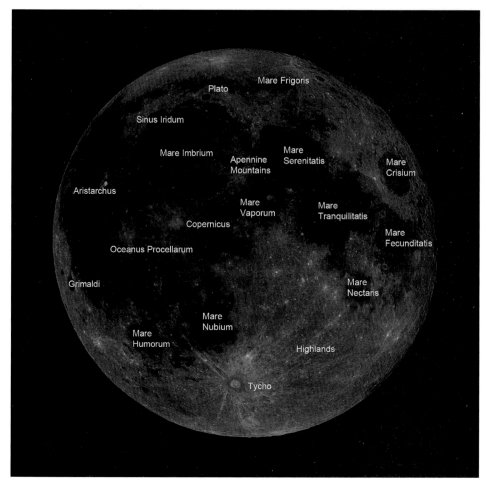

Figure 10.1 The lunar maria and highlands.

The Highlands and Mare

The lunar surface as seen from the Earth (the near side) is formed of two distinct regions, the lighter highland regions and the far darker mare regions. The amount of cratering is an indication of the age of the surface, and the heavily cratered highlands are the oldest parts of the surface. In contrast, the mare regions have relatively few craters; they are thus younger in origin and the majority date from about 3 billion years ago. They are formed from basaltic lava and cover about 16% of the lunar surface – mostly on the near side. They are easily seen with binoculars and are even visible with the unaided eye. Figure 10.1 is a full Moon image showing the main lunar features.

Earthshine

When the Moon is seen as a thin crescent near new Moon, the dark part can often be seen by light that has been reflected from the Earth – particularly when the side facing the Moon is well cloud covered. The dark part of the Moon is then said to be illuminated by earthshine.

Copernicus

Named after the astronomer Nicolaus Copernicus, Copernicus is a lunar impact crater, located in the Oceanus Procellarum. It is estimated to be about 800 million years old and has a prominent ray system. The circular rim has a terraced inner wall, outside of which is a gently sloping rampart that descends to the surrounding mare. Its floor, which has not been flooded by lava, has three central peaks that that are more than 1 km high. Ejecta from the impact have spread 800 km across the surrounding mare but are less distinct than the long, linear rays extending from the crater Tycho.

Plato

The lava-filled crater Plato is located on the north-eastern shore of the Mare Imbrium, at the western extremity of the Apennine Mountains. Though circular, it appears oval due to its position towards the pole and thus appears foreshortened. It is thought to be about 3.8 billion years old. Having a relatively low albedo, the crater floor looks quite dark and is free of any significant impact craters but does contain a number of small craterlets, which are quite a challenge to observe, requiring very good seeing.

Tycho

Tycho is a prominent lunar impact crater located in the southern lunar highlands, named after the Danish astronomer Tycho Brahe. With an estimated age of 108 million years, Tycho is one of the younger lunar craters and is thought to have been caused by the impact of a member of the Baptistina family of asteroids. Due to its young age and hence a lack of subsequent impacts, the crater is well defined. Its floor, in contrast to that of Plato, has a high albedo, or reflectivity, and becomes very prominent around full Moon when a ray system extending more than 1,500 km across the lunar surface becomes visible. They can even be seen when the Moon is illuminated only by earthshine. Its terraced walls slope down to a very rough floor, at the centre of which lies a peak some 1.6 km high with a lesser peak to its northeast. Tycho has appeared on lunar maps since 1645, when Antonius de Rheita depicted the bright ray system.

The Apennine Mountains

These are a rugged mountain range, named after the Apennine Mountains in Italy, that form the south-eastern flanks of Mare Imbrium; they form an arc from the crater Eratosthenes which gradually bends from east to north-east and which contains several prominent mountains. One of the most northerly is Mount Hadley, seen in images taken by the *Apollo 15* astronauts who landed close to Hadley Rille. The range extends for some 600 km, with individual peaks rising as high as 5 km.

The Alpine Valley

The Alpine Valley is a lunar valley feature that bisects the Apennines, discovered in 1727 by Francesco Bianchini. It stretches 166 km east–north-east from the Mare Imbrium basin to the edge of the Mare Frigoris. The valley is about 10 km wide at its greatest extent in the middle, narrowing at both ends. The flat, lava-flooded surface of the valley floor is bisected by a slender, cleft-like rille, which is a challenging target, requiring very good seeing to observe. It is thought that the valley was flooded with magma from Mare Imbrium and Mare Frigoris.

Mons Piton

Mons Piton is an isolated lunar mountain located in the north-eastern part of the Mare Imbrium, north–north-west of the crater Aristillus. The 25-km-wide peak forms prominent shadows when illuminated by oblique sunlight at first and third quarters, and the length of the shadow can be used to calculate its height, which is estimated to be 2.3 km. Due east of Mons Piton is the flooded crater Cassini.

Hyginus and the Hyginus Rille

Hyginus is a small lunar caldera located at the east end of the Sinus Medii. Its rim is split by the long, linear Hyginus Rille that extends for a total length of 220 km to the north-west and to the east–south-east. Together, the crater and rille make up a prominent feature on the flat lunar surface. Unusually, as Hyginus lacks the raised outer rim that is typical of impact craters, it is one of the few craters on the Moon that are believed to be volcanic in origin.

The Straight Wall

Rupes Recta, more commonly called the 'Straight Wall', is a linear fault in the south-eastern part of the Mare Nubium. Its name is Latin for 'Straight Fault' and it is, in fact, a gentle escarpment 100 km long and 2–3 km wide, with an overall height of ~250 m. However, when the Sun illuminates the feature obliquely on day 8 of the lunar cycle, it casts a wide shadow, giving it the appearance of a steep cliff. The fault has a length of 110 km, a typical width of 2–3 km and a height of 240–300 m. Thus, although it appears to be a vertical cliff in the lunar surface, the slope is actually a gentle incline.

Schröter's Valley

Schröter's Valley is a sinuous valley or rille located on the Aristarchus plateau, surrounded by the Oceanus Procellarum to the south and west and the Mare Imbrium to the north-west. The rille begins at a 6-km diameter crater located 25 km to the north of Herodotus and, due to its resemblance to a snake, has been termed the 'Cobra's Head'. It then follows a meandering path, first north, then north-east, before finally returning south to a 1-km-high precipice at the edge of the Oceanus Procellarum. It is believed that the rille is volcanic in origin. It has a maximum width of about 10 km, which then gradually narrows to less than a kilometre towards its end.

Sinus Iridum

Sinus Iridum (not iridium!) is the Latin for 'Bay of Rainbows'. It is a plain of lava forming an extension to the north-west of Mare Imbrium. It is surrounded on three sides by the Jura Mountain range. Two promontories, Heraclides and Laplace, protrude at either end of the range. The bay and surrounding mountains form one of the most beautiful features on the Moon. When the terminator crosses the bay (around day 10 of the lunar cycle), the tops of the western mountains are lit by sunlight and appear to be part of a skeleton.

The Lunar 100

The April 2004 issue of *Sky and Telescope* published an article by Charles Wood listing 100 of the most interesting sights to observe on the Moon – including, of course, those that I have mentioned here. The Lunar 100 Observing Club provides an observing log for recording your observations, and an excellent Web site – www.astrospider.com/Lunar100list.htm – includes many images of the individual lunar regions and gives links to related Web sites.

10.2 The Inferior Planets (Those Between the Earth and Sun)

Mercury

With a peak magnitude of –2.6 Mercury can at times outshine every object other than the Moon but, due to its close proximity to the Sun (~0.34 AU), it is never far from the Sun's glare and so is not that obvious to the casual observer. It can usually be observed for only a few days around its greatest eastern or western elongations from the Sun, when it can be seen above the horizon at sunrise or sunset. These directions refer to Mercury's position relative to the Sun, so when at eastern elongation it will be seen for a short time after sunset above the western horizon. Conversely, at western elongation it will be seen just before dawn in the east. Mercury has an elliptical orbit, taking 88 days to orbit the Sun. It so happens that the points in the orbit when it lies at greatest eastern or western elongation correspond to its closest and farthest distances from the Sun, perihelion and aphelion respectively. This means that the angular separations from the Sun at the two elongations are quite different: just 18 degrees at greatest eastern elongation but 28 degrees at greatest western elongation. So we would be more likely to spot it at western elongation but for the fact that this will then be before dawn!

The elevation that Mercury reaches above the horizon depends on the angle that the ecliptic makes with the horizon. It will thus be seen better at tropical latitudes, where the Sun is seen to rise and descend at steeper angles, than at more northerly or southerly latitudes. At northerly latitudes, it is best to observe Mercury when at greatest elongation – either after sunset near the spring equinox or before dawn near the autumn equinox. There thus tend to be two favourable apparitions of Mercury per year, with that in the autumn apparition being somewhat better, as Mercury is then farther in angle from the Sun. The precise dates of greatest elongation do not correspond to it greatest brightness, as the planet's phase affects its apparent brightness, which becomes greater when the phase becomes gibbous. It will thus appear at its brightest when observed a few days before greatest eastern elongation or a few days after greatest western elongation, in the evening and morning respectively.

Very little, if any, detail can be seen on Mercury's disk, which will subtend about 7 arc seconds when easily visible. It passes through inferior conjunction (in front of the Sun) three times a year but, as its orbit has an inclination of 7 degrees to the ecliptic, transits do not occur that often, there being 13 or 14 per century. These are currently occurring in May or November, a few days either side of May 8, having an angular diameter of 12 arc seconds, and November 10, with an angular diameter of 10 arc seconds. Upcoming transits will occur on May 9, 2016, in late afternoon, on November 11, 2019, after midday and on November 13, 2032, in the morning.

Finally, details of when Mercury is best seen can be easily found in amateur astronomy magazines and on Web sites such as the author's 'Night Sky' page on the Jodrell Bank Web site.

Venus

In contrast to Mercury, Venus dominates the sky when seen before dawn or after sunset. It is the brightest body in the sky apart from the Sun and Moon and can even, when at its brightest, be spotted in daylight with the unaided eye. It shows phases like the Moon and Mercury, presenting fuller phases when on the far side of the Sun and crescent phases when on the near side. It was this observation by Galileo that proved that Venus orbits the Sun as, if it were moving on an epicycle between the Earth and the Sun (as in the Ptolemaic theory), it could never show full phases. Interestingly, its brightness stays almost constant at about –4 magnitudes for much of each apparition as, when it is farther away from us and thus would tend to appear fainter, more of the disk as seen by us is illuminated, so tending to make it appear brighter, with the two effects cancelling out.

Because Venus is totally cloud covered, there is virtually no visual detail to be seen on the planet, but images taken in the ultraviolet do show some structure. It is thus, perhaps, not the most rewarding planet to observe. Venus does, very rarely, transit the surface of the Sun, most recently in June 2012, following its transit in June 2004. In the UK, we had an excellent view of the 2004 transit, and I was lucky enough to observe the 2012 transit from southern New Zealand – even using my observations to calculate the distance of the Earth from the Sun. Combining my observations with a set from Alaska gave a value of the astronomical unit that was 3 million km too low.

Transits of Venus occur in pairs, eight years apart separated by more than 100 years so, sadly, the next transits will not take place until December 2017 and December 2125.

Wratten Colour Filters to Aid the Observation of Venus

46 and 47	Dark blue to help reveal very low contrast shadings in the atmosphere
25 or 29	Red to darken the blue sky and increase contrast during daytime viewing

10.3 The Exterior Planets

These will have a new apparition (appearance in the night sky) after a number of months dependent on the orbital period of the planet: the farther out the planet, the nearer to 1 year this will be. Neptune, which takes 164.79 years to orbit the Sun, will be seen at its best at nearly yearly intervals (in fact every 367.49 days – its synodic period), whereas Mars will be seen at its best after every 26 months (2.135 years). The best time to observe the planet is when it is at opposition, that is, in the opposite direction from the Sun, when it will be due south and highest in the sky around midnight (UT). It will then also be closest to us, so that its angular diameter will be greatest and so allow us to observe more detail on the surface.

There is a second factor that determines how well we see the exterior planets: their position on the ecliptic. If, at opposition, they lie in the most northerly part of the ecliptic, which includes the constellations Taurus and Gemini, their elevation (as seen from a northerly latitude such as +50 degrees) can be as high as 65 degrees, whereas at the lowest part of the ecliptic in Sagittarius their elevation can be as little as 17 degrees. The opposite will be true for observers in the Southern Hemisphere. In late 2012 and early 2014, Jupiter was at opposition in Taurus and Gemini respectively, both in the most northerly part of the ecliptic, but then at successive apparitions it will have lower elevations in the sky, so our view from northern climes will be less clear. This will continue until 2020, when by far the best views will be had in the Southern Hemisphere. Saturn was highest in northern skies back in 2003; it is now moving to the southerly part of the ecliptic and will be at its most southerly point during the opposition of July 2018. Mars moves through the ecliptic far more quickly: at opposition in April 2014 it will lie in Virgo, in May 2016 it will lie on the boundary between Libra and Scorpius and in August 2018 in Capricorn. Sadly for those of us at northern latitudes, it will then lie some 8 degrees *below* the ecliptic and so will be at an elevation of only 15 degrees as seen from a latitude of 50 degrees north.

10.4 Mars

Mars is at its best every 26 months, but in contrast to Jupiter and Saturn, its orbit has quite a high eccentricity and, coupled with the lesser eccentricity of our own orbit, its distance from us at closest approach can vary by nearly a factor of 2: from 54 to

102 million km. (The date of closest approach can differ from opposition by as much as 8 days.) This means that the angular size at closest approach varies from 13 to 25 arc seconds, so that the observed surface area will differ by a factor of ~4. This also effects the apparent magnitude, with can range from –3.0 to –1.4 magnitudes.

Mars's closest approach for nearly 60,000 years was on August 27, 2003, when it lay in Aquarius at a distance of 55,758,006 km, had an apparent magnitude of –2.88 and had an angular size of 25.1 arc seconds. Then for some years the distance at each closest approach increased, so its angular size dropped and close to the minimum angular size was observed (~14 arc seconds) at the oppositions of January 2010 and March 2012. Happily, in the oppositions of April 2014, May 2016 and July 2018 the distance falls so the apparent size will reach 15.16, 18.6 and 24.31 arc seconds respectively. Sadly, for those of us in the Northern Hemisphere its elevation in 2018 will be very low – but superb for those, say, in Sydney, when it will be at an elevation of 82 degrees in the north!

Mars rotates with a very similar period to that of our Earth: 24 hours, 37 minutes and 22.66 seconds. This is somewhat of a pity in that each evening our view of the planet will not change that much, but it does has the advantage that, coupled with the fact that Mars is about half the size of the Earth, longer AVI files may be captured by means of webcam imaging (see Chapter 11) without obscuring any detail.

Although there is usually much to observe at each opposition it can be possible to see virtually nothing! For example, a Hubble Space Telescope image of Mars taken in September 2001 showed virtually no detail, and when *Mariner 9* reached Mars in September 1971, the planet was totally shrouded in dust; it was not until January 1972 before the dust storm subsided, first showing the caldera of the extinct volcano Olympus Mons, the highest point on the Martian surface (see Figure 10.2). As one might expect, these dust storms occur when Mars is closest to the Sun so, paradoxically, a very close approach, as in 2018, may not actually prove to be the best time to observe Mars.

Mars has a very thin atmosphere and, dust storms permitting, the surface will reveal much detail given a night of good seeing. A very worthwhile freeware program called 'WinJUPOS' provides an image of Mars that one might see at any time, and the screen shot (Figure 10.3) shows the surface that would be visible at a particular time during the close approach of July/August 2018. The following subsections summarise the many features that may be seen.

Polar Caps

These are perhaps Mars's most prominent and distinctive features and are seen at both north and south poles. They are composed mostly of water ice overlaid with a layer of carbon dioxide ice that forms in winter and sublimates again in summer, so their size changes with the seasons; as one cap shrinks, the other grows. In winter, a layer of clouds can often form over them, called 'polar hoods'. As they shrink in spring, dark bands, named after Percival Lowell, appear around them – nicely increasing the surface contrast. At the same time, clouds of water vapour called 'orographic clouds' may appear in the volcanic region.

Figure 10.2 A topographic map of Mars. (Image NASA)

Figure 10.3 The WinJUPOS representation of Mars at 00:26 hours UT on August 9, 2018, when it will subtend an angular size of 23 arc seconds. Syrtis Major is the prominent feature.

Light and Dark Markings

The Martian plains are covered in sand given a reddish colour by the presence of iron oxide particles and appear lighter than the areas where winds have swept the dust away, revealing the underlying darker volcanic rocks. Due to the winds shifting the dust around, these surface features change with time, and the most prominent feature on the Martian surface, Syrtis Major, now appears significantly different than it did in the nineteenth century. It now has a broad triangular shape, whereas then it was thinner and had a distinct hook at its northern extreme. As the Martian disk increases in apparent size, this is likely the first feature to become visible, along with, perhaps, the polar caps.

Topography and Volcanic Regions

The two Martian hemispheres are very different: much of the northern hemisphere is at a far lower elevation than the southern hemisphere; the young northern lowlands are flat and contain few craters, whilst the far older southern highlands are heavily cratered.

Mars contains the largest (extinct) volcano known in the solar system, Olympus Mons, which reaches a height of 27 km above the surface. It is located along with four other large volcanoes in a region called Tharsis that lies along the Martian equator and covers about 25% of its surface. A giant crustal fracture called Vallis Marineris (as it was first seen well in images sent back to the Earth by the *Mariner* spacecraft) extends eastwards from Tharsis. There is also a second, smaller volcanic region called Elysium containing the volcano Elysium Mons.

Impact Basins

Several large impact basins are found in Mars's southern highlands, the most obvious of which is Hellas Planitia, some 1,800 km in diameter. Two other basins, Isidis and Argyre, have diameters of 1,000 and 800 km respectively.

Clouds

Dust clouds, often originating in the Hellas Basin and most often seen in the southern Martian summer, can cover a small area or even, at rare times (once every 13 years on average), the whole planet. Polar hoods are formed over the polar caps in autumn when, as the temperatures cool, water vapour condenses to form clouds or fog. Orographic clouds form over regions such as Tharsis and Elysium when the atmosphere cools as it rises to these higher regions of the surface. They are most often seen around Olympus Mons and the other volcanoes on the Tharsis plateau. There is an interesting 'blue cloud' that tends to form over Syrtis Major, giving the surface a bluish tint. Clouds or haze caused by scattering in the atmosphere may be seen over the eastern and western limbs and also have a bluish tinge.

Wratten Colour Filters to Aid the Observation of Mars

12 or 15	Yellow to brighten the plains and darken blue/brown features
21 or 23	Orange to increase contrast and detect dust clouds
25 or 29	Red to maximize contrast and enhance fine surface details
80	Magenta to enhance red/blue features and darken green features
38 or 38A	Blue to detect clouds and enhance the polar caps
46 or 47	Dark blue to detect clouds and enhance the polar caps
57	Green to darken red/blue features and enhance the polar regions
64	Blue-green to detect fogs and polar hazes

10.5 Jupiter

When Jupiter is near opposition every 13 months, when it has a magnitude of about −2.9 and presents a disk around 49 arc seconds across, which, given a night of good seeing, allows us to observe wonderful detail in its atmosphere. It rotates once every 9 hours, 55 minutes and 30 seconds, so during some winter nights we can observe one complete rotation. Jupiter's most interesting object, the Great Red Spot, is brought into view for around 2–3 hours each rotation as it crosses the face of Jupiter. Around opposition, astronomy magazines and my 'Night Sky' page will give the times when it will face the Earth.

Because Jupiter is a gas giant, we observe only the cloud tops in the atmosphere, with the most obvious features being the two prominent equatorial belts, north and south, one on either side of a central, brighter equatorial zone. The surface is fluid, however, and for some time in the recent past the south equatorial belt disappeared completely but returned to its normal wide state within a year. The two equatorial belts are flanked by further alternating dark belts and brighter zones – all running parallel to the equator as a result of very strong east–west winds in the upper atmosphere. Their patterns vary due to motion within the atmosphere.

The most interesting single feature is the Great Red Spot lying within the south equatorial belt – though its colour was more of an orange pink as this was being written during the opposition of 2012. The colour can even fade completely, leaving what is called the 'Red Spot Hollow'. Having an oblong shape typically 20,000 km × 12,000 km and observed for hundreds of years, it is a storm system rotating with a period of 6–7 days. There can be other red spots too, with one, called 'Red Spot Junior', seen close by in 2012 and shown in Figure 10.4. During the 2011 opposition several very prominent dark brown ovals, called 'barges', were seen along the edge of the northern equatorial belt, though these were not present during the 2012 opposition. Small white ovals are also seen within the atmosphere. These and the bright zones are cooler clouds lying higher in the atmosphere than the darker, warmer bands lying lower down in the atmosphere.

Figure 10.4 Diagram showing the main features of the Jovian disk.

Other features that are present in the atmosphere are the bluish 'festoons' which generally protrude at a shallow angle from the northern equatorial belt into the equatorial zone, as well as white and dark spots that are smaller than ovals or barges.

It may be possible to observe the Galilean moons transiting the surface and, rather more obviously, their shadows falling on the cloud tops. Very occasionally, scars may be seen on the surface where a comet has impacted the planet, as was famously observed in July 1994, when, in the course of a week, fragments of the comet Shoemaker-Levy 9 crashed into the surface, leaving scars that were easily visible through amateur telescopes and that persisted for many months. The WinJUPOS representation of Jupiter during the opposition of 2015 (Figure 10.5) shows two of the moons and also the shadow of one on the surface. These 'shadow transits' are usually more obvious than the moons themselves.

The Moons of Jupiter

With a low-power eyepiece, up to four of Jupiter's moons will be observed as they weave their way around the planet. Discovered in 1610 by Galileo Galilei and so called the Galilean moons, they were independently discovered some days later by Simon Marius, who gave them their names: Io, Europa, Ganymede and Callisto. The moons have magnitudes ranging from 5.4 to 4.4 and so could, in principle, be observed with the unaided eye if they were not lost in Jupiter's glare. They can be easily seen with binoculars. Similar in size to our Moon, their angular sizes at opposition span from 1.0 to 1.7 arc seconds across, so only amateur telescopes with apertures greater than ~300 mm under excellent seeing conditions can allow any details to be seen or imaged. However, Io, which is pot marked with volcanoes resulting

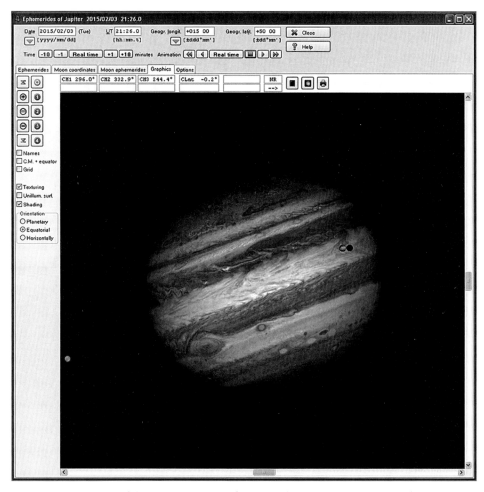

Figure 10.5 WinJUPOS representation of Jupiter during its opposition of 2015 showing the Great Red Spot and two moons: Europa (*left*) and Io with its shadow (*right*).

from the tidal heating due to its proximity with Jupiter, does appear to have an orange colour.

Wratten Colour Filters to Aid the Observation of Jupiter

38A	Light blue to enhance the contrast within the bright zones
57 and 47	Green and blue to darken the (brown) belts and make them stand out
12 and 15	Yellow to darken the blue festoons that appear below the north equatorial belt
30	Magenta to enhance the white ovals seen in the south temperate belts

Figure 10.6 WinJUPOS representation of Saturn in 2018 showing the ring system.

10.6 Saturn

With good reason, Saturn is regarded as the most beautiful of all the planets in the Solar System, and one's first view of it through a telescope is awe inspiring. A gas giant like Jupiter, it does not have a solid surface, and cloud bands exist in the atmosphere, though they are considerably less prominent than those we see on Jupiter. As is the case with Jupiter, the higher clouds are white and composed of ammonia ice crystals, whilst the yellow middle-level clouds contain water and ammonium hydrogen sulphide. In the lower levels are regions of water droplets containing an aqueous solution of ammonia.

Like Jupiter, the surface can also exhibit white ovals and, roughly every 30 years a Great White Spot – a large but short-lived storm – tends to occur near Saturn's equator. Last seen and imaged by the Hubble Space Telescope in 1990 and with similar storms having been seen in 1903, 1933 and 1960, around the time of the Northern Hemisphere's summer solstice, one might expect another around 2020.

Saturn's Rings

Saturn comes into opposition around 13 days later each year: in May 2014 and 2015, June 2016 and 2017 and July 2018 and 2019. Due to the 27-degree tilt of Saturn's rotation axis to the plane of the ecliptic, our view of its magnificent ring system varies from opposition to opposition (see Figure 10.6). The rings were edge-on in 2009, when, due to their thinness, they would have disappeared if Saturn was viewed through a small telescope. The ring system will reach a near maximum angle to the line of sight of 26.6 degrees in 2017. At opposition, the angular size of the disk reaches ~20 arc seconds, with the size of the ring system being approximately double that value.

Three main rings are visible with small telescopes: the outermost is the A ring, within which are the B, or bright, ring and the inner and fainter C, or crepe, ring. The rings are separated by gaps, the most visible of which is Cassini's Division, which is easily seen in a small telescope assuming that the ring plane is not too close to the line of sight. This is interesting, as its width of 0.8 arc second is less than the angular resolution (1.25 arc seconds for a 100-mm aperture) of a small telescope. The eye can detect linear features which are narrower than theoretically resolvable!

There is a second gap, called 'Encke's Gap', in the outer part of the A ring; however, it subtends only 0.06 arc second at opposition and so is virtually impossible to see and exceedingly difficult to image unless you have access to the Hubble Space Telescope. This rather begs the question as to how Johann Encke was able to observe it, and it is now generally believed that he had actually seen a wider darkening within the A ring called 'Encke's Minima'.

The ring system is made up of individual particle of water ice ranging in size from 1 cm up to several metres, each individually orbiting the planet – so the inner parts of the ring system are rotating faster. They are very thin; some parts might have a thickness of just 10 m, whilst other parts can be as thick as 1 km. Seen by reflected light from the Sun, there is an interesting effect when Saturn is close to opposition called the 'Seelinger effect' where the rings become significantly brighter than at other times, thus making the faint C ring more visible.

Saturn's Moons

Of Saturn's 53 moons, 7 are sufficiently bright to be seen with a small telescope. Titan is the largest and brightest, having a magnitude of 8.3 at opposition, and so is easily visible under dark skies with a pair of binoculars. The other 6 range in brightness from Rhea at 9.7, through Tethys at 10.2, Dione at 10.4, Iapetus at 11.1, Enceladus at 11.7 to Mimas at 12.9.

The *Huygens* probe landed on Titan in 2005 and found that it has a thick nitrogen atmosphere containing clouds of methane and ethane, as well as a nitrogen-rich organic smog. The geologically young surface composed of rock and water ice contains hydrocarbon lakes and has mountains that can reach 1 km in height. Enceladus is also very interesting and appears to have an ocean of water underlying a water ice crust through which plums of hydrocarbons vent into space.

Wratten Colour Filters to Aid the Observation Saturn

38A	Light blue to enhance the contrast within the bright zones
57 and 47	Light green and blue to darken the belts and make them stand out
57 and 30	Light green and magenta to highlight the ring system

10.7 Uranus

Although Uranus and Neptune are often called 'gas giants' like Jupiter and Saturn, they both have a rather different chemical composition and are now more properly termed 'ice giants'. Uranus takes 84.3 years to orbit the Sun and, in the years up to 2020, it will return to opposition in September – just a few days later each year. It rotates on its axis every 17 hours and 14 minutes but, interestingly, its rotation axis is tipped onto its side at an angle of slightly more than 2 degrees from the plane of the ecliptic so it actually appears to 'roll' around the Sun!

Due to its great distance from the Sun its angular size does not vary that much as seen from the Earth: from a minimum of 3.3 arc seconds when on the far side of the Sun to a maximum of 4.1 arc seconds at opposition, when its magnitude reaches 5.3, so Uranus becomes just visible to the unaided eye under dark, Moon-less skies. Like Jupiter and Saturn, its atmosphere is made mostly of hydrogen and helium but contains more methane and ammonia than they do. It is the methane that gives Uranus its rather beautiful blue-green colour. With the Hubble Space Telescope, some banding and surface storms are visible, but no surface details can be seen with amateur telescopes. It has a ring system too but, again, this is not visible with amateur telescopes.

Uranus's Moons

Uranus has five main moons, with the brightest, Titania, shining at magnitude 14.1 and the faintest, Miranda, at 16.7. They are thus challenging visual objects requiring dark, transparent skies and a telescope of 200 mm or more to even spot Titania.

Wratten Colour Filters to Aid the Observation of Uranus

12, 57 and 30   Yellow-green, light green and magenta can increase the contrast in a light-polluted sky

10.8 Neptune

The first observation and drawing of Neptune were made by Galileo on December 28, 1612, when, as shown in Figure 10.7, the planet was very close to Jupiter. He did not, sadly, recognise it as a planet even though he may have noted that it had moved from one observation to another. This is not too surprising, as he was observing its position relative to that of Jupiter, which would, itself, have been moving relative to the far more distant (and effectively stationary) Neptune. Now classified as an ice giant and similar in size to Uranus, it was finally discovered at the Berlin Observatory by Galle and d'Arrest, who had been given its predicted position by the French astronomer and mathematician Urbain Le Verrier based on the observations of a second French astronomer, Alexis Bouvard.

Neptune takes 164.8 Earth years to orbit the Sun and, in 2012, returned for the first time to the position in the heavens where it had been discovered. It rotates once every 16 hours, 6 minutes and 36 seconds and orbits the Sun at a distance of about 4.5 billion km. It is currently coming to opposition in September, about 2 days later each year, lying in Aquarius and shining at a magnitude ~7.9. This is too faint to be observed with the unaided eye, but it will be visible in binoculars.

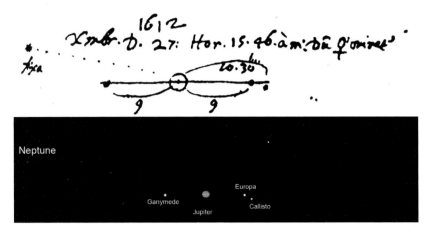

Figure 10.7 Galileo's drawing of Jupiter and Neptune on the night of December 27, 1612, along with a Stellarium plot at the same time.

As seen from the latitude of +50 degrees, it reaches an altitude of ~30 degrees when at opposition.

Due to its great distance, Neptune's apparent diameter varies little, from 2.2 to 2.4 arc seconds, and appears as a small blue disk, its colour caused by methane in the atmosphere. Its disk can be resolved with a telescope of 200 mm or more, but though there is some surface detail in the form of high white clouds and dark spots it is not possible to observe any details with amateur equipment. Neptune has a ring system but, like Jupiter and Uranus, it is too faint to be seen.

Neptune's Moons

Neptune has 13 known moons, of which only one, Triton at magnitude 13.5, can be seen in amateur telescopes. Its position relative to Neptune can be found using *Sky & Telescope's* Triton Tracker. The other moons are all at magnitude 19 or fainter.

Wratten Colour Filters to Aid the Observation of Neptune

12, 57 and 30   Yellow-green, light green and magenta can increase the contrast in a light-polluted sky

11

Imaging the Moon and Planets

Imaging is now becoming a major branch of the hobby, and a good feature of lunar and planetary imaging is that it can be done under light-polluted skies. It does not have to be expensive, as DSLRs and even iPhones can be used to image the Moon and, perhaps surprisingly, DSLRs can even be used to image the planets. Webcams are now routinely used for planetary imaging and can also be used to produce stunning lunar images. All of these possibilities are fully covered in this chapter.

11.1 Image Processing Programs

Virtually all images can be improved with the use of some post-processing in an image manipulation program such as Adobe Photoshop Elements or Adobe Photoshop and a freeware program such as GIMP (GNU Image Manipulation Program).

It has to be said that the majority of astro-imagers use Adobe Photoshop, and packages of 'actions' specifically designed to help process astro-images can be used with it. It is expensive, but bona fide students can purchase it for significantly less. Might it be worth signing on at a local college for an evening course in Photoshop use? Adobe Photoshop Elements is considerably cheaper and can carry out some of the image manipulation functions that are used, but it works only in 8-bit mode (whereas Photoshop can handle 16-bit images) and it does not have the curves function, which can be very useful for 'stretching' the brightness of an image to bring out detail in, for example, the spiral arms of a galaxy, which are far less bright than the central core.

GIMP is maturing and now provides a user interface very like Photoshop. It handles layers and does have the curves function, but the version (2.8.4) that was available as this book was being written could not handle 16-bit data. It may well be that, before long, GIMP will be able to do so, and it is thus well worth investigating. At present, a 16-bit TIFF file is converted to 8-bit as it is imported into GIMP, which means that 'stretching' of the data cannot be achieved so well.

But for both Adobe Elements and GIMP there is a (somewhat circuitous) way round this. A free program called 'FITS Liberator 3' can handle 16-bit data and provides several modes to stretch the data, so this part of an image processing work flow can be achieved outside of GIMP or Elements. However, the images it processes must be in the particular FITS format used by professional astronomers. Happily, another free program called 'IRIS' can convert JPEG and TIFF files into FITS format. These can then be imported into FITS Liberator, stretched as desired and exported as a TIFF file.

In all the image processing examples, I have given very detailed instructions as to how to use Adobe Photoshop, but though the specific ways of carrying out tasks are somewhat different in GIMP (less so in Adobe Elements) it is not difficult to achieve the same results. I purchased Photoshop some years ago before Elements and GIMP had reached their current functionality. I suspect that now I might use GIMP along with IRIS and FITS Liberator 3 (which did not become available until February 2013) to achieve pretty much the same results. It is well worth a try.

11.2 Imaging the Moon with a Compact Camera or iPhone

This is perhaps the first imaging project that an amateur can undertake, and it can be achieved with almost any telescope – even a Dobsonian. The technique is to use what is called 'eyepiece projection' coupling a compact camera to the rear of a low-power eyepiece. This can be done simply by holding the camera up to the eyepiece and taking an image. It may be necessary to zoom in somewhat to avoid vignetting and adjust the exposure until you get a reasonably bright image on the LCD screen. As exposures are generally very short, this can achieve excellent results. To be a little more professional, you can mount the camera on a moveable stage that is attached to the rim of the eyepiece. By movement of the stage, the camera lens can be made to align exactly along the optical axis of the eyepiece, making good results easier to achieve. One well-known example is the Baader Microstage II. Figure 11.1 shows my first-generation Baader Microstage holding a Canon G7 7-megapixel camera with an image of the third-quarter Moon taken with this set-up mounted on a 6-inch Maksutov-Newtonian. If you are using such a camera mount, set a timer delay of 2 or 5 seconds so that any vibration induced when you press the shutter has died out before the image is taken. It may be that, depending on the focal length of the telescope, more than one image may be required to fully cover the Moon, in which case two or more images are combined into one, as described later.

Mounts, such as the Orion SteadyPix Telescope Afocal Adaptor, are now available to mount even an iPhone or smartphone onto an eyepiece to image the Moon using eyepiece projection. A camera app will have to be downloaded that can provide control of the phone's camera. It should be able to provide an exposure delay to reduce the effects of camera shake. If a tracking mount is used, smartphones can even be used to image some deep-sky objects. The cameras in modern phones are really very good!

11.3 Imaging the Moon Using a DSLR

If you have a DSLR, you need to buy two items to link the camera to the scope. The first is a barrel that fits into the focuser. Buy one with a 2-inch diameter if your focuser will accept 2-inch eyepieces, but it is possible to get them in a 1.25-inch barrel

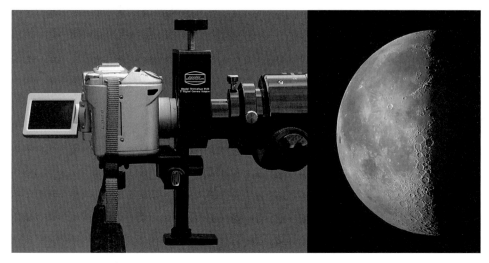

Figure 11.1 A Baader Microstage used to mount a Canon G7 compact camera onto a camera eyepiece with an image of the third-quarter Moon taken when used with a 6-inch Maksutov-Newtonian.

size. This barrel screws into a bayonet mount, which fits into your particular make of DSLR in place of a lens. Together these are called a T-mount. In principle one simply replaces an eyepiece with the T-mounted camera and focuses the Moon image onto the CCD (or CMOS) array. However, you may well find that you cannot bring it to focus and so need to buy a 2- or 1.25-inch extension barrel. With my 80-mm refractor I cannot extend the focuser sufficiently to get to focus without a star diagonal, nor can I retract it sufficiently if one is used! With some Newtonians, one might need to move the primary mirror slightly nearer to the secondary to bring the image plane far enough away from the side of the telescope tube; however, this problem can usually be overcome with the use of a Barlow lens, though this will increase the image scale. Figure 11.2 shows the items that I use with my DSLR cameras.

Ideally one would like the lunar image to fit nicely within the frame, but with a refractor of short focal length, say 500–700 mm, the lunar image will be rather small. A 2-inch Barlow lens, also shown in Figure 11.2, can help. If it is used as supplied, the image scale may be too big; however, the lens assembly of many 2-inch Barlows can be unscrewed from their mounts and screwed into the end of the T-mount barrel – increasing the image scale by 1.5 rather than 2, so that the Moon's image nicely sits within the frame. With a focal length greater than, say, 1,500 mm, it may well be necessary to take more than one image and combine images later, as will be described.

Manual mode should be used and a range of exposures tried using the histogram display to help find the correct exposure. To allow one useful post-processing technique to be used, it is best to under-expose slightly. If the shutter is to be fired manually, it is sensible to set a timer delay of 2 or 5 seconds to eliminate any resultant vibration. A second source of vibration is the lifting of the internal mirror, and many DSLRs allow for this to be lifted followed by a delay of 1 second before the image is taken. Thus, if a 2-second

Figure 11.2 Items used to link a DSLR to a 2-inch focuser: bayonet and 2-inch barrel forming a T-mount, 2-inch extender if required to gain focus and a 2-inch Barlow lens to increase image scale.

delay is used, the mirror will lift up after 2 seconds and the exposure is taken 1 second later. Canon provides the software to remotely control the camera using a laptop, so a delay is not necessary, but it may still be worth including the mirror lift-up delay.

A Single-Frame Lunar Image

Figure 11.3 and Colour Plate 11.3 are a single exposure of the Moon taken using a Nikon D7000 DSLR coupled to my CFF telescope's 127-mm, f7 apochromat refractor. An exposure of 1/250 second was given at ISO 100. The saturation in the colour image has been enhanced to show that the Moon is not just shades of grey and gives some indications of the chemical composition of the basalt lavas making up the mare floors. The orange-red colours indicate iron content, whilst the blue colour of Mare Tranquillitatis indicates the presence of titanium. The 'seeing' at dusk on the night of April 21, 2013, was superb and I have never previously seen such detail on a single-frame image. The image did not cover the full frame and ~1,800 pixels of the 16-megapixel sensor were placed across the Moon's image. This corresponds to 1 pixel per arc second, which 'sampled' the image well, given that the likely seeing disk would be about 2 arc seconds across and the theoretical resolution of the telescope is slightly more than 1 arc second. The noise level was creditably low (the D7000 has a superb sensor) and only a single pass of the 'Despeckle' filter (Filter > Noise > Despeckle) in Photoshop was used. I think that the image also attests to the superb contrast of the telescope.

Noise Reduction

A single image may be rather 'noisy'. A noise reduction program such as 'Picture Cooler' can leave the detail but smooth the maria regions. The despeckle function in Photoshop also works quite well. An alternative (and better) way of reducing noise

Figure 11.3 Single image of the Moon taken with a Nikon D7000 camera mounted on a 127-mm, f7 apochromat refractor.

is to stack several images together. This can be done manually with Photoshop. Take four images, say a, b, c and d. Open a and b, select a (Select > All or Ctrl A), copy it (Edit > Copy or Ctrl C). Select b and paste a over it (Edit > Paste or Ctrl V). This will add a layer containing image a above that containing image b, and both can be seen by opening the 'Layers' panel (Window > Layers or F7). In normal (the default) mode set the opacity to 50% and use the move tool and cursor keys to bring one precisely over the other. Flatten the pair (Layers > Flatten Image) to produce image ab. Do the same with c and d to make cd and then merge ab and cd to produce abcd. This is the average of the four and will show significantly less noise.

It is even possible to get a good image when the Moon is photographed in day-light or under heavily light-polluted skies. Take a 'sky' image close to the Moon with the same exposure. Open up both images in Photoshop, click on the sky image and smooth it using the 'Gaussian Blur' filter (Filter > Blur > Gaussian Blur) with a 200-pixel radius. Then select it using Select > All or Ctrl A, copy it with Edit > Copy or Ctrl C, click back on the Moon image and paste the sky image onto it using Edit > Paste or Ctrl V to produce two layers. Set the layer mode to 'Difference' and flatten the two layers (Layers > Flatten Image) to find that the sky light will have disappeared, leaving the Moon against a jet black sky. Magic!

The series of four images in Figure 11.4 shows how a reasonable image of the Moon was taken when at only 11 degrees of elevation and whilst the sky was still quite bright. Four lunar images were combined to reduce the noise and, as already described, a sky image (at left) was taken and differenced with the lunar image. Due to the low elevation, this produced a very red image. A monochrome image was then made using Image > Mode > Greyscale. Then the 'Levels' command, Image > Adjustments > Levels, was used to adjust the contrast by bringing the sliders into the edge of the histogram. This gave an image which was then enhanced somewhat using the 'Unsharp Mask' filter to enhance the local contrast of the image: Filter > Sharpen > Unsharp Mask. The radius is set to the maximum and the amount is set to a low value that is found by watching its effect with the preview box checked. This procedure works exceedingly well with lunar images, but it does have the effect of brightening the brighter parts of the image, which is why I suggested that the initial image be slightly under-exposed.

The 'Unsharp Mask' can also be used with a radius of a few pixels and a medium amount to sharpen the image, but alternative and better sharpening techniques can be used, such as those that employ the 'Smart Sharpen' and 'High Pass' filters in Photoshop, which are covered in Chapter 18. The final, sharpened image is shown at right.

Combing Two or More Images to Build Up a Lunar Image

Given a focal length of 1,800 mm or more, the Moon may well not fit onto an APSC-sized sensor – particularly near full Moon. Two (or possibly more) images have then to be composited into one final image. There is an easy and a hard way. The easy way is to download the free program 'Microsoft ICE', where ICE stands for 'Image Composite Editor'. The two or more images are selected in the image folder and dragged onto the central area of the ICE panel. Without any further effort it combines them into one image, which one saves in the image folder. Should one of the images be somewhat brighter than the other, Microsoft ICE matches the brightness of the two images automatically.

The harder way is to open the two images into separate windows in Photoshop. Extend the canvas size of one of them to be sufficiently large to hold the full lunar image. Click on the other image, Select > Select All or Ctrl A, and use the move tool to drag it onto the second image. In the 'Layers' window, which shows the two images as two layers, set the opacity in *normal mode* to 50%. It is then possible, first using the

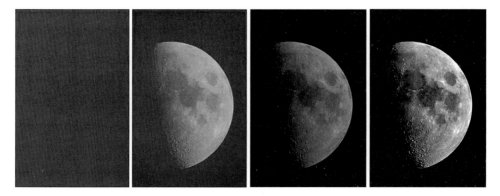

Figure 11.4 Processing a low-elevation lunar image as described in the text.

mouse and then the four cursor keys, to overlay the second image precisely over the first. The opacity is set back to 100% and, if the brightness of one part does not quite match that of the other, you can select one of them and then use the 'Levels' command (Image Adjustments > Levels) to make the transition from one layer to the other disappear. The two layers are then merged into one using the 'Flatten Image' command in 'Layers': Layer > Flatten Image.

Reducing the Effects of Curvature of Field

It is sometimes worth taking three images even if the Moon just fills the frame. If the telescope has a low f ratio it is likely to suffer from curvature of field so that, if the central part of the Moon is perfectly sharp, those parts imaged farthest from the centre of the field may not be. The solution is to take three images with the top, centre and bottom of the Moon across the middle of the frame, where field curvature effects will be least. The central parts of each image can then be cropped (making sure that there is some overlap) and then the three are combined, as previously described.

Lunar Webcam Imaging

This technique, which makes possible very high quality lunar images, uses the same equipment and software as that used for planetary imaging but, as many individual segments of the Moon will be imaged and then combined into a final image, some additional processing is required. The use of webcams or DSLRs for planetary imaging will thus be discussed first and a section on lunar imaging will follow.

11.4 Planetary Imaging with a Webcam, DSLR in Video Mode or Dedicated Planetary Video Camera

A major technique for imaging the planets, but also very useful for producing the very best lunar images, entails the use of webcams, DSLRs in video mode or dedicated planetary video cameras to capture video sequences of hundreds or even thousands of individual short exposures. Considerably more effort is required, so why do it? The main answer is that it becomes possible to ameliorate the effects of the atmosphere, which usually limits the resolution of a single image taken with a DSLR.

This technique became popular when amateur astro-imagers discovered a Phillips webcam, specifically the Toucam Pro II. It contained a surprisingly sensitive sensor and enabled videos to be taken with perhaps 1,000 or more individual frames where each frame was taken with a very short exposure to 'freeze' the seeing. Some of these frames will be significantly sharper than others, whilst all will be somewhat noisy. The object is then to extract the sharpest frames and 'stack' them to provide a final image which is both sharp and, due to the averaging of many frames, contains far less noise. Professionally this is called 'lucky imaging'.

Sadly, the Toucam Pro II is no longer available but, at greater cost, specialised webcams are available from companies such as Imaging Source, Point Grey and iNova. They are available in both monochrome and colour versions. In principle, the mono versions coupled with a set of RGB filters will give a better result, if with somewhat more effort. There is, however, one point to note regarding Jupiter and Saturn – prime imaging targets: they rotate quite rapidly so, unless the three imaging sequences are carried out quickly with the use of a filter wheel to quickly switch between filters, the image resolution could be compromised, as the three images will not register perfectly. Depending on the hoped-for resolution in the image, exposure sequences of no more than 60–90 seconds should be used. As this limits the number of frames that can be stacked, noise in the resulting image may be greater than desired. This is one reason that beginners in planetary webcam imaging should perhaps start by using a colour webcam to more easily achieve their first results. I have to say that when I went to the expense of acquiring a mono webcam, filter wheel and filters, I expected the results to be significantly better and was somewhat disappointed that, though the results were somewhat better, the improvement was not as great as I had hoped. I suspect that this was due to the seeing, which puts a limit on what can be achieved no matter what equipment is used.

For experienced practitioners in the art, there is now available a freeware program called 'WinJUPOS' which allows for a number of stacked images taken over a somewhat longer time to be derotated and combined to reduce the noise in the resulting image. This then allows for greater sharpening to be applied and hence shows more detail in the resulting image.

Dedicated Lunar and Planetary Webcams

Imaging Source produces a wide range of professional webcams in both mono and colour versions which would seem ideal for those wanting to make a start in planetary imaging. For some years I have been using the Imaging Source DFK 21AU04.AS, a colour camera with a sensor of 640 × 480 pixels and a maximum useable frame rate of 30 frames per second. (It can record at 60 fps, but artefacts may appear.) It is equipped with an infrared filter that cuts off at 7,000 angstroms, so giving an image similar to

185

that seen by the human eye. A newer colour camera, the DFK 21AU618.AS, at ~25% greater cost, uses a higher-sensitivity CCD sensor, so reducing the individual exposure times and thus helping to cut through the atmospheric seeing. Monochrome versions of both cameras are available with which filter wheels are used to switch between the red, green and blue filters. With no filter in place, a luminance image can also be taken using shorter exposures, which will thus tend to have a higher resolution and which can be combined with the colour image, as described later. I have recently acquired the DMK 21AU618.AS, a mono version using the more sensitive CCD sensor, partly for lunar imaging but also for planetary imaging with the use of a filter wheel. Both these cameras use a USB2 interface, but Firewire and Ethernet interfaces are also available.

These cameras, which use 8-bit digitisation, can be bought with sensors having differing sensitivities. That in my colour camera has a quantum efficiency of 38%, but the newer chip (Sony ICX618ALA) in the mono camera has a quantum efficiency of 68%. They also come in a number of sensor sizes. Point Grey with its Flea3 cameras and iNova with its PLA and PLB series also provide high-quality colour and monochrome webcams which use 12-bit digitisation. The iNova series are interfaced through USB2, whilst the Point Grey can be interfaced through Firewire or Ethernet; some now come with USB3 interfaces to allow even faster data transfers – which could well be needed if the up to 130 frames per second imaging speeds are used! At the time of writing, the DMK, Point Grey and iNova monochrome cameras using the Sony ICX618ALA sensor are amongst the very best for planetary imaging.

One might first think that 12-bit digitisation will give a better result than only 8 bits. This is, of course, true when one is taking a single image but not necessarily so when stacking a sequence containing several hundred images. Provided that there is some noise present – there will be – adding images together gives the effect of increasing the bit depth, so I am not convinced that this will make a major difference in the final, stacked image.

Registax

Analysing perhaps 2,000 or more images to select those of the highest quality and then combining them would be impossibly time-consuming if done manually, and the technique became viable only with the introduction of software programs such as the free program 'Registax'. Having imported a video sequence into Registax, one can scan along the individual frames, and one of good quality is selected as a reference frame. Registax then analyses each frame in turn, rating it in order of goodness. The program also measures the offset in position of each frame relative to the reference frame, so it is then able to 'stack' the images to produce the final image. These offsets can be random due to the seeing (the planetary image dances about somewhat) or show a steady drift if the tracking of the mount is not quite perfect. I am an advocate of allowing a small drift as, with the image moving across the sensor, the effects of any dust spots will be greatly reduced. If a colour webcam is used, it will also mean that the array of micro-filters above the sensor also shifts relative to the object and so, I suspect, a slightly truer colour will result.

There is one aspect of Registax that requires some judgement, and that is how many of the images to use in the stack. The program gives options as to how to do this; for example, one can select, say, the best 100 frames or use those with a goodness parameter of greater than, say, 90%. Using a few of the very best individual images would give the sharpest image, but it would tend to be rather noisy. Putting more lower-quality images into the stack will remove the noise but there will most likely be some loss of sharpness. Finding an appropriate compromise will vary depending on the seeing when the video sequence is taken. The consensus is to err towards using more frames rather than fewer.

A word of warning: Registax is very computing intensive and will run very slowly on a single-core processor; a laptop with a four-core processor will be a useful asset – not only for carrying out the image processing but also for data acquisition. The data files can be very large and an external USB-powered hard drive can also be a very useful accessory.

The latest versions of Registax (Registax V6 onwards) allow for a large number of alignment points to be set across the image, and the procedure is described in detail in the subsection on lunar imaging by webcam. I have not been too happy using this for planetary imaging and I, along with some other (and far more experienced) imagers, still use Registax V4 for processing planetary images, which gives a little more hands-on control. In this version, a 'box' of appropriate size is placed around the planetary image. Registax then performs a two-dimensional Fourier transform of the image, with is then used in aligning each other frame in the sequence. I like the way one can 'see' the registering take place and feel that I have somewhat more control over the process than with Registax V6.

There can be another problem in using Registax. AVI sequences can be uncompressed or compressed by a codec so that they use up less disk space. The effect can be quite considerable, as planetary images will probably not cover a major portion of the frame, the remainder being black. As an example, a video sequence that was 233 Mbytes in compressed form was 1.76 Gbytes in uncompressed form! Registax 6 can decompress compressed files, but earlier versions may not – giving rise to an error box 'Cannot decompress AVI file'. There are two solutions to this problem: one is to save uncompressed files in the first place, and the other is to use a free program called 'Virtual Dub' to read the compressed file and export it as a temporary uncompressed AVI file.

The way to save planetary AVIs is to use an uncompressed 'raw' file. If the camera is a mono version, each pixel will have one 8- or 12-bit number assigned to it. In this case, an uncompressed raw data file will take up two-thirds less storage space than a full colour uncompressed AVI file, as there is only one number assigned to each pixel rather than three. Above each pixel in a colour camera will be placed a red, green or blue filter. Provided that the processing software knows the specification of filter layout of what is called a 'Bayer mask', it can generate each frame's colour image. To achieve this, the raw data must be 'debayered'. In Registax V4 there is an 'Additional Options' tab above the image and within this are the 'Debayer' options which, for use with the Imaging Source cameras, require the 'Debayer' box to be ticked and the GB button selected. In Registax V6 a box at the top right – 'Prefilter' – has to be ticked and then 'Debayer' box ticked along with GB. (I have included these details as, when one is reading the forums, these aspects of Registax can cause quite a bit of anguish – for myself included.) An example of the use of Registax V4 for use in imaging Jupiter will follow after a discussion of more general aspects of planetary imaging.

11.5 Planetary Imaging Problems and Some Solutions

You may well find, as I have, that planetary webcam imaging can be a slightly frustrating and sometimes dispiriting activity. The latter is because, no matter how well you prepare to carry it out, it is unlikely that you will be able to match the incredible quality of some of the images seen in the astronomical press and on the Internet, such as those by Damian Peach (www.damianpeach.com/), who has kindly allowed me to include one of his images in this book. The problem is the atmosphere. Unless the seeing is really very good, even when thousands of individual images are taken, the final result may still have a resolution of only 1 arc second or more, and although this is sufficient to show detail, such as the Great Red Spot on Jupiter, the results will often be somewhat disappointing. You simply have to hope that, having taken all the appropriate steps to get the very best out of your mount and telescope, one night when, say, Jupiter is high in the south, the seeing will be excellent and a first-class image will result. Sadly, this was never the case when I imaged Jupiter in its apparition at the end of 2012. However, the example that will be described later does indicate what one might achieve under more typical conditions.

One aspect of planetary imaging is of prime importance: the use of a very stable mount with excellent tracking. In order for the planetary image to be sufficiently well sampled, a long effective focal length will be required, usually achieved by using a telescope in conjunction with a Barlow lens. The sensor size of these webcams is very small, typically 4.6 × 3.8 mm, so unless the mount is stable it will be almost impossible to place the planet on the sensor and then, unless the tracking is good, keep it within the sensor whilst the imaging takes place.

A very helpful device that can remove some of these frustrations is called a 'flip mirror' – as is produced, for example, by Vixen (Figure 11.5). In this case the webcam and Barlow, if used, are placed in the straight-through port and an eyepiece, made par-focal, is placed in the other. An eyepiece with an illuminated reticule would be a good choice. One can then easily 'flip' between the eyepiece and webcam without affecting the pointing and, of course, as the weight remains constant the position of the telescope is unlikely to shift.

It does not matter if the tracking is not quite perfect as long as the planet stays well within the frame whilst the exposure sequence is carried out. As already mentioned, some movement across the frame can be helpful but it is also important that, at the end of an imaging sequence, the planet is still within the frame so that slight manual adjustments can bring it back into the frame. If the planetary image moves out of the sensor, it can be annoyingly difficult to bring it back. (This is again where a flip mirror would be very useful!)

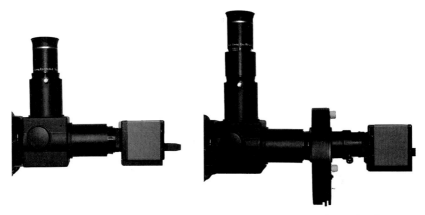

Figure 11.5 A Vixen flip mirror and colour webcam (*left*) and a flip mirror with manual filter wheel and monochrome webcam (*right*).

The image quality will be determined by the fundamental resolution of the telescope, the seeing on the night in question and the size of the image on the CCD chip. By selecting the best frames in the video sequence, one might hope that the seeing, if good to excellent, will not significantly limit the resolution, in which case it is the telescope that will limit the fundamental resolution. As an example, the resolution of my 102-mm refractor is theoretically 1.25 arc seconds. It is then important that the image be 'well sampled' by the pixels in the CCD sensor. Ideally, there should be at least 2 CCD pixels for each resolved point so, in this case, each pixel should not subtend more than ~0.5 arc second in the image plane. Thus, if one were imaging Jupiter having a 45 arc second angular diameter, there should be at least 90 pixels across the image. The pixel size on a typical webcam sensor is about 5 microns so, with my refractor of focal length 800 mm, each pixel will subtend 1.3 arc seconds across – rather more than one might like. When a ×2.5 TeleVue Powermate is used before the webcam, this is reduced to 0.5 arc second and the image will be well sampled.

I would not wish you to assume that I set up my 800-mm f8 refractor with a ×2.5 Powermate as a result of the reasoning just given – which I only carried out as I was writing this chapter. I simply found that the use of a ×2.5 TeleVue Powermate seemed to give the best results. It was thus somewhat gratifying to find that when I measured the number of pixels across the image of Jupiter the total came to 93 – very close indeed to the 'theoretical value' of 90!

It is not sensible to try to increase the magnification further to produce a larger image on the CCD sensor, as the brightness will be reduced, so requiring longer individual exposures when the seeing may well reduce the image quality. When I tried using a ×5 effective Barlow with my refractor, so doubling the image scale on the CCD sensor over that obtained with my ×2.5 Powermate, the exposures had to be increased by a factor of 4 and the resulting image quality was not as good.

It is a happy consequence that larger-aperture telescopes which have higher theoretical resolution also collect more light so that the exposures can be kept short. They also tend to have greater focal lengths, automatically increasing the image scale on the

CCD sensor, ensuring that the image is still well sampled. There is a very simple way to calculate what equipment to use with a webcam. Given a pixel size of ~5 microns it turns out that the effective focal ratio of the telescope should be ~f20, so that if a telescope of f8 is used a ×2.5 Barlow would be ideal, whilst if a Schmidt-Cassegrain of f10 is used a ×2 Barlow is about right. Obviously, if the pixel size is greater than 5 microns then the required effective focal ratio will have to be increased pro rata. The fact that the pixels are larger and so more sensitive should mean that, even though the image is larger, the exposures can remain similar. If the seeing is exceptionally good it can be worth increasing the image scale on the CCD sensor; then the fact that longer exposures are needed will not adversely affect the image quality and over-sampling may well increase the resolution of the final result.

From the preceding discussion it should not be surprising that larger telescopes will provide higher-resolution images, as testified by the fact that leading planetary imagers are using telescopes such as Schmidt-Cassegrains of 11-, 12- and 14-inch aperture.

11.6 Processing a Planetary Colour Image in Registax V4

The first step is to import the AVI files into Registax, taking note of the points made earlier about compressed video files. Registax V4 may not be able to open some AVI files, in which case a free program called 'VirtualDub' can usually be used to convert the files into a form that can be read by Registax. I believe that the ideal file type would contain uncompressed raw data. In this case one must 'debayer' the data by clicking on 'Additional Options' and selecting the appropriate debayer algorithm for the colour webcam. The program should detect that it is in colour, but it can be worth ticking the 'LRGB' box, which helps with noise suppression and preventing highlights from becoming over-exposed. Movement of the bottom slider to the right shows each frame in turn, and one of the sharper and rounder ones (seeing often 'squashes' the planetary disk) should be selected. Then, when one moves and clicks the cursor over the image, a box will be drawn. The box size can be selected within the 'Alignment Options' so that the planet is suitably surrounded – larger but not too much larger. At this point an FFT window will appear at whose bottom right is a box with a number. This should be increased until there are no colours at the edges of the FFT window, with the central bright spot about one-fifth the width of the box. This parameter is usually between 5 and 9. When one is analysing a Jupiter image it is good to see some faint linear features occurring in the FFT which are due to the equatorial belts (Figure 11.6, top).

One then selects the percentage of images to be stacked in the 'Quality Estimate' box. The default 'Gradient' method works well for planets. As the number of frames that will actually be stacked can best be decided later, this can be set at a very low value, say, 10%. The 'Align' button is clicked and one can watch the alignment box follow the planet. The 'Limit' box is then clicked and then the alignment optimisation process is entered, which will attempt to optimise the offset information relating to each frame. The first thing to do is to create a reference frame by clicking on the

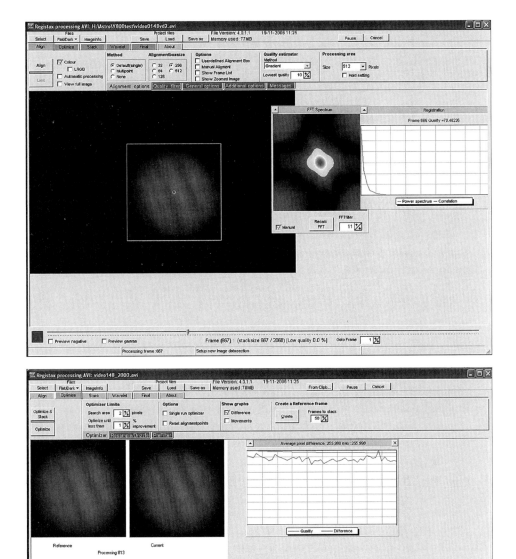

Figure 11.6 Registax V4 screens. *Top*: Setting the alignment box and Fourier transform parameter. *Bottom*: Having made a reference frame with which to optimise the alignment of each video frame.

'Create' button at the top right. This produces a stack of some of the best images to give a better reference against with each frame is compared in order to make fine adjustments to its offset as compared with the reference frame (Figure 11.6, bottom). If the 'Single Run Optimiser' is unchecked, more than one (as one might expect) optimising run will take place – taking longer to process but producing a somewhat better result. Clicking on 'Continue' and then 'Optimise' – *not* 'Optimise and Stack' – allows one to initiate the optimisation process in which one sees the reference frame along with each frame in turn as the alignment is optimised in one or more steps. The green plot that is produced shows the offset and, after each optimisation run, the offset, average and maximum, in number of pixels is indicated, which should be reduced at the end of each optimisation run. This should make the stacking process more accurate and so is well worth the time taken.

Next click on the small 'Stack' button in the line including 'Align, Optimise, Stack …', not the large button! Then check the 'Show Stackgraph' box within the options section. Moving the slider at the bottom of this box will control the number of frames that will actually be stacked out of those aligned. The slider will typically be moved over to the left to eliminate the worst of the images from the stack. (This is why the initial alignment percentage or number control is not too important.) By moving the 'difference' slider down, one can reject those frames where the alignment was significantly off. This is one aspect of using Registax V4 where some judgement is required. The general consensus is that using more frames rather than fewer tends to give better results. Then one can click on the bigger 'Stack' button to initiate the stacking process of those frames selected. Figure 11.7 (top) shows how the frames to be selected were chosen, with those on the right deselected using the bottom slider and some of those with greater offsets also removed by bringing down the left slider.

After the stacking is completed (Figure 11.7, top) the 'Save Image' button can be clicked to save the stacked image for further processing in Photoshop. There is, however, one process that may be useful, and that is to align the red, green and blue images. If the planetary image was taken at a relatively low elevation, one effect of the atmosphere is to act like a prism, so that the red, green and blue images of the planet will be at slightly different positions and it may appear that one edge of the image will look a touch red and the opposite edge a touch blue. When one clicks on the 'RGB Align' tab a set of buttons are presented which allow the red and blue images to be moved relative to the green image, so aligning them and providing a cleaner overall image.

At this point I will also apply some additional processing to sharpen the image and then export a second version. The sharpening process is opened by clicking on the 'Wavelets' box. This brings up a set of six sliders which affect the sharpness of the image. The image can be observed as the effect of each slider on the image is seen. If, whilst a slider is being adjusted, the noise level in the image increases, it is time to back that slider off. One then clicks on the 'Do All' box before saving the image. This stage is shown in Figure 11.7 (bottom). Once a set of slider positions that work well have been found, they can be saved for use on later images. It is important not to over-sharpen an image, as artefacts will appear. Clicking on the 'Final' tab brings up a number of further

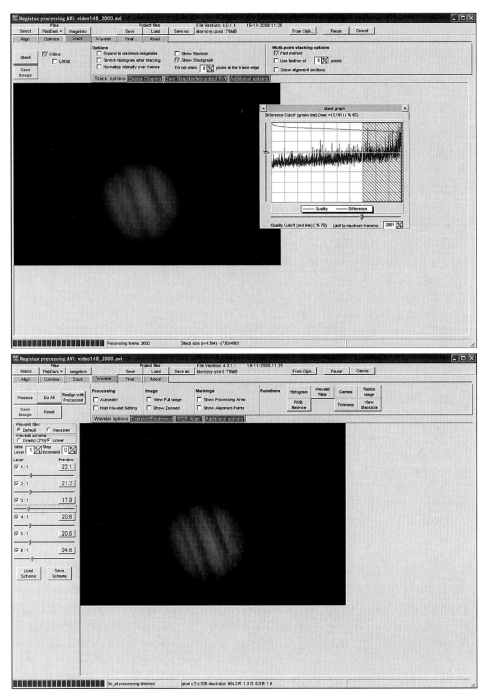

Figure 11.7 Registax V4 screens. *Top*: Having stacked 1,394 frames out of 2,001.
Bottom: A sharpened image, having used the wavelet mode.

options: the image can be flipped horizontally or vertically and rotated by an arbitrary angle, and the hue, saturation and lightness of the image can be adjusted before the final image is saved. The initial, unsharpened stacked image can always be imported back into Registax to make use of the 'Wavelets' and 'Final Processing' options.

In Photoshop, I may use the 'Smart Sharpen' filter to apply some further sharpening to the image and use the lab colour mode to increase the saturation if necessary. Details of these sharpening and saturation enhancement techniques are provided in Chapter 18.

R-Colour Imaging

With Photoshop, it may be possible to improve the image by using one of the channels as a luminance layer which can then have colour added to it, as is usually done in CCD imaging. One first saves a second version of the colour image '... _chan'. By bringing up this image in 'Channels' (Window > Channels) it is possible to inspect the individual red, green and blue channels by turning off the two that one does *not* wish to inspect. Theoretically, the blue channel should be sharper than the green and red as the resolution of the telescope will be higher, but this is not usually the case, as the atmosphere affects the red image least and this may well appear to be the sharpest. This channel is then used as the luminance image, which is then coloured using the original colour image to give a sharper overall result. This technique is thus called RRGB imaging. In 'Channels', enter Image > Adjustments > Levels, select the green channel and move the left slider across to the extreme right. Do the same with the blue channel. The image will now look red! Convert to greyscale, Image > Mode > Greyscale, and then back to colour: Image > Mode > RGB (it will still be monochrome). Now bring back the colour image and apply a Gaussian blur of, say, 2 pixels, select it, Select > All or Ctrl A, and copy it, Edit > Copy or Ctrl C. Finally, select the mono image and paste the coloured image onto it, Edit > Paste or Ctrl V. Open the 'Layers' box, Window > Layers or F7, and, with the blending mode set to 'Colour', flatten the two layers after adjusting the opacity to give the best result.

RGB and LRGB Imaging

This requires the use of a mono camera such as the DMK 21AU618.AS or Point Grey Flea3 and a filter wheel to take the red, green and blue filters along with a clear filter to provide the luminance image if one is used. In principle, only the three RGB images are required to produce a colour image, so why is a further mode listed in the heading? In a technique called 'LRGB imaging', widely used by deep-sky astro-imagers, the three RGB channels are used to make a colour image which provides the colour to add to a monochrome luminance image. As no filters are in the imaging chain (except, perhaps, an infrared cut-off filter), the luminance exposures will then be shorter and so will be better able to cut through the seeing.

Due to the rapid rotation of Jupiter and Saturn, care must be taken in producing the image. To achieve the full resolution that a large telescope is capable of, the luminance image should be limited to perhaps 90 seconds, which will provide 2,700 frames at a frame rate of 30 frames per second. On either side of this luminance exposure the filtered images should be arranged in the order blue, red, luminance and green. As the blue image provides the lowest resolution, it is imaged first. The three colour images are used to produce the colour image. As this is often smoothed (which reduces the noise in the image), the exposures can be somewhat shorter and 1-minute AVI sequences might well be suitable. These will, of course, require greater exposures than the luminance images.

There are two types of RGB filters. The cheapest filters use glass doped to absorb the unwanted two colours and are essentially the same as the RGB filters used in DSLR Bayer matrix. Their pass bands will be broad and will overlap somewhat. They do not incorporate an infrared cut-off filter, so one is needed in front of the CCD chip and is screwed into the camera's 1.25-inch barrel. This is certainly true when a refractor is used but may not be needed with a reflecting telescope, when the luminance image can then be taken without a filter in the optical path. This can mean that the focus will be slightly different, and it can be useful to include a plain glass filter to reduce this effect. A system using these filters is probably quite sufficient for planetary imaging, and I have a set of absorption filters mounted in a manual filter wheel for this purpose. RGB filters that have much tighter pass bands are available at higher cost; they are well worth having for deep-sky imaging and can obviously be equally well used, but are not essential, for planetary webcam imaging. They are discussed in detail in the section on CCD imaging in Chapter 15.

Producing a Colour Image from the Red, Green and Blue Channels

Once the AVI files in a program such as Registax have been processed as described earlier, the red, green and blue images are combined into a colour image which, due to the rotation, will be theoretically less sharp than the luminance layer – both because the exposures have to be longer and because they will have been taken over a period of 4.5 minutes, not 1. This does not matter and the colour image can be made even less sharp, but with less noise, by applying a Gaussian blur of, say, 2 pixels, assuming the image is of the order of 200 pixels across. The three red, green and blue mono images must then be combined to form a colour image. Each of the three images must be in 8-bit greyscale mode. At this point the relative brightness of the three images should be checked using the 'Levels' box, Image > Adjustments > Levels. It may well be that the blue image is not as bright as the other two and, if so, the right-hand slider should be moved to the left to increase its brightness. It is very likely that the three planetary images will not be in precisely the same location within the frames and, if so, the three can be combined and aligned using the following method.

First open a new image whose size is equal to or larger than the RGB images. If the 'Channels' window is then opened, three empty R, G and B channels will be seen. First select the green mono image, select it (Select > All or Ctrl A), copy it (Edit > Copy or Ctrl C) and then click on the appropriate channel box and paste the image into it (Edit > Paste or Ctrl V). Now repeat with the red image. The result will look

6" Maksutov   Christopher Hill   11" Schmidt-Cassegrain   Damian Peach

Figure 11.8 Images of Jupiter and Saturn.

pretty awful, as it is unlikely that the two will be aligned. Then select the move tool and, by initially using the mouse and then, for fine adjustments, the cursor keys, you can accurately align the two. (This assumes that an equatorial mount has been used so there is no frame rotation between the three RGB images. If not, two of the images may have to be rotated before they are combined, as described in Chapter 15.) Finally add and align the blue image, and the final colour image should appear.

LRGB Imaging

In this imaging mode, the colour image produced as already described is combined with a luminance image. This is a mono image taken with just an infrared cut-off filter prior to the CCD sensor. As the exposure used to take the luminance image is likely to be one-third that required for the RGB images, the luminance image may well be sharper. Both the luminance and colour images are opened in Photoshop. The colour image is selected (Select > All or Ctrl A), copied (Edit > Copy or Ctrl C) and pasted over the luminance image by clicking on that image and applying Edit > Paste or Ctrl V. The move tool is selected to align them, the blending mode is set to colour and its opacity adjusted to give the best result before the two layers are flattened (Layers > Flatten) to give the final result.

The Jupiter image by Christopher Hill shown in Figure 11.8 and Plate 11.8 was taken during the apparition of 2012 when Jupiter was at high elevation, whilst the Saturn image by Damian Peach (also in the Figure and Plate) was taken back in 2003 when Saturn was far higher in the sky as seen from northern latitudes than in the apparition of 2013.

11.7 Planetary Imaging with a DSLR Camera

Until I read an excellent article by Jerry Lodrigess in *Sky and Telescope* I had not thought that planetary imaging with a DSLR camera was sensible or even possible. His CD-based 'book' – *A Guide to DSLR Planetary Imaging* – is available on the Internet and is a superb introduction to planetary imaging, whether with a webcam or DSLR, and it includes excellent tutorials on using the alignment and stacking programs. Not only can many DSLRs be used for planetary imaging, they can do it well, and if you have a suitable camera it would be well worth trying the technique out before purchasing a webcam.

The majority of new DSLRs have the ability to take video sequences, and so it is not unreasonable that one could be used to undertake planetary imaging. However, just taking a video sequence as one would normally do will not work well! The basic problem is that each frame of the video will likely have an image size of 1,280 × 720, but this image has been taken with a sensor which might have, as in the case of my Canon 1100D, a sensor size of 4,272 × 2,848. Either the image from a group of adjacent pixels (roughly 3–4 pixels wide) will be averaged to produce each output 'pixel' or a grid of pixels within the sensor will be used. In either case the AVI frames would provide a heavily under-sampled image of the planet.

Happily, there can be ways round this which are somewhat dependent on the camera. The easiest solution results if the camera is a Canon DSLR which has a live view mode – as virtually all now do. You will, of course, also need a T-mount to attach the camera to the telescope, as described earlier. All that is then necessary is to download the free program 'EOS Camera Movie Record' and attach the camera to the computer with a USB cable. When the program is opened, the live view is seen on the computer screen with a white rectangle superposed in the centre. In the control panel is a box containing '5×' beside the word 'zoom'. If this is pressed, this part becomes the full frame so that all pixels in this area will be utilised, exactly as we want. A button on the extreme left enables a folder to be selected in which to save the video files and then, by clicking on the red 'Write' button, the capturing of the movie sequence is initiated. This program records files that can be immediately opened by Registax V6 or AutoStakkert!2. A program called 'Images Plus' will also capture live view images from a Nikon DSLR, but it is rather expensive and it might actually be cheaper to purchase a second-hand Canon body. It records files in a custom format which then have to be converted into bit-mapped images for use by the stacking programs.

Some cameras, for example, the Canon 60D, have a 640 × 480 pixel 'movie crop mode' that can be accessed from the video recording menu. This uses all the pixels from the central part of the sensor – just as required – at rates of up to 60 frames per second. In this case, the video file is recorded onto the SD card within the camera. (You will need a high-capacity card!) These files are compressed and a program, such as the free program 'SUPER', will be needed to convert these files into uncompressed AVI files for use by the stacking programs.

11.8 Processing Planetary Images in AutoStakkert!2

AutoStakkert!2 is a relatively new program that is also free – although a donation would be appreciated – and is now used by many of the world's leading planetary imagers. It currently lacks the ability to process most compressed AVI files, so a free program such as VirtualDub must be used to produce an uncompressed file first.

Neither does it include a wavelet sharpening tool, but it does provide two resulting images, one unsharpened and one sharpened. For planets, this latter image seems to provide a very good, not over-sharpened result.

The processing sequence is very similar to that of Registax. The uncompressed AVI file is opened by clicking on the 'Open' tab in the control window. The first image of the sequence then appears in the 'Frames' window. It appears that the program automatically searches through the frames to find one of good quality. As with Registax, I prefer to use a single alignment box for planetary images and, if so, the 'Single' button is clicked in the 'Alignment Points' box and, by means of the mouse, a box is made to just surround the planet's disk. To estimate the image quality, a gradient method tends to work well with planetary images, so the 'Gradient' button in the 'Quality Estimator' box is clicked.

The 'Analyse' tab is then clicked on and a progress bar is seen to move across until the process is completed, when a quality graph appears. The green bar can be moved across to view individual images that are ranked from left to right and can then be set so that a suitable number of frames will be stacked. Setting the 'cut-off' at the 50% level seems a good compromise. In the 'Stack Options' box, I would click on the 'TIF' button and click the 'Sharpened Images' box. The final tab to click on is 'Stack', when, again, a bar shows the progress in first aligning and then stacking the selected frames. When this is completed, two new sub-folders containing the resulting images appear in the folder in which the AVI file was selected.

11.9 Lunar Imaging with a Webcam or DSLR

In contrast to the sensor size of a typical DSLR, 23×15 mm, the sensor size in a typical webcam is just 5×4 mm. This means that an individual image will cover a relatively small area of the Moon's surface. The result is that one needs to take quite a number of images and then combine them into one composite image. By hand this would be an incredibly time-consuming job but, as previously mentioned, there are programs that will do it for you. I am now using the free program Microsoft ICE, which, in this application, is well-nigh miraculous. One needs to make sure that there is plenty of overlap between the individual frames that will make up the final image and, for one particular lunar image, I used 22 individual frames when ~14 would have just covered the Moon.

The area of the Moon covered by an individual image will obviously depend on the telescope's focal length. The numbers that I have just given relate to my Takahashi FS102 refractor of focal length 800 mm. If I were using images obtained with my Schmidt-Cassegrain of focal length 2,350 mm, they would cover about one-ninth the area – so requiring well more than 100 individual images to be processed!

So let's get to some specifics. In November 2012, Jupiter came quite close to a full Moon, and I felt that it would be nice to image the two together in the sky. At the time I was using an Imaging Source colour webcam type DFK 21AU04.AS having a 640 × 480 pixel array. It comes with a 1.25-inch eyepiece adapter and image acquisition software. When connected through a USB port on my laptop, the software provides a live view display to enable easy focusing of the image. Within the device properties, one can select either 'Auto Exposure' or set a specified exposure. For this particular imaging exercise I was recording compressed AVIs to minimise the amount of memory storage required on the controlling laptop's hard disk.

Moving over the Moon's surface, I selected an exposure of 1/1,500 second so as not to over-expose the brightest regions of the surface. (This is quite important, so one should do an initial scan of the whole surface before selecting a suitable exposure.) The FS102 was mounted on my iOptron Minitower Alt/Az mount. This is not ideal due to 'frame rotation' so that later images will be rotated somewhat relative to that taken first; however, Microsoft ICE appears to be able to correct for this. The camera was imaging 30 frames per second, so each 1,000-frame sequence took about 33 seconds. The telescope was simply driven over each part of the lunar surface and an AVI sequence taken, ensuring that there would be plenty of overlap so that Microsoft ICE could composite the images into a final image. Once the Moon was covered, the telescope was moved over to Jupiter and a single 1,000-frame AVI sequence was taken with Jupiter suitably exposed. The Jovian moons were too faint to appear in this image, so the exposure was increased so that the moons were visible (but Jupiter was heavily over-exposed) and a final AVI taken. This was the easy part – the image processing to be described later took considerably longer!

As described in the section on planetary imaging, I prefer to use Registax V4 for analysing planetary AVIs, but for lunar imaging I use Registax V6. Each of the .avi files were opened up in Registax V6, which displayed the first image of each sequence. The first task is to set 'align points' across the surface. This is first done automatically and one can determine the number spread across the image (I was using around 130), as shown in Figure 11.9 (top). One can add additional points if the automatic process has missed them by placing the curser over the point and left-clicking. Once this is done, the alignment routine is initiated which, for every image in the sequence, finds out how much each specified point in the image shifted from the base image. The next task is to limit the number of frames that will be used. I was simply using those within 90% quality of the best. The display then shows lines indicating how much the points have moved during the 40-second period of the exposure. If all lines are aligned in the same direction, it is a good sign and is simply a measure of how well the mount has tracked during the exposure. Lines moving in all directions indicates bad seeing! Finally the selected images are stacked to give a smoothed image, which one then sharpens by clicking on the 'Wavelets' tab. For these images I brought each of the six sliders to the 9.7 position and saved this wavelet mode for use on all images. Once the 'Do All' tab was clicked on, the final image (Figure 11.9, bottom) was saved. Microsoft ICE was then opened and the 14 individual images were selected as a group and 'dragged' onto the working area to merge the images into one composite image. Microsoft ICE had distorted the overall image somewhat in the fitting process and so a compensating adjustment was applied in Photoshop by means of the transform tool.

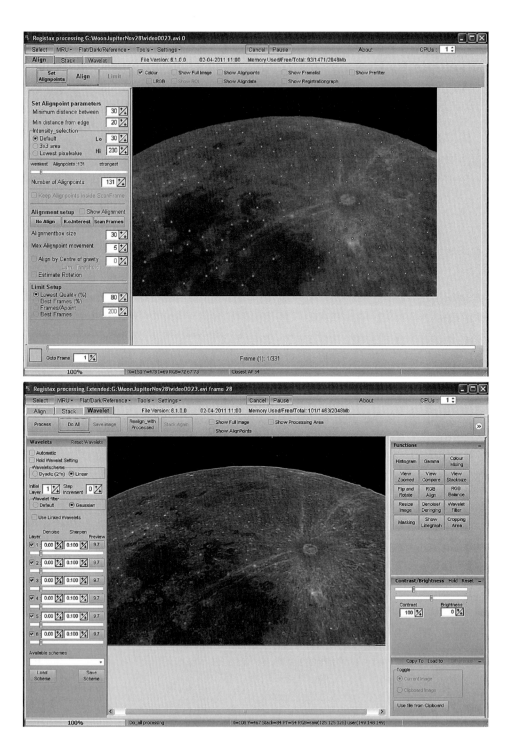

Figure 11.9 The alignment points (white dots) applied to the video frame by Registax V6 (*top*) and the resultant stacked and sharpened image (*bottom*).

Figure 11.10 The full Moon and Jupiter; the image produced as described in Section 11.9. (To reduce the image size, Jupiter has been brought closer to the Moon than was, in fact, the case.)

Enhancing the Lunar Image

This image may well look rather flat with somewhat low contrast, particularly if a reflecting telescope has been used to make the image. It is surprising what can be achieved with a few procedures. One method is simply to use the 'Levels' command in Photoshop. The first thing is to check the histogram in the (Image Adjust > Levels) box. Hopefully the initial image was not over-exposed, and so there will be a gap to the right of the histogram. It is best to leave this for the time being but, the 'Preview' box ticked, it may be worth adjusting the left and middle sliders to improve the image. It is important to be careful; if the left slider is moved too far to the right, detail at the limb of the Moon will begin to disappear.

Another procedure that is less obvious and not well known can make an amazing difference. For this to be used, there should be a gap between the right-hand side of the histogram and the peak white (so the initial images must be a little under-exposed), as the process tends to increase the brightness of the brighter parts of the image. Select the 'Unsharp Mask' filter (Filter > Sharpen >Unsharp Mask), and in the box move the radius slider over to the right (it may then be set to 250 pixels). Then bring the amount slider down to between 10 and 30% and observe the result with the 'Preview' box ticked. Adjust the slider to give an enhanced, but still natural image. I think that you will be impressed.

You can then apply some sharpening to the image, and a number of methods are described in Chapter 18.

Some Lunar Imaging Projects

Why not try to image the Moon on every day of its cycle except very close to new Moon? Be warned, though: this will require some early-morning observations. You could also use a webcam to image the lunar 100 list. Also, look out for a total lunar eclipse, when you can get a beautiful image of the Moon. This will require quite long exposures, so the telescope must be mounted on a tracking mount.

12

Observing and Imaging the Sun

12.1 A Warning

Before describing how to observe and image the Sun, I must give the standard warning. The Sun is the only astronomical object that could cause harm to an observer. If any observing aid, binocular or telescope is used to directly observe the Sun without the use of suitable solar filters, the retina can be irreparably damaged, leaving a blind spot or worse. It is not so much the visible light as the infrared radiation that is the problem. So if a filter is used, perhaps to observe a partial solar eclipse, it is vital that it be opaque to infrared. Any filter used to observe the Sun must be specifically designed for this use and it must be one that will totally block the infrared emission whilst bringing the visible light down to safe levels.

When the Sun is observed (and very great care is taken to use appropriate filters!) it appears to have a sharp edge but there is, of course, no actual surface. We are, in fact, just seeing down through the solar atmosphere to a depth where the gas becomes what is called 'optically thick'. This deepest visible layer of the atmosphere is called the 'photosphere' (as this is where the photons that we see originate) and is about 500 km thick (Figure 12.1). The effective temperature of the photosphere is ~5,800 K. The convective transport of energy from below gives rise to a mottling of the surface – solar granulations that are about 1,000 km across. Each granulation cell lasts about 5–10 minutes as hot gas, having risen from below the surface, radiates energy away, cools and sinks down again.

The region, about 2,000 km thick, above the photosphere is called the 'chromosphere'. The gas density in this region falls by a factor of about 10,000, and the temperature increases from ~4,400 K at the top of the photosphere to about 25,000 K. Above this is the 'transition region', in which the temperature rises very rapidly over a distance of a few hundred kilometres to ~1 million K. The transition region leads into the outer region of the Sun, called the 'solar corona', where temperatures reach in excess of 2 million K. Its form and extent depend strongly on the solar activity, which varies through what is called the 'sun spot cycle' but typically extends for several solar radii into what is called the 'heliosphere'. At the time of solar minima, when activity is low, it usually extends farther from the Sun at its equator and the pattern of

Figure 12.1 The photosphere of the Sun showing limb darkening, granulations and sunspot groups. (Image: Paul Cannon, using a Celestron 102-mm refractor equipped with a Baader Solar filter and imaged with a Canon EOS 1000D camera).

the Sun's magnetic field is often well delineated near its poles. At solar maxima, the overall shape is more uniform and has a complex structure.

The density of the corona is very low, ~10^{14} times less than that of the atmosphere at the Earth's surface, and its brightness at visible wavelengths is about a million times less than that of the photosphere. It can thus be observed only during a total eclipse of the Sun or by means of a special type of telescope, called a 'coronagraph', that can block out the light from the solar disk. How the corona can reach such high temperatures is still somewhat of a mystery, but it is thought that energy might be transported into it by magnetic fields. The million-degree temperatures give rise to X-ray emission that can be observed from space.

Sun Spots

A photograph of the Sun's surface will usually show some darker regions on the surface. These are called 'sunspots' and often appear in pairs or groups – a sunspot group. Each spot has a central 'umbra' surrounded by a lighter 'penumbra'. Sunspots

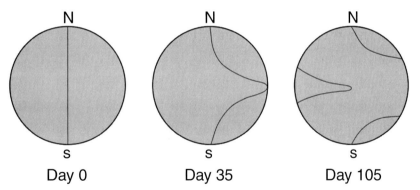

Figure 12.2 The winding of the Sun's magnetic field due to the differential rotation of the Sun.

appear dark because they are cooler than the average surface temperature. The umbra has a typical temperature 1,000 K less than its surroundings. Around the outside of a sunspot group may be seen an area which is brighter than the normal surface. This is called a 'plage' – the French word for 'beach'. (French beaches often have white sand!)

The passage of sunspots across the Sun's surface has been used to measure its rotation period. It has been found that at the equator the period is ~28 days, but this increases with increasing latitude and is ~35 days near the Sun's poles, an effect called 'differential rotation'. Sunspots are intimately linked to this differential rotation and how it affects the Sun's magnetic field. Imagine that at one particular moment the Sun has a uniform bipolar field just like that shown by iron filings under the influence of a bar magnet. Looking at the Sun's face we could imagine a field line just below the photosphere directly down the centre of the Sun's disk. The field and surrounding material remain locked together in what follows, as shown in Figure 12.2. Now move forwards in time by 35 days. The field close to the poles will have made one complete rotation and will be in the same position as seen from Earth, but close to the equator it will have rotated an additional amount, which is (35 – 28)/28 of one rotation, or 1/4 × 360 degrees and so 90 degrees. After three further rotations (as measured near the poles) the field near the equator will have made one additional rotation. The field is thus being 'wound up' and becomes more intense. It gains 'buoyancy' and, in places, rises in a loop above the surface. Where it passes through the surface, the magnetic field inhibits the convective flow of energy to the surface, so the localized region will be cooler than the surface in general – a sunspot appears. The energy that is inhibited from reaching the surface here will tend to reach the surface in the area surrounding a sunspot group, making this region hotter and thus brighter – a plage.

The twisting of the magnetic field from its initial state gradually produces an increase in the number of sunspots observed across the Sun's surface. This determines what is termed the 'sunspot number', which reaches a peak after three to four years – called 'sunspot maximum'. The strength of the field then begins to diminish,

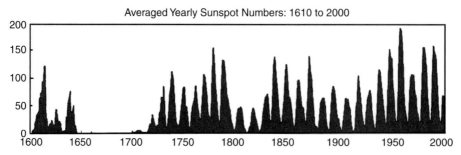

Figure 12.3 Sunspot numbers.

and the sunspot number declines for a further seven to eight years to a point where the Sun's face can be totally devoid of spots – called 'sunspot minimum'. (This was the case during a very prolonged solar minimum around 2009–11.)

The Sunspot Cycle

The whole process then starts over again and is called the 'sunspot cycle'. It is often said to be an 11-year cycle, though it can vary in length somewhat, and the average length of the cycle over recent decades is 10.5 years. The following cycle has, however, one important difference; the field has the opposite polarity. So perhaps we should call it the 21-year, not 11-year, solar cycle. Figure 12.3 shows a plot of the sunspot numbers since 1600. The cyclical variation is very apparent. The plot shows two interesting features: a complete lack of sunspots during the late 1600s, called the 'Maunder minimum', and an increase of solar activity during the past 50 years.

The Sun can be viewed either in white light or, using specially designed solar telescopes, in the light of the hydrogen alpha (H-alpha) or calcium K lines to allow details in the solar chromosphere to be seen.

12.2 Observing the Sun in Visible Light

This allows the observer to view the photosphere and thus to observe sunspot groups on the surface and (less obviously) the solar granulations that form a 'honeycomb pattern' on the Sun's surface.

Nowadays, this tends to be achieved by the use of full-aperture solar filters or a 'Herschel wedge' to allow direct viewing of the Sun but, until the advent of suitable solar filters, the most common method entailed eyepiece projection, and this will be covered first.

Eyepiece Projection

The basic idea is to use the telescope's eyepiece to project an image of the Sun onto a piece of white card, which may be mounted within a box to prevent direct sunlight

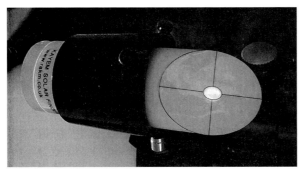

Figure 12.4 The KAYEM solar finder and Baader Planetarium Herschel wedge.

from falling upon it and so improve the contrast of the image. The advantage is that it is then easy to place a piece of paper over the card and easily sketch the sunspot groups. (Pre-prepared sheets are often made with a circular outline of the Sun already drawn.)

There are several potential dangers. The finder scope objective must be covered so one cannot inadvertently observe the Sun though it. The projection should be done using a star diagonal in front of the eyepiece so as to project the Sun's image downwards. The Sun's heat can damage both the telescope and the eyepiece, and so its intensity must be decreased by the use of a small-aperture mask, perhaps 1 inch across. Many telescope covers have such a mask aperture built in. If this is not done, the Sun's heat has been known to damage the adhesive that is used to hold the secondary mirror in a Schmidt-Cassegrain and split up the doublet lens elements found in all modern eyepieces. It might well be worth having a simple Ramsden or Huygens eyepiece with two singlet lenses for this use.

Aligning the Scope

It is not sensible to look along the telescope tube to align it on the Sun. Instead, one can just observe the telescope's shadow on the ground behind (or onto a piece of white cardboard) and minimise its projected area. It is also possible to buy or make a solar finder where a pinhole is used to project a tiny image of the Sun on a screen (having a central dot or cross) a couple of inches behind. A rather neat KAYEM solar finder, which fits directly into the new design of an 8 × 50 finder mount, can be bought at low cost directly from www.rakm.co.uk (Figure 12.4).

Using a Full- or Part-Aperture Solar Filter

The second method is to place a solar filter in front of the telescope aperture. Again, the finder scope objective must be covered so that one cannot inadvertently observe the Sun though it. Aperture filters come in two kinds. Many observers use a filter made of metallized Mylar plastic, called Baader Solar Film, and sold either as A4 sheets or protected by glass in mounts for specific apertures. This type of filter gives the Sun a bluish hue. If you use solar film, do make sure that it is taped securely over the telescope aperture so that it could not possibly fall off. Wrinkles do not matter and should not be removed by stretching the Mylar. Before being mounted over the telescope aperture, the filter should be held up to the light to check for any pinholes in the film. If only a few are visible, Tippex can be used to block them, but if there are many it is time to reject it. It is possible to buy ready-made filters, such as those made by Kendrick and AstroZap, that mount this film into a suitable holder for placing over the telescope aperture.

A second type of filter, made in sheets or mounted in glass, is marketed by Thousand Oaks and Seymour Solar Filters. This is a black polymer and its use gives the Sun a very attractive yellow-orange hue. There appears to be a consensus that the Baader Film–based filters give a slightly sharper image.

Using a Herschel Wedge

A Herschel wedge (Figure 12.4) is a prism within a mount looking somewhat like a star diagonal, which reflects a very small portion, about 4.6%, of the Sun's light at right angles whilst passing the remaining 95.4% through to be reflected by a mirror downwards to the ground or caught within a heat trap (as in the Lunt and Baader Planetarium versions). In the latest Baader version, the Sun's image falls on a translucent ceramic plate at the rear of the unit, giving an image of the Sun and so also acting as a solar finder. The idea was first proposed and implemented by the astronomer John Herschel in the 1830s. The light reflected towards the eyepiece is still too bright, so that those currently on the market also incorporate a neutral density filter to reduce the light to safe levels. In addition, as the light reflected by the prism is linearly polarized, a polarizing filter can be added in the light path after the neutral density filter, which allows the Sun's brightness to be finely controlled.

Herschel wedges must be used only with refractors of up to 6 inches in aperture, as otherwise the heat being absorbed within the wedge will be too great, and they must not be used with reflecting telescopes (as the heat could damage the secondary reflector) or refractors with small correcting elements near the focus (as in the Petzval designs). It is said that the use of these wedges gives the highest definition and contrast of any method of observing the Sun in visible light.

12.3 What Does One See?

When observing in white light one can observe darkening of the solar limb, sunspots, faculae and solar granulation.

Sunspots and the Sun's Rotation

As previously described, sunspots are cooler, relatively dark areas seen on the Sun's photosphere. Most have a central dark umbra surrounded by a lighter grey penumbra. If the seeing is good, and high magnification is used, thin fibrils may be seen radiating from the umbra into the penumbra. However, the very smallest sunspots, known as 'pores', lack penumbras. The appearance of a pore in a clear area of the Sun's surface is a sign that a sunspot group may later appear. Some pores die out in a couple of days or so but, if not, they might grow and develop a penumbra. Other spots may appear in the same region, so that after perhaps 10 days an active region may develop between two major spots (where the Sun's field leaves and enters the surface) orientated in an east–west direction. Producing daily solar drawings (clouds willing!) makes the Sun's rotation become readily apparent, with a sunspot group taking about 2 weeks to traverse the Sun's face. However, as described earlier, the rotation period as seen from the earth varies with solar latitude. At the solar equator the 'synodic' rotation period is 27.25 days, but this lengthens at higher latitudes, being the order of 30 days at solar latitudes of ~40 degrees.

Sunspot Number

The 'international sunspot number' is a measure of solar activity and varies from zero around the time of solar minimum to more than 200 at the time of solar maximum. Also known as the Wolf number, it does not relate particularly well to the number of sunspots that might be seen with small telescopes or binoculars. The daily 'Boulder sunspot number' is computed by the NOAA Space Environment Center. In 1960 the sunspot number reached 250 at the peak of the cycle. Following a lesser peak 11 years later, those in 1980 and 1991 reached ~180. The solar maximum in 2001 reached ~130 and, following a very drawn out solar minimum, the solar maximum of early 2013 reached a maximum sunspot number of only ~80. Indeed, it appeared that the rather unimpressive maximum was reached on one hemisphere in late 2012. It could be that we are in for a period of reduced solar activity.

Limb Darkening

Proof that the Sun has an atmosphere is the fact that the limb of the Sun is darker than the centre. When viewing towards the centre, we see down to a deeper and hotter layer of the photosphere, whereas towards the limb and so the surface is seen obliquely, our vision does not penetrate so deeply and the light is coming from a higher and cooler layer of the photosphere.

Faculae

Faculae are bright areas, usually seen most easily near the limb, associated with concentrations of magnetic field. They are often seen above active regions and may persist after a sunspot group has died. If faculae are seen towards the eastern limb, a sunspot group will usually make an appearance within a day or so. During solar maximum the lower-temperature sunspot regions will tend to reduce the Sun's brightness, but the faculae, hotter than the average photospheric temperature, actually win and the Sun is about 0.1% brighter at solar maximum as a result.

Solar Granulation

Granulation is the name given to the textured appearance of the photosphere due to the convective cells that bring heat to the surface. Each cell (which persists for 10–20 minutes) is about 1,000 km across, so subtends an angle of close to 1 arc second, and a telescope of at least 4-inch aperture coupled with excellent seeing is thus required for granulation to be visible.

12.4 Imaging the Sun in White Light

If eyepiece projection has been used, a DSLR or compact camera can be employed to photograph the projected image. The image is bound to be distorted, as the camera cannot be on axis. If the image is imported into Photoshop, the image can be selected (Select > Select all) and the image corrected using Edit > Transform > Distort.

DSLR and webcams are used in exactly the same way as for imaging the Moon. With a refractor of short focal length and a DSLR, a Barlow lens may be used, so that the image can be made to fill the frame, and with a telescope of long focal length, two or more images might be needed to cover the whole disk. If there is little detail on the Sun apart from a few sunspots, it may not be possible to use a compositing program, and manual overlay of the images in Photoshop may well be required. Webcams can be used to take higher-resolution images of the individual sunspot groups, and they can be processed in Registax exactly as in the case of lunar webcam images. I make AVI sequences of ~2,000 frames and, as for planetary imaging, prefer to process them in Registax V4. When the sequence is opened in Registax, a suitable-sized box is placed around one of the sunspots in one of the sharper images in the AVI sequence and the alignment process started. The procedure for setting the Fourier transform parameter, producing a reference frame, optimizing and finally stacking the selected frames is described in Chapter 11, Section 11.6 to produce images such as those in Figure 12.5.

One trick that I have used to give the effect of a higher-resolution full solar image is to overlay the detailed webcam images of the sunspot groups onto the full disk DSLR image. The DSLR's image is increased (Image> Image Size) to have the same scale as detailed images of the sunspot regions taken by a webcam. The webcam images can then be rotated (Image > Image Rotation > Arbitrary) to have the same orientations as the sunspot groups on the DSLR image. With both images open, the higher-resolution webcam image of the sunspot group can be copied using the clone stamp tool onto the DSLR image.

Adding Colour to a Solar Image

The images taken with either a DSLR or webcam are monochrome, but may, if a Baader filter has been used, show a bluish colouration which is not particularly attractive. A

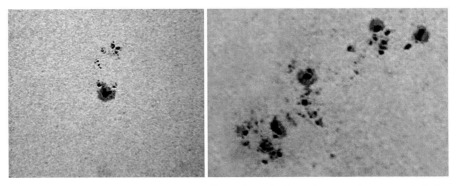

Figure 12.5 Sunspot groups imaged in white light using a webcam and Celestron SkyProdigy 130-mm Newtonian equipped with a Baader Solar filter.

more appropriate colour may be added if desired. One first makes the monochrome by changing the colour mode to greyscale (Image > Mode > Greyscale) and then back into the RGB colour mode. Having found that the colour of the SOHO satellite images was very appropriate, I decided to apply this colour to the image. The solar image is first duplicated (Layer > Duplicate Layer) and then the 'Foreground Colour' square box is clicked on to bring up the 'Colour Picker (Foreground Colour)' box. Within the small boxes to the right of the R, G and B symbols the numbers 252, 193 and 86 are typed. The foreground colour will now appear orange-yellow and could be changed as desired by altering these numbers. The paintbrush tool is now selected and the duplicate image over-painted to give an orange-yellow rectangle. If the 'Layers' box is now opened (Window > Layers), the blending mode set to 'colour' and the two layers flattened (Layer > Flatten Image), the solar image will appear in colour.

Photographing the Sun During a Solar Eclipse

My honest advice is not to try photographing the Sun during totality – there will be many superb images available on the Web soon after. Just enjoy the wonderful spectacle. The partial phases can be imaged by the use of a solar filter, perhaps made out of Baader Solar Film. A very nice image results from photographing the effect of the leaves on a tree forming multiple 'pinhole' cameras projecting images onto the ground. Perhaps if one were observing in a desert it would be worth taking a small twig of foliage to make such an image possible.

The interesting thing is that, during totality, no one exposure will capture the full majesty of the event. To image the outer parts of the corona, a far longer exposure will be needed than for the inner parts, so the correct procedure is to take a sequence of images, doubling the exposure each time. The longest exposures can even show the surface of the Moon illuminated by earthshine! This is the problem: unless one can pre-program a sequence of different exposures, this takes up precious time – which is fine if totality lasts several minutes but not so good if, as often, totality lasts for only a minute or so.

12.5 Observing the Sun in the Light of H-Alpha and Calcium II K Emission

This shows far greater detail in the solar surface than does observing in visible light and allows additional features to be seen.

Prominences and Filaments

Prominences and filaments are dense, somewhat cooler clouds of material suspended above the surface of the Sun by loops of magnetic field. Filaments are dark, thread-like features seen as darker features against the solar surface, whilst prominences are actually the same thing, except that they are seen as light features projecting out above the limb, or edge, of the Sun. Prominences are often visible during a total solar eclipse. Images taken over the several hours that the eclipse track passes over the Earth's surface can show them rising into the Sun's atmosphere and then falling back.

Observing with H-Alpha and Calcium II K Solar Telescopes

This requires the use of a specially designed telescope. At its heart will be a Fabry-Perot etalon which allows a 'comb' of wavelengths to pass through, typically less than 1 angstrom wide and spaced every 10 angstroms across the visible band. One of these narrow pass bands will be centred on the H-alpha line (or Ca K line). The etalon will reject ~90% of the total sunlight, but this is not sufficient to provide a safe viewing level so that there will also be an energy rejection filter to bring the light down to a suitable level. This will also block all the infrared light, which is most damaging to our eyes. The final element is a 'blocking filter' placed close to the eyepiece, which is an H-alpha filter that passes only a very narrow band of light about 8 angstroms wide, centred on the H-alpha line at 6,562.81 angstroms, so removing all but the H-alpha emission.

It is important that the light passing through the etalon be a collimated parallel beam at right angles to the optical axis. The obvious way to achieve this is to place the etalon at the front of the telescope before the converging lens, and this is done in many solar telescopes. However, etalons are very expensive, and an alternative approach which has lowered the cost of solar telescopes is to place the etalon within the optical tube when the light cone has become smaller. The light passing through it must still be in the form of a plane wave, so a diverging (concave) lens is placed in front to provide a collimated beam through the etalon with a second (convex) lens following it to bring the sunlight to a focus. For example, the Coronado SolarMax II solar scope of 60-mm aperture uses an internal etalon ~40 mm across, reducing its cost by almost 50%.

The bandwidth of the etalon has a bearing on the resolution of the image, as the gas in the Sun's photosphere and chromosphere is in motion, and the differing Doppler shifts will broaden the bandwidth of the spectral line. If a broader bandwidth system is used, the effective image resolution on the surface will be less than with a narrower band system. On the other hand, to observe the prominences erupting from the photosphere (which have a greater spread in velocity and hence spread in wavelength), a broader bandwidth tends to work better. If what is called a 'single stack etalon' is to be used for all H-alpha solar viewing, a good compromise is a bandwidth of 0.7 angstrom. A bandwidth of 1 angstrom may show the prominences better, and one of 0.5 angstrom will give somewhat better resolution of surface details. This narrower bandwidth can be obtained by the use of a 'double stack etalon', but it does increase the cost substantially. The second etalon is normally a full-aperture etalon placed at the front of the telescope. By offsetting the two band passes slightly, one produces a slightly dimmer but narrower combined pass band. Using two stacked 0.7-angstrom etalons together typically provides a bandwidth of 0.5 angstrom.

Rather than purchasing a dedicated solar telescope, it is possible to buy an H-alpha observing kit to convert a normal refracting telescope into an H-alpha telescope. All the major manufacturers provide etalons with apertures ranging from 40 mm in clear aperture upwards – with larger apertures giving the potential for higher resolution – along with the required blocking filters. A point to note is that, as only a single wavelength is being imaged, the telescope does not even have to be an achromat; it is important only that spherical aberration be well corrected.

The vast majority of observers will, however, purchase a dedicated solar telescope, and such telescopes can be bought for quite reasonable sums. Coronado produces the PST (personal solar telescope), which limits the cost by the use of a telescope of 40-mm aperture. The less expensive version has a bandwidth of 1 angstrom, whilst a more expensive version using a double stack etalon gives a bandwidth of 0.5 angstrom. Lunt Solar Systems produces a solar telescope of 35-mm aperture with a bandwidth of less than 0.75 angstrom.

Coronado, Lunt and Solarscope all produce dedicated solar telescopes with greater apertures (see Figure 12.6). Each aperture size has several options: they can be single or double stack and can be provided with up to three sizes of blocking filter. As pointed out earlier, a double stacked version will give a narrower bandwidth. The smallest blocking filters (5 or 6 mm in diameter) are said to be suitable only for visual observing, as they will vignette the image taken by a DSLR or CCD camera. (But see my thoughts about the use of webcams later in the chapter.) The medium-sized blocking filters will be suitable for both visual and imaging use, with the largest best for large-sensor imaging such as when a full-frame DSLR or binoviewer is used.

An important point is related to 'focusing' the image. It is necessary to first adjust the focus to give a sharp focus and then 'tune' the etalon to give the clearest image. There are two reasons the etalon has to be tuned. The first is that the effective spacing between the two glass plates making up the etalon depends on the atmospheric pressure, as air at higher pressures has a higher refractive index. Thus, on a mountaintop, where the air is thinner than at sea level, the etalon separation would have to be increased. The second is that the surface of the Sun is dynamic, so that the wavelength of light emanating from, say, a prominence will be altered by the Doppler effect, so requiring the observed wavelength band to be adjusted.

As just stated, the band of H-alpha that is passed through the etalon is determined by the *effective* distance between the two parallel glass surfaces that make up the etalon.

Figure 12.6 Lunt pressure-tuned 60-mm H-alpha telescope with TeleVue solar finder and Meade Coronado SolarMax 60-mm H-alpha telescope with a Coronado solar finder.

There are three ways in which this can be adjusted so that the etalon can be tuned to give the sharpest images. One method is to finely adjust the actual separation between the glass surfaces. It is possible that this is what is done in the Coronado 'Richview' tuning system, but Coronado gives no details as to how this patented system works. Its etalon does, however, have a pad resting on it, which is perhaps used to adjust the separation. Lunt Solar Systems (for some of its telescopes) and Solarscope both use tilting of the etalon to alter the effective separation as, when tilted, the effective path length between the two glass surfaces is increased, as the light is passing through the etalon at an angle rather than straight across (just as the route one takes when walking across a road at an angle will be longer than that when one goes straight across). Because the etalon is then no longer perfectly at right angles to the light path, this could degrade the image quality very slightly, but I have seen no evidence of this.

The fact that the effective spacing of the etalon depends on the air pressure allows for a further method of tuning the etalon – that of altering the air pressure within it. Lunt Solar Systems now also provides solar scopes which are pressure-tuned by adjusting the air pressure within the etalon to provide a tuning range of ±0.4 angstrom. Thus the etalon always remains perfectly at right angles to the light path, so this highly elegant system is theoretically the best for etalon tuning.

Coronado, in its SolarMax range, provides telescopes of 60- and 90-mm aperture. The single stack versions have a bandwidth of 0.7 angstrom, with the double stack versions reducing this to 0.5 angstrom. Each telescope can be provided with blocking filters of 5-, 10- or 15-mm diameters, the models being designated as BF5, BF10 or BF15 respectively. It is worth explaining why these three versions are available. The focal length of the 60-mm SolarMax II is 400 mm, which means that the image of the Sun it produces is about 3.5 mm in diameter. If a blocking filter 5 mm in diameter is placed close to the image plane, all the light will pass through it and the image will be unvignetted. This will be the case when a simple eyepiece is used (one is provided with the telescope), and so this size is all that is needed for visual observing. If, however, it is necessary to move the blocking filter away from the focal plane to allow the use of a DSLR camera, this will not be wide enough to allow all the light to pass through and the image will be vignetted. A filter 10 or 15 mm in diameter will then be better and, for use with a binoviewer, 15 mm will be best.

I really believe (with others) that an excellent way to obtain solar images is to use a webcam (as will be described later), and I suspected that the small webcam sensor could be close enough to the blocking filter that a 5-mm blocking filter would not cause vignetting. I thus carried out an (expensive) experiment to try this out by acquiring a Coronado SolarMax II 60-mm, BF5 solar scope. I was somewhat relieved to find that my supposition was correct!

Lunt Solar Systems produces solar telescopes with apertures of 60, 80, 100 and 150 mm. They incorporate blocking filters 6 or 12 mm in diameter for the 60-mm version, with 12- or 18-mm blocking filters for those of larger aperture. These have a 0.7-angstrom pass band etalon but can be provided with a second etalon to 'double stack' the pair, thus reducing the bandwidth to 0.5 angstrom. Solarscope produces 50- and 60-mm H-alpha telescopes with 0.7 angstrom pass bands; these can have a second etalon attached to reduce the pass band to 0.5 angstrom.

Calcium K Solar Scopes

Solar telescopes can also be provided with an etalon tuned to the wavelength of the calcium K line at 3,933 angstroms. This gives a rather different view of the Sun, with much of the CaII-K line emission coming from a lower part of the photosphere than the H-alpha emission. The CaII-K line is very sensitive to the presence of magnetic fields. Regions where the field is weak show up as darker areas, whilst regions of moderately strong magnetic field show up as bright regions. However, regions of very strong magnetic fields, such as that in a sunspot, appear very dark. This effect makes sunspot regions show very high contrast as compared with imaging in H-alpha. A calcium K image produced by Greg Piepol is shown along with an H-alpha image in Figure 12.7 and in colour in Plate 12.7.

12.6 Drawing and Imaging the Sun in H-Alpha and Calcium K

Given an H-alpha telescope, one can get real satisfaction from sketching the full solar disk showing the sunspot groups, filaments and prominences, such as those drawn by my colleague Paul Cannon. The lower pastel drawing in Figure 12.8 is of the prominence group seen on the lower right of the solar sketch. The SOHO satellite Web site provides the designated numbers for the sunspot groups.

A DSLR can be attached to a solar scope using a T-mount to image the whole disk with one exposure, or a compact camera can be used with eyepiece projection, but often more detail will be captured by imaging segments of the Sun using a webcam. As the CCD sensor of a webcam is only a few millimetres in size and its sensor can be placed closer to the blocking filter than with a DSLR camera, a large blocking filter will not be required and the cost of the solar scope will be reduced. When one is using a webcam, reaching focus can be aided if necessary by the use of a Barlow

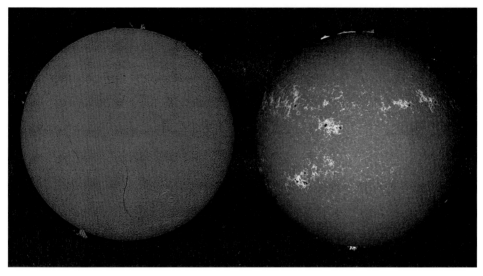

Figure 12.7 The solar disk imaged in the light of H-alpha (Ian Morison) and calcium K (Greg Piepol).

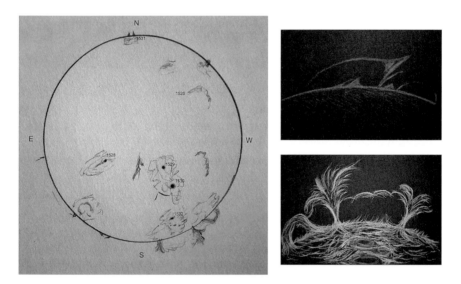

Figure 12.8 A sketch of the Sun observed in H-alpha and pastel drawings of prominences made by Paul Cannon.

lens, removed from its mount and screwed into the front of the camera's 1.25-inch nosepiece. This will also increase the image scale of the Sun, giving the potential for higher-resolution images but requiring more individual images to be combined to give a full disk image.

The image is essentially monochrome – though it appears red to our eyes – and thus a monochrome webcam will be ideal. This will be more sensitive than a colour

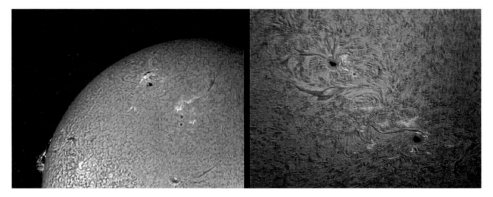

Figure 12.9 H-Alpha images produced using a monochrome webcam mounted on a Coronado SolarMax II 60-mm telescope without (*left*) and with (*right*) the use of a ×2.5 TeleVue Powermate.

version, so allowing shorter exposures. I thus set out to image the Sun using the Imaging Source DMK 21AU16.AS, which has a CCD chip with an excellent ~68% quantum efficiency. This was initially coupled to a Solarscope SA60, which had been kindly loaned to me. The Sun barely shone when I had the telescope, but one morning I was able to take a number of video sequences as gaps appeared in the clouds. Because the Sun's image in the image plane is not that large, only six individual AVI sequences were required to cover it, whilst providing good overlap between them. This technique is exactly that used for imaging the Moon with a webcam. The individual AVI sequences were processed in Registax V4 and the resulting frames composited into a single image shown in Figure 12.7 (left).

The prominences are not as bright, so a further set of AVI sequences were taken to correctly expose them. The prominence and solar disk images then need to be combined. One method is to have the two images open side by side and simply 'clone' the prominences from the brighter image over the just-visible prominences in the image exposed correctly for the disk.

Just as a 'monochrome' image of the visible Sun can be coloured, so can that of an H-alpha image, but this time the colour layer must be the red colour of H-alpha. This can be achieved simply by making sure that the initial image is put into RGB 8-bit mode and then duplicated (Layer > Duplicate Layer). The 'Colour' box is clicked on and a set of suitable R, G and B numbers placed in the panel that opens up. A good starting point would be red = 231, green = 33 and blue = 11. The paintbrush is then used to paint over the duplicate layer to give a completely red image and, in 'Layers', the blending mode is changed to 'Colour'. The layers are then flattened (Layer > Flatten Image), and the resulting image saved. Plate 12.7 (top) is the final result of this solar imaging exercise using the excellent Solarscope SA60.

The Coronado SolarMax II gives a somewhat brighter image than the Solarscope SA60, and I found that I had to use the minimum exposure of 1/10,000th of a second to image with the DMK 21AU16.AS. If, when one is using a larger-aperture solar telescope, there is a danger of over-exposing parts of the image, one can use the concave element of a 1.25-inch Barlow screwed into the webcam barrel to increase the image scale somewhat and so increase the exposure required. Rather than producing full disk images, I have tended to image interesting areas of the Sun, and Figure 12.9 shows two of these. That on the left is with the webcam mounted directly into the telescope eyepiece, whilst that on the right is with the insertion of a ×2.5 TeleVue Powermate to give a detailed image of a particularly interesting region.

In all, solar observing can be quite a rewarding aspect of the hobby which, almost by definition, can usually be carried out when it is warm. Why not give it a try?

13
Observing and Imaging with an Astro-Video Camera

This is an interesting, if not common, branch of amateur astronomy which bridges the gap between visual observation and imaging. One problem with imaging is that to produce an image that stands comparison with those seen in magazines and on the Web will require many hours of work, so that only one or two images might be taken during one observing session. Unless one lives where clear skies are common, the number of celestial objects that might be imaged in a year will be quite small. But, of course, imaging enables us to eke out faint nebulosity that our eyes cannot see unless we have access to a very large telescope. Could there be a middle way to allow us to 'see' faint details even with a relatively small telescope? An astro-video camera gives us that ability.

For some years, a number of companies, including Watec, have been making video cameras containing very sensitive CCD chips for surveillance purposes under conditions of low light. Watec realized that these might have a useful role in astronomy, and the Watec 120N video camera was the result. The camera uses a 752×582 pixel array to provide a completely standard video output stream that can be observed on any monochrome video monitor. It is provided with a 1.25-inch adaptor that fits in the telescope focuser in place of an eyepiece. The monitor display is a very useful feature that enables quite a number of people to 'see' what is being observed at one time – very useful at star parties. (Incidentally, some portable DVD players have an external jack input that allows them to be used as a monitor.) The video output can also be 'frame-grabbed' and so imported into a laptop for display (in place of a video monitor) and for storage if future processing if desired. I use an EasyCAP USB2 video capture dongle for this use.

The basic exposure time of the 120N is 1/25th of a second but with this exposure the sensitivity is less than that of the eye – so why use it? The trick is that it is able to internally integrate up to 256 sequential images, giving a total exposure time of up to 8 (NTSC) or 10 (PAL) seconds. The output image is updated at the completion of each integration. The control unit for the camera is equipped with a dial which can set the number of integrated exposures from 1 to 256 in powers of 2, that is, 1, 2, 4, 8, ..., 256. In the author's experience, with a setting of 8, equivalent to an exposure time

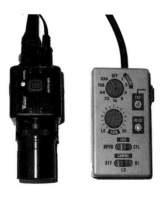

Figure 13.1 The Watec 120N with its control unit and mounted on a Celestron 130 SLT Newtonian with a TeleVue ×2.5 Powermate used to image Saturn and its moons.

of ~0.3 second, the monitor output is comparable to that seen by the eye. The point is, of course, that as the exposure is increased up to 10 seconds, the monitor or laptop display shows fainter and fainter detail in the object well beyond what our eyes can see. It is equivalent to having a telescope whose aperture is at least twice the size of that in use! Figure 13.1 shows the 120N camera and 1.25-inch eyepiece adaptor with its control unit. As well as adjusting the integration time it is possible to adjust the gain (this increases sensitivity but also noise) and the gamma (which controls contrast). It is also possible to freeze a frame on the display. A green LED flashes as each new integration is completed.

With all imaging systems Newtonians can prove a problem in that there may not be enough 'back-focus' to reach focus. A Barlow lens will usually solve this problem but with the result that the field of view is reduced. Refractors usually have sufficient back-focus and there should never be a problem with Schmidt-Cassegrains and Maksutov-Cassegrains given their very wide focus range.

The sensor sizes of these astro-video cameras are pretty small, so to encompass many objects a relatively short focal length is required: 500 to 1,000 mm is about optimum. This can be a problem with Schmidt-Cassegrains and Maksutovs, which tend to have quite long focal lengths and so a focal reducer could be useful. My 200-mm, f4 Meade Schmidt-Newtonian is ideal for use with an astro-video camera with its focal length of 800 mm.

I have owned a Watec 120N for some years and used it with a number of telescopes, including the Celestron 130 SLT (Newtonian of 130-mm aperture). Two images were taken of Saturn with the 120N mounted on the 130 SLT through a TeleVue ×2.5 Powermate. The first, short exposure was exposed for Saturn, with the second having the maximum exposure of 10 seconds. These were combined to show Saturn and six of its moons (Figure 13.2, left). The faintest of these is only 11.7 magnitudes. In fact, under typical conditions, using a 120N with a 5-inch scope should enable one to easily 'see' stars down to about 13th magnitude. With a 200-mm Newtonian, the

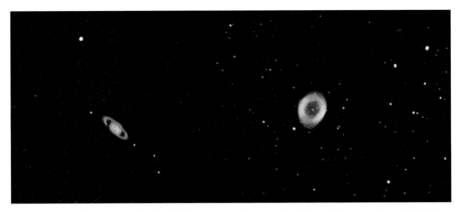

Figure 13.2 Saturn and six of its moons imaged using a Newtonian of 130-mm aperture (*left*) and the Ring Nebula, M57, imaged with a 200-mm Newtonian (*right*).

spiral arms of the Whirlpool Galaxy, M51, were easily seen, as was the central star of the Ring Nebula, M57, which has a magnitude of 14.8. To produce the image of Figure 13.2 (right) ten 10-second-exposure images of M57 were frame-grabbed and stacked in Deep Sky Stacker.

As will be discussed in the next chapter, it is now possible to buy a modern SLR camera body such as the Nikon D5100 for less than an integrating CCD video camera, and such cameras are very suitable for astro-photography. However, SLR cameras which use high-resolution colour CCDs, now typically 12–18 megapixels, have two disadvantages for imaging faint objects. Firstly, each pixel is covered with a colour filter, red, green or blue, and this reduces the light input into the pixel by more than a factor of 3. Secondly, the high nominal resolution of the SLR means that pixels must be small, and light input is proportional to pixel area. The area of the pixels used in the Watec 120N is 3.5 times that of the Nikon D5100 so, overall, the 120N is about 10 times more sensitive. So, to image faint objects, a low-resolution monochrome CCD with a large pixel area, rather than a high-resolution colour camera, is to be preferred. SLR cameras are also relatively large and heavy, so a sturdy telescope tube and mount are required. It would be impossible to mount a DSLR camera on the Celestron 130 SLT!

The Watec 120N has now been updated to the 120N+. This features both additional control over exposure to allow both very short exposures so it can be used for planetary imaging (good) and a 'bulb' setting to allow exposures of longer than 10 seconds. Astrovid (Adirondack Video Astronomy) sells the StellaCam3 Monochrome CCD, which appears to be essentially identical to the Watec 120N+, along with a Peltier-cooled version to give lower noise levels when used with longer exposures (Figure 13.3). Astrovid also produces a wireless controller used in place of the control unit to allow remote operation.

At somewhat lower cost, Mintron produces the 12V6HC-EX Mono Camera, which can integrate for up to 2.5 seconds. This has a metal ring around the base of the barrel to help dissipate heat from the CCD chip. It may be remotely controlled from a PC

Figure 13.3 The Mallincam Hyper Plus and Astrovid StellaCam3 Peltier-cooled astro-video cameras.

or laptop. With very similar back panels (so possibly a development of the Mintron), Mallingcam produces three handcrafted models: the Xtreme, VSS and Hyper Plus, all of which are Peltier cooled and, like the Mintron, can be remotely controlled from a PC or laptop. The VSS is available with only a colour sensor, whilst the other two can be obtained in both mono and colour versions. Due to the Bayer colour matrix in front of the sensor, the colour versions will be less sensitive and are recommended only for larger-aperture telescopes but, for example, with an 11-inch Celestron and an exposure of 56 seconds, the Horsehead Nebula can be easily seen against the red glow of the background nebula IC434 with the nearby reflection nebula, NGC2023, highlighted in blue.

The price of these cameras is not insignificant, but they can often be found second-hand at reasonable cost. If you do acquire one, they can be a great way to show your-self, and others, many of the most interesting objects in our night sky.

14
DSLR Deep-Sky Imaging

As the cost of large-sensor cooled CCD imaging systems has, until recently, been very high, many astro-imagers wanting to make wide-field images of the sky have taken to using DSLR cameras. With their inbuilt colour filters forming a Bayer matrix above the sensor, colour images are easily produced without the cost of the filters and filter wheels that are used with the majority of CCD cameras – not to mention the increased image processing time that is then required afterwards.

There is no doubt that Canon DSLRs have, up to now, been the choice of virtually all astro-imagers, with Canon providing free software to allow computers to remotely control their cameras and download the resulting images. Two more sophisticated programs to control Canon cameras are 'Astro Photographic Tool' (APT), which is freeware, though a small payment is (rightly) requested, and 'Backyard EOS', which is somewhat more expensive. Both allow sequences of exposures to be made so that the imaging system can be left unattended. They also record the sensor temperature for each exposure, which can be very useful if 'dark frames' are to be subtracted in the image processing software rather than in the camera, as will be discussed later. For example, APT can add the sensor temperature to the file name as in 'L_3745_29C', where L indicates a light frame.

The equivalent Nikon software has to be bought though some freeware programs are available. I was using a Nikon camera in the days of film and had acquired some superb prime lenses to go with it so, when finally moving to digital, I first bought a Nikon D80 and then, in March 2010, a Nikon D7000 with a Sony 16-Mpixel sensor, which has greatly impressed reviewers and which, at that time, was reckoned to be the best APSC sensor that had ever been made. However, more recently, I have also acquired a Canon EOS 1100D (Rebel T3), which, as described later, has been modified to capture more of the H-alpha emission that gives the pink-red colour to many astro-images.

A useful feature of the Canon EOS cameras is that the flange-to-sensor distance is less than for many other DSLRs. This means that adapters can be bought so that prime lenses from other camera systems, such as Olympus, Pentax and Nikon, can be used, and so I have been able to use my Nikon prime lenses with the EOS 1100D. An

M42-to-Canon adapter is also available, so it is possible to acquire high-quality prime lenses for wide-field astro-imaging at relatively low cost.

14.1 Making a Start in Deep-Sky Astro-Photography

Taking your own astronomical images can be very rewarding and it needn't be very expensive, as you can make a start using just a DSLR and tripod. Prime lenses probably give better image quality than zooms, but if stopped down somewhat they can also be used. Quite an interesting thing to try first is to simply photograph the constellations using a relatively short focal length. When a fixed tripod is used, the exposure times are limited, as the stars move across the sky surprisingly quickly. The closer to the celestial equator you are imaging and the greater the focal length of the camera lens, the shorter the allowed exposure time. It is worth remembering that the effective focal length of a lens is increased by ~1.5 when used with the common APSC sensor-sized DSLRs. Thirty seconds is about the maximum with a wide-angle lens near the equator. Lenses of greater focal length will require shorter exposures, but as one moves towards the North Celestial Pole, exposures can be made longer without the stars trailing. One advantage of using a digital camera is that the images can be inspected very quickly, so you can find out what is the longest reasonable exposure. The great thing is that with a free program such as 'Deep Sky Stacker', described later, it is possible to combine a number of short exposures and so get the effect of one longer exposure.

Choice of ISO Rating

One choice to be made is to select the ISO rating used for the image capture. Low-ISO (80 or 100) images will be less noisy but longer exposures will be necessary. Often a mid-range ISO, say 400 or 800, will work best. Up to, typically, 800 ISO the gain of the amplifiers reading out the data from the CCD pixels is increased to give a brighter image for a given exposure, but an interesting fact is that with many DSLRs the amplifier gain and hence the raw data read-out from the CCD array with an ISO of 800 is identical to one captured at higher ISO values. The values read out of the CCD array are simply multiplied by 2 for an ISO of 1,600, by 4 for an ISO of 3,200 and so on. There is simply no point in using ISO values higher than 800. If you are stacking a number of images, I would suggest ISO 800 since the stacking will reduce the increased noise level, as illustrated in Figure 14.8.

Focusing

Being able to view the image in 'live view', now present on virtually all new cameras on the rear LCD screen, is a great help in getting the focusing right. If there are stars in the field then, when in focus, the tiny stellar images on the screen will excite only 1 pixel – which could be red, blue or green – so a perfectly focused image may well show brightly coloured stars! (But don't worry, they will be correctly coloured in the output image.) Given a laptop, the camera can be coupled and driven through its USB port by an image acquisition program such as 'EOS Utility', provided free with Canon DSLRs. Again, using the live view mode, the image can be viewed on the laptop screen at a larger image scale than that on the camera's LCD screen, so making focusing easier.

Raw Mode

Ideally the camera should be used in raw mode, as the number of bits used to record each pixel value is increased from 8 bits in JPEG mode to 12 or even 14 bits in raw mode. When processed in the camera's raw developer, the images should be output as 16-bit TIFF images to retain their inherent quality. When the images have, as often, been stacked and stretched as described later, the final image may also be saved as a JPEG to reduce the image data size. I save both raw and JPEG fine modes so that the JPEGs can be quickly viewed to decide which images to use for processing from raw.

Having said that, I have processed the same 20 short-exposure images captured in both raw and JPEG modes and found surprisingly little difference in the final result. Stacking a number of 8-bit images will increase the effective bit depth, but the JPEG result was definitely far better than I had expected.

Auto 'Dark Frame Subtraction'

When long exposures are made with a DSLR, two problems arise. The first is that some 'hot' pixels will appear across the image as white speckles. The second is that edges or corners of the frame may become bright because of heat from the amplifiers that read out the data from the CCD array and, dependent on the temperature of the sensor, noise due to 'dark current' will become more apparent with long exposures. Most DSLRs will (when activated) automatically follow a long exposure with a second exposure of similar length with the shutter closed to produce what is called a 'dark frame'. This is then subtracted from the 'light frame' to remove the hot pixels and amplifier and dark current noise from the main image. The hot pixels and amplifier noise become worse with increasing sensor temperature and longer exposures.

In principle, one might take an image with the lens cap on to produce a single dark frame which is subtracted in the imaging software from all the light frames. This will reduce the total imaging time by half. This is the technique used for imaging with cooled CCD cameras when the imaging chip is kept to a constant temperature, but it will not work as well with a DSLR, as the temperature of the chip will increase with time during an imaging session. It is not unusual for the sensor temperature to rise from an initial 20 C to 30 C or more within 1 hour of use. One reason for heat generation within the camera is the battery, and this can be eliminated by using an external power source. This involves the purchase of two parts (often as a package); the first part is a dummy battery placed within the camera which has a short cable and connector to couple with the second part, which is a mains adaptor to provide current at the appropriate voltage for the camera. Another advantage of using an external power source is that camera batteries, when cold, do not last long.

Given the sensor temperature of each 'light frame' provided by, for example, ATP and Backyard EOS, it becomes possible to produce a set of dark frames at a range of temperatures that can be matched to the light frames and used to carry out dark frame subtraction later in the image processing procedure. This is discussed in more detail in the following chapter. The use of a cooling system, as described later in this chapter, can keep the sensor temperature more constant, making the process rather easier.

A Constellation Image Using a Fixed Tripod and Prime Lens

The image of the Plough shown in Figure 14.1 and Plate 14.1 was taken with a 55-mm Micro-Nikkor – a very sharp lens. A Nikon D7000 camera was mounted on a very sturdy tripod with a ball and socket joint in order for the camera orientation to be aligned with the constellation being imaged. The exposure time of each image was 8 seconds following a 2-second delay (to prevent camera shake from pressing the shutter) and a 1-second 'mirror up' delay. The set of 10 images were simply imported into the free program 'Deep Sky Stacker' and the 'align and stack' process initiated. The program looks for quite a number of stars in the image to carry this out and will give error messages if it cannot find them. However, the number of stars it attempts to find can be reduced by the use of a sub-menu in the 'Register Checked Pictures' command box; this is opened by clicking on the command in the menu and opening the 'Advanced Settings' tab, which shows a slider that determines the number of stars needed to be detected.

Several techniques, described in Chapter 18, were used to refine and enhance the image: the sky glow from light pollution was removed and, even though the Plough is relatively near the Pole Star, a little star trailing was visible, which was also removed. A further technique puts some of the colour back into the stars whose centres are often burnt out.

However, I find that constellation images taken with a DSLR are not as good as those taken with film cameras: the brighter stars do not appear significantly brighter than the fainter ones so, in particular, the constellation pattern that is made up from the brighter stars is not that obvious. Ironically, this is due to the fact that a digital camera images the sky more accurately! Stars are point sources of light and so should have no 'size' in the image. In a film camera, 'halation' scatters the light from the brighter stars into the surrounding emulsion, so increasing their apparent size in the image. It is the size of stellar images that enables us to perceive their apparent brightness both in images and also, of course, in star charts. A DSLR sensor does increase the size of the brighter stellar images to some extent, but not nearly so much as film. One trick is to stretch part of a black nylon stocking over the camera lens; its diffraction effects increase the size of the brighter stars. In Chapter 18 a technique is described to make a digital image appear more 'film-like' and so, perhaps, making a more pleasing image, as demonstrated in Figure 14.1.

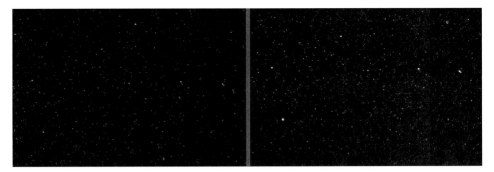

Figure 14.1 An 80-second total exposure image of the Plough (*left*) an enhanced image using a technique described in Chapter 18 (*right*).

14.2 Long-Exposure Imaging

In order to take longer exposures some form of tracking must be used. At very low cost one can make a 'barn door' mount made up of two hinged pieces of wood with the hinge axis pointing towards the Pole Star. A rotating worm thread is used to move the 'door' open at just the right rate. The instructions as to how to make one can be easily found by searching for 'Home-Made barn door Mount'.

If one has a driven equatorial mount, with a little ingenuity one can mount a camera on it – indeed, two of my telescopes already have camera mounting screws in place, and the clamshell support for one of my refractor tubes has one mounted on top. With a wide-angle lens, the telescope body may actually get in the way if the camera was aligned along the axis of the telescope, but the camera does *not* have to be pointing in the same direction as the telescope, so a sturdy ball and socket joint can be placed between the mount and the camera to offset the camera by any desired amount.

Equatorial mounts tend to be heavy, and now several companies produce tracking mounts for use with DSLRs that are light so that they can easily be carried on observing trips, perhaps by air, to dark-sky locations. AstroTrac produces the TT320X-AG, which, once polar-aligned, can track accurately for nearly 2 hours. The 'AG' indicates that it is equipped with an ST4 auto-guiding port to allow guiding in right ascension. However, even without guiding, it is capable of making exposures of up to 5 minutes with a typical tracking error of ±5 arc seconds. This accuracy is due to the fact that the drive, resulting from the rotation of a precision screw under the control of a small computer, is at the base of a long arm. Thus the angular motion of the mounting head is far less than the motion of the screw, and the result is that any periodic error in the screw thread drive is greatly reduced. To achieve equivalent tracking accuracy a very expensive equatorial mount would be required. Though the mount weighs just 1 kg (2.2 pounds), given a suitably solid tripod or pillar it can support up to 15 kg (33 pounds), so even 80-mm refractors or a Celestron C6 Schmidt-Cassegrain could be used – not just a DSLR with attached lens. AstroTrac

Figure 14.2 An AstroTrac mounted on a tripod with pan and tilt head with the Vixen Polari Star Tracker and the iOptron SkyTracker.

provides a TH3010 counterbalanced head to allow such telescopes to be mounted on it. A sturdy tripod is required with a pan and tilt head to mount the AstroTrac so that it can be polar-aligned along with a sturdy ball and socket head to hold the camera. AstroTrac also provides a very neat and compact equatorial wedge to replace the pan and tilt head and an ultra-light pier, which, for transporting by car or air, can contain and protect the AstroTrac itself. My AstroTrac, mounted on a Manfrotto tripod, is shown in Figure 14.2. AstroTrac provides an optional illuminated polar scope to carry out the polar align – to which I have added a webcam so that the alignment can be made without having to lie on the ground! As an alternative, a friend has made me an adapter so that a laser pointer can be used to polar-align the system – pointing the beam 0.7 degree away from Polaris towards the star Kochab, as described in Chapter 8. The pointing accuracy obtained by this method is perfectly adequate for the 30-second exposures that I employ, using Deep Sky Stacker to give the equivalent of longer exposures.

Vixen has introduced its camera-sized Polari Star Tracker, and iOptron a similar-sized SkyTracker. The Vixen unit incorporates a small 'North Star alignment window' to polar-align its mount (a polar scope is available as an accessory), whilst the iOptron unit has an integral rotating azimuth base and latitude adjustment wedge as well as an illuminated polar scope. Their weight limits and tracking accuracy are, of course,

Figure 14.3 The anti-tail of comet PanSTARRS imaged using a 66-mm refractor mounted on an AstroTrac. (Image: Andrew Greenwood)

lower than those of the more expensive AstroTrac, but both will support cameras and lenses weighing up to a total of ~3 kg. Both are shown in Figure 14.2.

My colleague Andrew Greenwood (VisualSense.com) used an AstroTrac TT320X-AG at the end of May 2013 to image comet PanSTARRS sporting a magnificent anti-tail as the Earth approached the plane of its orbit (Figure 14.3). A total of sixty 20-second exposures were made at 1,600 ISO with a Nikon D5100 camera allied to a William Optics ZS66 refractor and Type 6 field flattener.

Using a DSLR and a Prime or Zoom Lens to Image Star Clusters and Nebulae

Perhaps the first question is what focal length is required of a lens or telescope to image a given object with a reasonable image size within the frame. Here is a list of the approximate angular fields of view given with lenses of different focal lengths when used with a full-frame DSLR:

35-mm lens:	40 degrees	Good for Milky Way vistas
100-mm lens:	14 degrees	Good for constellation images
200-mm lens:	7 degrees	Good for the Andromeda Galaxy
400-mm lens:	3.5 degrees	Good for the Beehive Cluster and Hyades Cluster

Note: When an APSC sensor DSLR is used, the effective focal length is increased, usually by a factor of ~1.5.

The technique of combining a number of short exposures to give the effect of a longer exposure minimizes the problem of tracking errors elongating the star images, but it also solves another problem if a tracking Alt/Az mount is used. With such a mount, the field observed by a fixed DSLR will rotate relative to the sky during the exposure. Supposing that a 20-second exposure does not show any tracking or rotation problems, then the Deep Sky Stacker program will automatically align and rotate the individual frames before combining them into a single image.

Employing a Camera Cooler to Reduce Thermal Noise

A DSLR camera cooler is now marketed by Orion, the Orion DSLR Camera Cooler 52095, which appears to be identical to the Teleskop-Service Geoptik 30B330. These coolers will encase a number of Canon DSLRs and cool them up to 24 degrees Celsius below ambient temperature. A 51-watt Peltier cooler (taking up to 4 amps at 12 volts) is used to provide the cooling, so a fair-sized battery or mains adapter will be required. As the unit weighs about 1 kg, a capable focuser will also be required. The cool-down time to drop the sensor by 16 degrees Celsius is about 30 minutes, which is significantly longer than the few minutes of cool-down time with a dedicated CCD camera. However, it may well mean that, once stabilized, a single dark frame could be used for a sequence of images, so reducing the overall imaging time. This, in combination with a Canon EOS 1100D (Rebel T3), could make an alternative to a cooled colour CCD camera at considerably lower cost.

Gary Honis has provided details of how to make a low-cost equivalent using the Peltier cooler from a thermoelectric travel cooler. The instructions can be found at ghonis2.ho8.com. One problem with the cooling of a DSLR camera is condensation, and it is useful to include some desiccant packs within the cooling box. The Dutch company JTW Astronomy takes the sensor and electronics of a Canon EOS 1100D, replaces the infrared filter with one that passes four times more H-alpha emission (as discussed later) and mounts the components within a custom metal case in which the cooling is applied directly to the sensor to reduce its temperature by up to 50 degrees Celsius below ambient. As the sensor in an uncooled camera can increase to 10 or more degrees Celsius above ambient this is very impressive. The Bayer filter matrix can also be removed to make a monochrome version. An image of M31 taken with a monochrome modified camera is shown in Figure 14.4.

A very low cost technique that can provide a useful amount of cooling in warm weather is to simply mount the camera in a plastic box sufficiently large (I used a 2.3-litre food box) so that the camera can be surrounded by small flexible ice packs and insulation. Loosely packed layers of kitchen roll work quite well. Again, some desiccant packs should be included to prevent moisture from condensing on the camera. The way to use the cool box is to switch the camera on, insert the final ice pack and insulation directly behind the camera back and then use the astro- photography tool to take one short exposure every 5 minutes (so keeping the camera on) whilst monitoring the sensor temperature (which is given in the real time log) as the sensor cools.

Figure 14.4 An Orion DSLR cooler (*top left*) and JTW Astronomy–cooled Canon 1100D (*bottom left*), along with a 180-second, 1600 ISO single image of M31 taken with a monochrome JTW modified camera by Dr Fritz Hemmerich.

In one test made at an ambient temperature of 24 °C, I allowed the sensor temperature to drop to 10 C and then took a continuous sequence of 30-second exposures, at which point the sensor temperature increased rapidly to 14 °C, where it stabilized and remained within 1 degree for more than an hour.

This would allow one to average a number of dark frames taken at the stabilized temperature and then carry out the dark frame calibration later, as described in Chapter 15. The result would be to effectively double the available imaging time as well as significantly reduce the sensor noise levels. In an identical test carried out without the cooling jacket, the sensor temperature rose gradually to 30 C, at which point it stabilized. As sensor noise levels approximately double for each 6 degree Celsius rise in temperature, the 16-degree reduction produced by the simple cooling jacket would reduce the noise level by a factor of ~6. The effect will, of course, be less the closer the ambient temperature is to 0 °C, and there would be no point in using one if the temperature were close to, or less than, this.

14.3 Using a DSLR with a Telescope

In Chapter 11 the T-mount adapters used to mount a DSLR onto a telescope were described, and exactly the same system is used for deep-sky imaging. As the chip size of DSLRs is quite large and the majority of telescopes will show some curvature of field, it might well be found that star images towards the edge of the imaged field are blurred. This is particularly noticeable with refractors of short focal ratio. The corners of the frame may well show elongated stars radiating away from the frame centre. By selecting each corner in turn and using the techniques described in Chapter 18 to improve their shape, one can often greatly improve the overall image.

Field Flatteners

As discussed in Chapter 1, manufacturers often provide dedicated field flatteners for their telescopes to overcome this problem. These often reduce the focal ratio somewhat, commonly by 0.8 times. Recently, the German company Teleskop-Service has introduced a field flattener of 2-inch diameter which has no focal ratio reduction and can be used on a very wide variety of refractors by adjusting the separation between the lens assembly and the camera sensor. This is typically 109 mm and so allows plenty of room for off-axis guiders and filter wheels when used with CCD cameras. I have obtained one of these for use with my refractors, and Plate 1.5 shows an image of the open cluster M35 taken using an 80-mm ED refractor, with insets showing a corner of the frame taken with and without the flattener in place. As this chapter was being finalised, I added a William Optics 72-mm, f6 Megrez FD refractor to my collection to use for wide-field imaging and also as a 'lens' for nature photography. Given the relatively low focal ratio it is perhaps not surprising that the image quality in the corners of an APSC sensor is pretty poor, and it was very pleasing to find that the incorporation of the TS field flattener produced an essentially perfect image right across the APSC sensor. I cannot but highly recommend this excellent imaging accessory.

Focal Reducers

Imagers using f10 Schmidt-Cassegrain telescopes can obtain a field flattener/focal reducer to widen the field of view and reduce exposure times. The standard reducer for both Meade and Celestron reduces the focal ratio from f10 down to f6.3. It will also help to remove field curvature. This can be employed for both visual and imaging use. There is also an f3.3 version that is recommended only for imaging use. The f6.3 versions have also been used with some success with refractors, and I made an adapter so that one could be used with my 127-mm f7 refractor, giving an effective focal ratio of ~f4.5. The image quality across the APSC-sized frame is very acceptable, though there is some vignetting towards the extreme corners. This can be removed by using a 'flat' test image, as described in Chapter 15 on deep-sky CCD imaging or using the vignette correction tool in Photoshop: Filter > Distort > Lens Correction.

Figure 14.5 The Bahtinov mask for a 9.25-inch Schmidt-Cassegrain and the star images seen as one passes through focus.

An Aid to Focusing a Telescope

Many astro-imagers now use a Bahtinov mask to aid focusing. Placed at the telescope's aperture, this produces a star image with diffraction spikes consisting of a pair of crossed lines separated by an angle of ~20 degrees together with one line which, when in focus, will bisect the other two. Figure 14.5 shows the mask for use with a 9.25-inch Celestron and the images seen in and out of focus.

14.4 Using a DSLR and an 80-mm Refractor on an iOptron Minitower Alt/Az Mount

Comet PanSTARRS

Early in 2013, comet PanSTARRS (named after the telescope with which it was discovered it) appeared low in the western sky after sunset. I set about attempting to photograph it with my Nikon D7000 attached to an 80-mm refractor. The telescope was mounted on my iOptron Minitower Alt/Az mount. I set an ISO of 800 and used 8-second exposures following a delay of 2 seconds to eliminate any camera shake due to my pressing the shutter and a 'mirror up' delay of 1 second. A trial image showed no obvious trailing of the comet's coma. Each of the raw images was converted into 16-bit TIFFs. To lower the noise level in a single image, a sequence of 24 images were stacked in Deep Sky Stacker. Some further noise reduction was carried out by importing the image into an excellent noise reduction program called 'Picture Cooler', which is very easy to use and can be downloaded from FreePhotoSoftware.shorturl.com. Figure 14.6 (left) shows a section of a single frame whose contrast was low due to sky glow from light pollution pervading the image; Figure 14.6 (right) shows the result of stacking the 40 frames and having removed the sky glow, as described in Chapter 18.

The Perseus Double Cluster, C14

The combination of my 80-mm ED semi-apo refractor and APSC sensor DSLR gives an excellent image scale for imaging the larger open clusters such as the M35 open

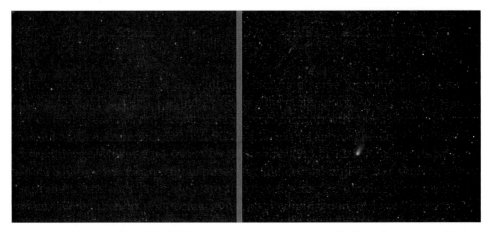

Figure 14.6 Comet PanSTARRS as seen in a single frame (*left*) and as stack of 40 frames, the sky glow having been removed.

cluster (shown in Plate 1.5) and the Perseus Double Cluster, C14. They were imaged from a relatively dark-sky site in Wales but with, sadly, a small town to the west, which was causing some light pollution. The iOptron mount was aligned and synchronised on Jupiter, which was also used to focus the Nikon D7000 and TS 2-inch field flattener combination. A total of forty 8-second exposures (with each exposure following a 2-second exposure delay and 1-second mirror up delay) were stacked in Deep Sky Stacker, and in both cases the simple light pollution removal procedure as described in Chapter 18 was used to remove the sky glow.

To be more specific, the image produced by Deep Sky Stacker was output as a 16-bit TIFF and imported into Photoshop. Once the sky glow was removed, the curves function was opened (Image > Adjustments > Curves or Ctrl M) and the straight line lifted towards the darker end, as shown in Figure 14.7. This left the brighter stars much as they were but brightened the fainter ones. This procedure can well be made an 'action' so that it can easily be applied several times until the background noise becomes too apparent. Some judgement is required here. Finally, the 'Levels' command is entered (Image > Adjustments > Levels or Ctrl L) and the black level slider moved to the right to remove the background noise.

An Image of M51 Showing the Noise Reduction Achieved by Stacking Frames in Deep Sky Stacker

Following the imaging of the Double Cluster, I thought it might be worthwhile attempting to image the Whirlpool Galaxy, M51, which lay almost overhead and away from the local light pollution. As I was using a refractor of short focal length, M51 was very small in the frame and could not even be seen in an individual image. I was pleased that, when the stacked image was stretched, it appeared very close to the frame centre. The iOptron mount had been aligned on Jupiter, well across the sky, some 2 hours previously, proving that its 'go-to' ability is pretty impressive. Having

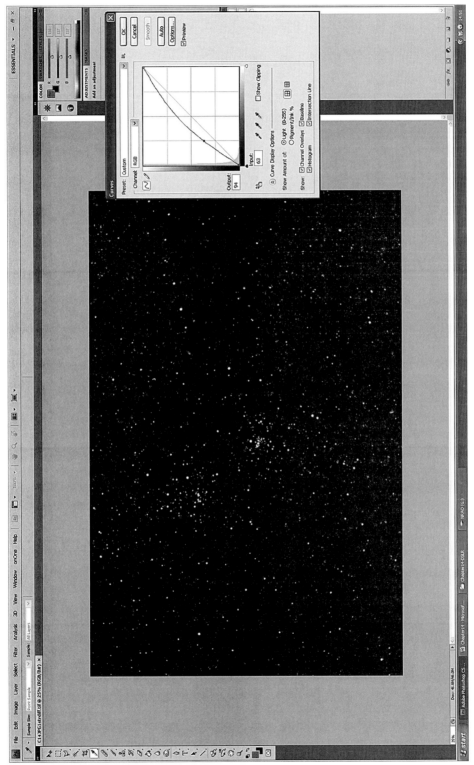

Figure 14.7 The Double Cluster in Perseus. The curves function is being used to 'lift' the fainter parts of the image without over-exposing the brighter stars. This is one of the later applications.

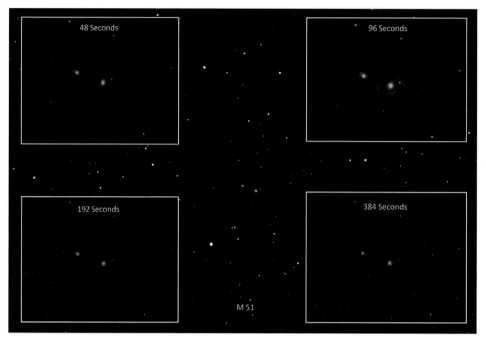

Figure 14.8 The Whirlpool Galaxy imaged with a Nikon D7000 mounted on an 80-mm ED refractor. The insets show the effect of increasing the total exposure time by stacking more individual frames.

produced the full-frame image of Figure 14.8 (which also had had a little noise reduction applied in 'Picture Cooler'), I stacked differing numbers of frames, whose results are shown in the four insets, to give some idea of how the total integration times will affect the final image.

14.5 A Night at the Isle of Wight Star Party

I hope that the following description of a night at the Isle of Wight star party will give an idea of how a DSLR coupled to a telescope can be used for imaging a variety of objects and further illustrate the techniques that are used to process the images captured by the camera.

My new Nikon D7000 DSLR arrived the day before I attended the Isle of Wight star party in March 2010 and I was eager to try it out. I had mounted the camera on my Takahashi FS102 fluorite refractor, which was carried by an iOptron Minitower Alt/Az mount. Each month I write a Web page describing what may be seen in the heavens (just Google 'Night Sky Jodrell Bank'), and I had written that that night there would be a thin crescent Moon hanging in the western sky, with Jupiter not far to its left along with Mercury low above the horizon. As the Sun set that evening, the Moon and Jupiter were easily seen but none of us were able to see Mercury, even using binoculars and knowing where to look. A colleague suggested that we try to photograph

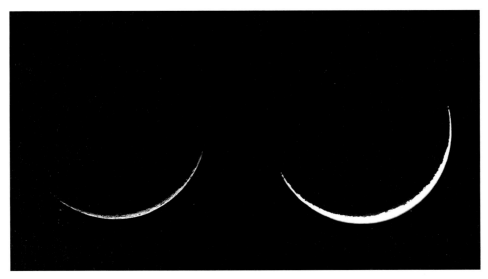

Figure 14.9 An image of the 46-hour-old Moon exposed (*left*) for the new Moon and (*right*) to show the earthshine.

it, so the telescope was aligned on Jupiter, slewed to the position of Mercury and a single image taken. To our great surprise Mercury was just visible, its image stretched into a small spectrum due to refraction in the atmosphere. The Nikon's first light image had been a success!

As darkness fell, the very thin crescent Moon became more prominent, and it was easy to see the earthshine illuminating its dark side. This was a second imaging challenge! I made images to correctly expose for the illuminated crescent (Figure 14.9, left) and then made a number of images to show the earthshine on the unilluminated part of the Moon. One might think that a considerably increased exposure should have been used, but the problem then is that the light from the over-exposed bright limb of the Moon would have 'bled' over the dark side. So I deliberately under-exposed these images. They would thus be noisy and so, to overcome this, a number of images were taken and stacked together, thus reducing the overall noise level. The Takahashi FS102 uses a fluorite doublet which scatters almost no light within it, so these fluorite refractors are reckoned to have the highest inherent contrast of virtually any telescope. Very little light was thus scattered into parts of the field of view away from the thin crescent. As a result, a very pleasing result, as seen in Figure 14.9, right, was obtained.

With the Sun well set, Orion was visible in the south-west, so the next target was the Orion Nebula region – perhaps one of the most photographed objects in the heavens. There is a real problem when one is photographing this region due to the great contrast in brightness between the region immediately surrounding the stars of the Trapezium and the far fainter nebulosity surrounding it. As a result, many images that show the nebulosity well have a burnt-out central region. There are two approaches to overcoming this problem. One is to take a number of images, some exposed correctly for the central region and some for the fainter nebulosity, and then combine them – perhaps by using the clone tool to 'paint over' the blown-out central region of the deeper image with the correctly exposed central region from a short-exposure image. There is an alternative approach, which is what I used for this image. This was to take a large number (20) of 12-second-exposure images at ISO 800 that were correctly exposed for the central region and showed the four Trapezium stars well. As these were short exposures, no guiding was required. All images were taken in raw format with 14-bit digitisation and converted to 16-bit TIFFs. A 2-second delay was introduced prior to a 1-second mirror up delay before each image was exposed. With a DSLR, the chip temperature will almost certainly rise during an observing session so, as previously described, with each light frame a dark frame was automatically taken, which was then subtracted from the light frame in camera.

As an Alt/Az mount had been used, there would have been some frame rotation between exposures but, providing that there are enough stars visible across the frame, the Deep Sky Stacker program can be used to rotate the images before combining them and producing a 16-bit TIFF. The combination of frames will improve the signal-to-noise ratio of the fainter parts of the image, so allowing the fainter parts of the image to be made visible using the levels or curves tools in Photoshop. (The free programs 'IRIS' and 'FITS Liberator 3' can achieve a similar result, as described in Chapter 11, Section 11.1.)

The quick way to 'stretch' the image is to use the 'Levels' control. The top slider is untouched, as the Trapezium stars and surrounding region were correctly exposed. The centre slider is then moved down to the left to bring up the fainter parts of the image, and finally the bottom slider might be brought up a little to darken the sky background. A slightly more time-consuming method which may produce better results is to repeatedly apply a gentle curve to the image, as described earlier. One weakness in the original image was that the fainter parts of the nebula showed rather a lot of noise. A noise reduction program such as 'Picture Cooler' can be used as well as a technique that is described in Chapter 18. This involves removing the stars from the image to make a 'star field' frame along with a second frame showing just the nebula. The nebula image is smoothed using the 'Gaussian Blur' filter and, finally, the star field is added back.

To be honest, I was pretty surprised as to how well this first deep-sky image using the D7000 came out – as were some very experienced astro-imagers at the star party – and it indicated that the Sony sensor in the D7000 was indeed rather special. It should be pointed out that it was a pretty cold night and one might well not get as good results on a warm one. However, the image showed rather less of the pink-red H-alpha emission than is seen in many images and the reason is discussed next.

14.6 DSLRs with Modified Infrared Filters

Standard DSLRs incorporate filters to remove the infrared light from the image. This is for a very good reason in that with lenses – or when used with a refractor – the infrared light comes to a different focus and so will produce a halo around bright objects such as

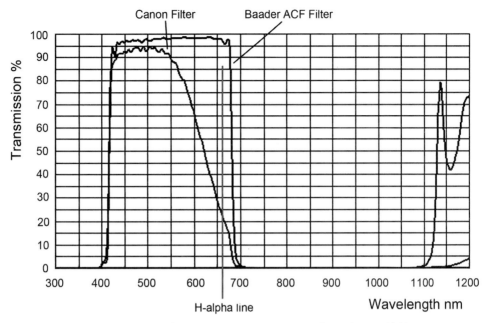

Figure 14.10 Comparison of the Canon and Baader infrared cut-off filters.

stars. The problem is that these filters have a very gentle cut-off across the red part of the spectrum to give a result similar to that seen by the eye, and the result is that the light from the H-alpha emission at 6,563 angstroms (that adds so much to images of nebulae) is attenuated by 75% or so. Canon has twice produced cameras specifically aimed at astro-imagers which incorporate a filter with a steeper cut-off, allowing greater sensitivity to H-alpha emission. In 2012 Canon brought out the 60Da, which incorporates an 18-megapixel sensor and increases the sensitivity at the H-alpha line by a factor of 3. It is equipped with a high-resolution tilting screen that can be used in live view mode to aid focusing – and so is essentially the perfect camera for DSLR imaging.

Because so many astro-imagers are using Canon cameras, an after-market industry has grown up to remove their infrared filters and (optionally) replace them with a filter that allows the H-alpha emission to pass through virtually unattenuated. The standard filter is made by Baader Planetarium. This allows more than 90% of the H-alpha emission to pass through and then attenuates very rapidly, as shown in Figure 14.10. I purchased a modified Canon EOS 1100D in order to exploit this difference, as described later.

Imaging the North America Nebula

A modified camera is an obvious choice for imaging nebula regions that emit a lot of H-alpha light, and so an excellent target is the North America Nebula (NGC 7000), which lies close in direction to the star Deneb in Cygnus. The first decision to make is what lens would be most suitable to cover the region including Deneb and the

Figure 14.11 An image of the North America Nebula region taken with a modified Canon EOS 1100D, as described in the text.

North America Nebula. Looking at a star chart indicated that the region would subtend about 1 hour in right ascension. At the celestial equator this corresponds to 15 degrees but, at the declination of the region of about 45 degrees, this reduces to 15 × sin(45 degrees), which is ~10 degrees. A very nice app for tablets or smartphones is 'SkySafari Plus'. This beautifully shows the North America and Pelican Nebulae and their location relative to Deneb in a field about 9 degrees across. A Web site, www.howardedin.com/articles/fov.html, allows a camera and lens combination to be entered and returns the field of view in both axes. This showed that the Canon EOS 1100D and a 135-mm, f3.5 Zeiss lens (using an M42-to-EOS adapter) would give a field of view of ~10 × 7 degrees and so appeared ideal.

The camera was mounted on an AstroTrac which was aligned using a laser pointer, and then the laser was laid along the lens barrel so the lens could be pointed towards the region close to Deneb. An ISO of 800 was used and, with an exposure of 30 seconds along with a subsequent 30-second dark frame subtracted in camera, each 'exposure' took 1 minute. A total of 17 frames were taken using the Canon EOS Utility and downloaded into the laptop, giving a total exposure of 8.5 minutes. The imaging conditions were not ideal as, though imaged from a dark site in mid Wales with Deneb overhead, there was an 11-day-old Moon low in the south-west.

Figure 14.12 An image of the Orion Nebula region combining an image taken with a Nikon D7000 with the red channel of an image taken with a modified Canon EOS 1100D added to enhance the H-alpha emission in the region.

Deep Sky Stacker could not correctly stack the .CR2 raw files produced by the Canon 1100D, so these were first converted to Adobe .dng files. Some of the techniques given in Chapter 18 were applied to enhance the image, which included the use of 'Unsharp Mask' to provide some local contrast enhancement along with the curves and levels to produce a reasonable colour image. Without colour to highlight the nebula, the monochrome version shown in Figure 14.11 lacked contrast and, so to produce the image shown, some further adjustments were made to the brightness and contrast of the image. To the right of the North America Nebula can just be seen the fainter Pelican Nebula.

Adding Some H-Alpha Light to the Orion Nebula Image

Given that I now had a camera that was more sensitive to H-alpha emission, a worthwhile experiment was to add some of this light to the image taken with the Nikon camera. Although the image was taken with the same 102-mm refractor, the Canon is only a 12-megapixel, rather than 16-megapixel, camera, and the sensor size is slightly different, so the Canon image had to be first scaled and slightly rotated to match that of the Nikon. The image then had the green and blue channels reduced to zero by means of the 'Levels' command: the three colour channels making up the image (Window > Channels) were opened up and then the green and blue channels individually selected. By entering 'Levels' (Image > Adjustments > Levels) and moving the black level to the extreme right these channels were, in effect, removed from the image, leaving just the red channel. The 'Remove Dust and Scratches' filter (Filter > Noise > Dust and Scratches) was used to remove the stars from the image, the nebulosity smoothed using the 'Gaussian Blur' filter and the image cropped down so that it contained only the nebula region. This was selected (Select > Select All or Ctrl A), copied (Edit > Copy or Ctrl C) and pasted and aligned on top of the Nikon image; the Nikon image was opened and the Canon red channel image pasted over it (Edit > Paste or Ctrl V) and the move tool used to give an accurate alignment. The colour blend mode was selected and the opacity slider adjusted to give the final result of combining the Nikon and Canon images shown in monochrome in Figure 14.12 and in colour in Plate 14.12. There are far, far better images of the Orion region, but at least the central part is not burnt out and it is the result of only 288 seconds of imaging time. I include it to show that quite reasonable results can be obtained with a DSLR, quite simply and without too much imaging time.

15
Imaging with Cooled CCD Cameras

The latest DSLR cameras can do a very good job of astro-imaging and can, of course, be used for general photography as well, so why go to the expense of buying a cooled CCD camera? The main reason lies in the word 'cooled'. All imaging chips produce dark current noise which increases with exposure time and is also highly dependent on its temperature, that of a typical chip dropping by half for each drop of 6 degrees Celsius in temperature. So, if the chip is cooled by 30 degrees below ambient temperature, the dark current noise will have dropped by about 5 times, so allowing longer exposures to be taken before dark current noise becomes a problem. Given dark skies that do not suffer from light pollution, this can allow images to reveal faint nebulosity that would otherwise be lost in the noise. When significant light pollution is present, the exposure times, and hence the dark current contribution, have to be less, before the skylight becomes obtrusive, and so cooling does not confer as great an advantage. The latest chips have very low dark currents, and it is rarely worth cooling them down below about −20 C. This temperature can normally be reached with the single-stage Peltier cooling employed in CCD cameras aimed at the amateur market.

15.1 Aspects of Cooled CCD Cameras

Chip Size

The dimensions of the CCD chip allied to the scope's focal length dictate the field of view of the resulting images. The chip dimensions in millimetres divided by the focal length in millimetres give the field of view in radians. Multiplying this by 57.3 gives the field of view in degrees. So, obviously, a bigger chip will provide a bigger field of view. For example, my 80-mm refractor has a focal length of 550 mm, and this is used with a CCD camera whose sensor has dimensions of 18 × 13 mm. This gives a field of view of 1.9 × 1.4 degrees.

However, the biggest chips can highlight telescope problems. If the scope does not have a sufficiently large flat field, star images will become blurred towards the edge and the image may also suffer from vignetting (darkening towards the corners).

However, the former problem can be overcome by means of field flatteners such as the Teleskop-Service 2-inch field flattener that I have acquired for use with my refracting telescopes, whilst the latter problem can be corrected by the use of 'flats', as will be described later. The camera supporting a larger chip is likely to be heavier and so will need to be very well supported to ensure that the CCD chip remains perfectly at right angles to the optical axis of the scope. It is thus important that the focuser be able to handle the weight, and often telescopes can be bought with upgraded focusers specifically for this reason.

Pixel Size

It is important that a CCD chip have an appropriate pixel size to match the focal length of the scope and the expected seeing conditions at the observing site. But it is worth pointing out that pixels can be binned to increase their effective area (and hence sensitivity). In most imaging software, pixels can be binned as 2 × 2 (reducing the resolution by a factor of 2 but increasing the sensitivity by a factor of 4) and 3 × 3 (3 and 9). As we will see, this can allow a chip with small pixels to be optimised for different scopes and conditions. Binning is often employed for focusing the camera, as shorter exposures can then be used to speed up the process.

The combination of pixel size and focal length gives rise to what is termed the 'image scale' of the system, which is expressed in arc seconds per pixel. To calculate this, divide the pixel size (in microns) by the focal length (in millimetres) and multiply by 206.3. For example, take my Takahashi FS102, which has a focal length of 820 mm and is used with a CCD camera which has a pixel size of 5.4 microns. The calculation then gives an image scale of 1.36 arc seconds per pixel.

How does this relate to the resolution that the telescope will achieve? In most locations the effective resolution is limited by the seeing. This could well give typical stellar images that have 'full width at half maximum' (FWHM) of 3.5 arc seconds or so. The FHWM is the width of the star profile measured when the brightness has dropped to half its maximum value. One might thus think that an image scale of 1.36 arc seconds is too small. However, sampling theory (the Nyquist theorem) implies that the image scale be at least half, and preferably a third, of the apparent stellar size. So with seeing of 3.5 arc seconds this would indicate an image scale of ~1.2 arc seconds. This suggests that FS102/5.4-micron pixel size is a pretty good combination unless the seeing happens to be particularly good. In fact, images will start to look blurred only when the image scale exceeds 7 arc seconds or so, and even an image scale of 3 arc seconds could produce images with quite reasonable detail.

What if the scope were my Celestron 9.25-inch Schmidt-Cassegrain? It can be used with its native focal length of 2,350 mm or with the f6.3 focal reducer/flattener to give a focal length of 1,480 mm. The image scales are then 0.47 and 0.75 arc second respectively. In both cases, under typical seeing conditions, the image will be over-sampled, with a 5.4-micron pixel size sensor and one could certainly employ 2 × 2 binning (or even 3 × 3 when using the longer focal length) and so significantly reduce the required exposure times. Thus a camera with ~5-micron pixels and capable of being used in a binned mode is a pretty good choice. So, if the seeing conditions at your site are average, consider cameras which will give image scales in the range 1.0–3.0 arc seconds per pixel. It should be pointed out that the fundamental resolution of the telescope has some input into the calculations as, no matter what the seeing, the resolution will be limited by the telescope aperture. This will not usually affect the conclusions, but if you are using a 14-inch Celestron telescope (with a 3,910-mm focal length) having a theoretical resolution of ~0.4 arc second under skies having exceptional seeing (Damian Peach's images of the planets taken in Barbados come to mind), then far smaller imaging scales will be appropriate. Using a camera with a pixel size of 7.5 microns would give an image scale of 0.4 arc second. If a 2× Barlow were employed this would halve to 0.2 arc second and so be capable of extracting exceptionally detailed images.

Anti-Blooming Measures

CCD chips can have a problem if very bright stars are included within an image. The electron wells of pixels where the light falls can overflow and bleed into the surrounding pixels, most noticeably into a vertical or horizontal streak along a row of pixels. This effect is called 'blooming' and would, for example, seriously detract from an image of the Pleiades Cluster when one is trying to detect the faint nebulosity surrounding the very bright stars. Blooming will effect CCD chips which have what are called 'non-anti-blooming gates' (NAGB). (Interestingly, this effect can be used to focus an NAGB CCD camera. Simply observe a bright star and adjust the focus until the streak is at its longest and thinnest. This means that the star's light is largely concentrated onto 1 or 2 pixels and will then be in perfect focus.)

CCD chips can also be constructed with 'anti-blooming gates' (ABG). In this case the electron well is surrounded by circuitry which bleeds off the excess electrons before they can spill into adjacent pixels. Such chips thus seem to be the obvious choice; however, this circuitry takes up space, so the effective collecting area of each pixel will be less and hence the quantum efficiency of the chip will be lower, thus requiring longer exposures. There used to be quite a big difference in quantum efficiency, and so for some applications NAGB chips would be the preferred choice. More recently, CCD chips have become available which use micro-lenses above each pixel to concentrate the light from the whole pixel area into the electron well. This naturally increases the quantum efficiency, which may well have a peak of 56%, as opposed to an NAGB peak efficiency of ~80% – not as good, but as the difference is not now so marked a chip utilising micro-lenses may well be the best choice.

Dynamic Range

CCD cameras usually employ a 16-bit analogue-to-digital converter (A-D) in the chip read-out electronics, as opposed to 12- or 14-bit converters in DSLR cameras. This too can make a slight improvement in image quality. However, the number of bits is not quite as important as one might think. Many images are now made by stacking a number of sub-exposures. Not only does this have the effect of reducing the noise in the image, it also increases the effective number of bits in the conversion process. Consider a simple A-D that gives a value of 1 for the range 0.5–1.5, a value of 2 for 1.5–2.5 and so on. Suppose there is no noise in the system and the input has an actual value of 1.4. This will always be given the value of 1. Now suppose that there is random noise added which spreads the input value over 1 unit – so that it will vary from 0.9 to 1.9. Successive values given by the A-D will be either 1 or 2, but over time there will be slightly more 1s than 2s. If one averaged 10 readings the value would come to ~1.4. If, however, the input value was 1.0, all measured values would be 1, so the average would be 1.0. So if there is noise in a system – as there is bound to be in a CCD image – the averaging of many sub-exposures will give the effect of increasing the resolution of the A-D.

Monochrome or Colour?

Many CCD cameras are now available with either a monochrome or colour CCD. In the latter case the pixels are covered by a 'Bayer mask' of red, green and blue filters. Each block of 4 pixels will have two covered by a green filter, one by a blue and one by a red. Thus the colour resolution will be somewhat reduced, though, to be honest, this is not usually a problem. If one wishes to do colour imaging, a colour CCD chip saves the cost of both a set of filters and a filter wheel and so is a somewhat cheaper option. This is a very simple system to use and can give some surprisingly good results, but it should be pointed out that, as the light falling on each pixel is reduced by around one-third due to the filter above, the effective sensitivity will be less and longer exposures will be needed.

In contrast, when one is using a mono CCD to give a colour image, individual images are sequentially shot through red, green and blue filters but also, usually, through a clear glass filter to image the full spectrum in order to make what is called a 'luminance image'. Many imagers will take a long monochrome exposure (which will usually be the result of stacking many shorter exposures) to provide a deep, high-resolution luminance image. They will then take shorter-binned, (hence) lower-resolution images through the three colour filters. These may have some noise reduction applied to give three very smooth colour images. As will be shown in a later example, these are then combined with the high-resolution luminance image to give the final result. The overall imaging time taken will be comparable to that using a colour CCD chip, with maybe an edge given to using a monochrome CCD.

Under very dark skies with little light pollution, both types of camera have been shown to give very comparable results. As described later, under light-polluted skies astro-imagers are now tending to image with narrowband filters, and in this case colour CCD cameras are at major disadvantage when imaging the light from a particular emission line such as the H-alpha line at 6,563 angstroms. It is this line that produces the pink-red colour from excited hydrogen in emission nebulae such as the Orion and Eagle Nebulae. Often, imagers will add an H-alpha layer into the overall image, which can provide added contrast and definition and show up faint nebulosity. It is possible to use a narrowband filter with a colour CCD but, in the case of H-alpha, only 1 in 4 pixels will be utilised, reducing both the resolution and sensitivity.

Combining images taken through narrowband filters enables high-quality results to be obtained even under light-polluted skies or when the Moon is brightening the sky. Often three filters are used: that for the two sulphur II lines in the deep red at ~7,620 angstroms, the H-alpha line at 6,563 angstroms in the red and the two OIII lines at ~5,000 angstroms in the green. The three images (which will, of course, be monochrome if a monochrome CCD is used) are then allocated individual colours. In the so-called Hubble Palette, the deep red sulphur line is assigned to red, the red H-alpha assigned to green and the green oxygen lines assigned to blue. The false colour images that result can look very beautiful, as in the famous 'Pillars of Creation' image taken by the Hubble Space Telescope. A further colour palette giving more natural colours assigns the H-alpha line to red, the OIII lines to the green and the SII lines to the blue. One palette that I like assigns H-alpha to the red and OIII to the green but uses a mix of the two to provide the blue layer. This reduces the overall imaging time, as the SII lines are weak and so long exposures are needed. To equalise the brightness in the three images, the SII layer will usually have to be 'stretched' more than the other two, and this tends to increase the star sizes in the image, giving a coloured 'halo' about the resulting stars. The way in which the two, three, four or even five (red, green, blue, luminance and H-alpha) individual images are combined will be covered with actual examples.

So, for the many of us who live under light-polluted skies, a mono CCD camera is probably best.

15.2 Choosing a CCD Camera

To some extent the choice of a CCD camera depends on what you wish to use it for, but in general a camera with a larger CCD chip and more pixels will be more versatile. Small pixels will give a good imaging scale when used with short-focus refractors for wide-field imaging, and their pixels can be binned for use with telescopes of long focal length. For general use an ABG chip that eliminates blooming is a sensible choice and, if it employs micro-lenses, not that much less sensitive.

A First Cooled CCD Camera

It is now possible to buy a cooled CCD camera at not too great expense. The Atik Titan and the QHY QHY6 are two examples (Figure 15.1). As one might expect, the sensors are not large and are of the order of 6 × 5 mm. The Atik Titan employs a 325 K, 659 × 494 array of 7.4-micron square pixels, whilst the QHY6 uses a 752 × 582 array giving 440 K pixels of 6.5 × 6.25 microns. The other cost-saving aspect is that, though the sensor is cooled with a single-stage Peltier cooler and can reach 25–30

Figure 15.1 The Atik Titan (*left*) and QHY QHY6 (*right*). Each has USB2 and ST-4 guiding ports along with a 12-volt power input for Peltier cooling when required.

degrees Celsius below ambient temperature, there is no temperature control system to keep the sensor at a specific temperature and no way of knowing what that temperature is. This means that dark frames (described later), if used, have to be taken at the same time as the light frames so that the sensor temperature does not change significantly between the two. (With a stable, regulated sensor temperature, fewer dark frames need be taken.) However, the cooling will both significantly reduce the number of hot pixels present and reduce the dark current, so is well worth having. The procedure is then to take many exposures which are short enough that the dark current is insignificant and then stack them to produce the effect of a long-exposure image.

Should one wish to move on to a CCD camera with a larger sensor such as those described later, then this first CCD camera will not be wasted. Such cameras make excellent guide cameras, and both of those just mentioned are equipped with the standard ST4 guide port which makes them easy to use for this purpose, as will be described in Chapter 16.

LRGB Filters and Filter Wheels

When using a monochrome CCD camera for colour imaging, a filter wheel (not absolutely necessary but highly convenient) is used to select the colour being imaged. The filter sets available until relatively recently used glass doped to transmit a particular colour range. Their bands overlap somewhat, and the transmission efficiency is perhaps of the order of 70–80%. Filter sets are now available, such as those manufactured by Baader Planetarium, which use dichroic filters, offering many advantages: their individual pass bands cut off very steeply and their transmission efficiency in band is more than 95%; the H-beta line is passed only by the blue filter, giving it the correct blue colour; but the OIII line is passed by both blue and green filters, both giving it its correct colour when the RGB image is produced (blue-green or teal) and increasing the overall sensitivity at that wavelength. The red filter accepts both the (highly important) H-alpha and SII lines with near 100% transmission but then efficiently cuts off the infrared. There is a narrow gap at ~5,800 angstroms between the green

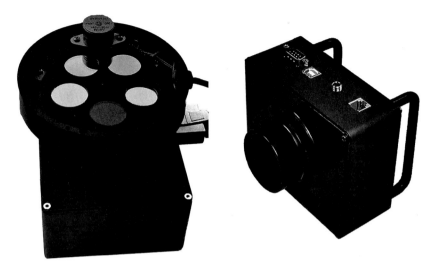

Figure 15.2 The author's SBIG ST-8300M used for monochrome imaging and coupled to the SBIG CFW9 filter wheel. This is opened to show the red, green, blue, and H-alpha dichroic filters along with an infrared cut-off filter through which to take the luminance images. The filter wheel is coupled to the camera through a nine-pin D-type connector.

and red pass bands to help suppress the emission from sodium streetlamps. The luminance filter simply cuts out the infrared emission. All three colour bands have equal sensitivity, so that the same exposures can be made for each, which greatly simplifies automated imaging. Very similar LRGB filter sets are manufactured by Astronomik with its 'LRGB type 2 filters' and Astroden with its 'True Balance filters'. Given the expense of the CCD camera and filter wheel it is simply not worth using filters that cost less than these, even though they are somewhat expensive.

With a typical five-slot filter wheel, the LRGB set could be supplemented with an H-alpha filter to add an H-alpha layer to the image. With seven-slot filter wheels both SII (deep red) and OIII (blue-green) narrowband filters can be included to allow both LRGB and narrowband imaging to take place easily.

The filter wheels can be manual or controlled from the imaging computer either directly or via a coupled CCD camera. This latter approach helps to reduce the number of cables between the computer and telescope, which is good and which can also allow for filter changes in imaging sequences to be directly controlled by the imaging program. Many CCD cameras have dedicated filter wheels which can have five or seven slots to hold 1.25-inch or 50-mm-diameter filters. The larger (and more expensive) filters are required with large (such as APSC size) CCD sensors and even with the somewhat smaller 8.3-megapixel chip in my SBIG ST-8300 camera when used with telescopes of short focal length (Figure 15.2). I am using the SBIG CFW9 filter wheel in conjunction with the ST-8300 camera. This incorporates 1.2-inch mounted filters which are sufficiently large in diameter to give unvignetted images when used with my 102-mm, f8 refractor, but when used with my 80-mm, f6.8 refractor, I have to accept some minor vignetting towards the corners of the frame. This can, however, be pretty easily corrected for by taking 'flats' in the imaging process or, later, in post-processing. SBIG now provides a filter wheel that takes a new size of filter, 36 mm in diameter. This provides unvignetted imaging with the ST-8300 series with a telescope of any focal ratio. The new filter is cheaper than the 50-mm-diameter filters that would otherwise have been needed.

Large-Sensor Cooled CCD Cameras

The problem with cooled CCD cameras, particularly those with large chips for use in wide-field imaging, is that, until recently, they have been very expensive. For several years the cost deterred me from buying one; I was happy to use my Watec 120N camera for taking images of small deep-sky objects and my DSLRs for wide-field imaging.

More recently a new Kodak KAF-8300 CCD chip has become available in both mono and colour versions. This utilises 8.3 million 5.4 micron square pixels in a 3,326 × 2,504 array across an 18 × 13.5 mm chip. It utilises anti-blooming gates (which would tend to reduce its sensitivity) and micro-lenses (which increase it) and has a peak quantum efficiency of around 56% in its mono version.

This chip has been incorporated into a number of CCD cameras, for example, the SBIG ST-8300 in mono and colour versions, the Starlight Express SXVR-18 mono camera, the Alta® U8300 camera in mono and colour versions, the QSI 583 mono camera (which can include an integral filter wheel), the Moravian Instruments G3 8300 camera incorporating an integral filter wheel, the Atik 383L+ mono camera and the QHY QHY9 in mono and colour versions. Celestron has introduced its Nightscape colour CCD camera using the colour version of the same CCD sensor. They all represent excellent value for the money and have made large-size, multi-mega-pixel imaging much more accessible. Finally tempted, I was able to purchase the SBIG mono version second-hand.

It is very likely that, when large-sensor CCD cameras are used, field curvature may well become a problem, with star images at the corners of the field becoming blurred and distorted. As discussed in Chapters 1 and 14, a field flattener (often also acting as a focal reducer) will almost certainly be used unless the telescope already incorporates correcting optics in its light path, as in the Celestron Edge HD telescopes and the Vixen VC200L.

15.3 Light, Dark, Bias and Flat Frames

At this point a few imaging terms should be defined and explained. Imaging software refers to three types of 'frame' which relates to one exposure. The first is a 'light frame', which is a frame taken of the object to be imaged. The second is a 'bias frame', which has had zero exposure. It gives a measure of the read-out noise or 'bias current' contributed by the chip electronics each time the image data is downloaded into the computer. This may vary slightly with temperature but, at any given temperature, will be constant – and hopefully small. The third is a 'dark frame', which is a measure of the total noise due to the bias and dark currents when an exposure is made with the camera shutter closed. It will also expose 'hot pixels' that become more prominent the longer the exposure. If a dark frame has been made with the same exposure time and at the same temperature, this will be subtracted from the light frame images in the first stage of the image processing. In a cooled CCD camera the temperature control electronics will normally stabilize the chip temperature to about half a degree, which means that the dark and bias current noise contribution will remain constant. This means that dark frames (which will be averaged by the imaging software) taken at the start (and perhaps end) of an imaging sequence can be used to correct for all the light images.

As it is obvious that the dark frame will contain the bias current, I was confused for some time as to why any bias frames have to be taken. It turns out that they are needed only if the light frame exposure time is different from the dark frame exposure – in which case they allow the imaging software to estimate what a dark frame with the same exposure as the light frame would be. This can work, as the bias current contribution remains constant and the dark current contribution increases linearly with exposure time. Let us suppose that the light frame exposure was 10 minutes, but there is only a dark frame of exposure 5 minutes along with a bias current frame. (They must all be at the same temperature.) The software will take the 5-minute dark frame and subtract the bias frame to give only the dark current noise contribution to the dark frame. The noise in this frame is then doubled to give an estimate of the dark current contribution that there would be in a 10-minute dark frame. The noise from the bias frame is then added back to give an estimate of the total noise in a 10-minute dark frame. The point is that if you make dark frames of the same exposure time as your light frames – which is always best – there is no need to take any bias frames.

By taking bias frames and short dark frames, one allows the imaging software to estimate what the dark frame would be for a longer exposure and so reduce the total imaging time for an observing project, but the result will never be as good as dark frames taken with the same exposure time. A number of dark frames, perhaps ~20, will be taken and then averaged in the image processing software. The image acquisition software may well allow a suitable number of dark frames to be taken unattended after the observing session has ended.

The fourth type of frame is a 'flat' frame. Particularly if a focal reducer has been used in the imaging chain, the image may suffer from vignetting in the corners of the field. Photoshop has a filter (Filter > Distort > Lens Correction) which allows for vignetting to be corrected in post-processing, but a better alternative is to take and average some 'flat' frames before or after the imaging session. A flat frame is simply an image of a uniformly illuminated field. There are several ways of achieving this. One way is to observe a twilight or pre-dawn sky (when the stars are invisible) at an altitude of about 30 degrees towards the south-east or south-west. A second way is to image a matt-white area which is uniformly illuminated. This could be painted on the inside of an observatory dome or a white sheet held in place some distance from the telescope. A third way is to make a light box with a uniformly lit background, whilst a fourth, similar approach is to stretch a tight and wrinkle-free sheet or white T-shirt over the telescope aperture and point the telescope at a diffuse light source. The first owner of my 350-mm Maksutov made a telescope aperture cover out of white translucent plastic which is used in just this way. Image processing software will use flat frames to correct for vignetting, but this will also reduce the effect of dust or hairs on the sensor. Again it is recommended that one take quite a number of flat frames, which are averaged by the imaging software. It is important that no part of the image be over-exposed. A peak brightness of ~40,000 would work well when taken with a 16-bit camera which has a range of 0–65,535.

15.4 Monochrome CCD Imaging

First Light with an 8.3-MPixel Camera Imaging M31, the Andromeda Galaxy, in Monochrome

The fact that an 8.3-Mpixel SBIG CCD camera that I had acquired boasted a relatively large and high-resolution CCD array meant that it could be used for quite wide-field imaging given a telescope of short focal length such as my semi-apo refractor of 80-mm aperture and 550-mm focal length. The sensor size is 18×13.5 mm. Dividing these values by 550 gives the chip field of view in radians, and multiplying these values by 57.3 gives the field of view (1.9×1.4) in degrees. The combination would thus be able to encompass much of the Andromeda Galaxy, M31, so this seemed a good choice for the camera's first light observation.

The first step is to choose an appropriate exposure, which is determined by two factors. The first is the tracking ability of your mount. If the mount is auto-guided, this is not a significant determining factor, but if not, it will depend on how well the mount tracks unguided. A well-set-up equatorial mount or computerised Alt/Az mount with a short-focal-length telescope should be capable of several tens of seconds, as determined by a number of test exposures. The second factor is that *important* parts of the image not be over-exposed. If individual exposures are made to bring out the (relatively faint) spiral arms of M31, then the core of the galaxy is very likely to be over-exposed (as can be seen in many images of M31 found on the Internet). To get this aspect of the imaging right is not quite so easy. Figure 15.3 shows the on-screen image of the 'first light', 20-second-exposure image of M31 produced by the ST-8300 using the SBIG imaging software 'CCDops'.

You will see that a large area of the central core appears to be burnt out. But note the box entitled 'contrast' appearing in the frame. The image that is presented to you by CCDops is 'stretched' to show faint detail in the image: the two numbers, 172 and 526, show the range of values (out of 65,536) that have been displayed, so any actual values below 172 will appear black and any values greater than 526 will appear white. If the image is exported as a TIFF file and viewed in Photoshop, hardly anything is seen – just one bright star and a very faint fuzzy region, which is

Figure 15.3 A 20-second exposure made of M31 by an SBIG ST-8300 using CCDops. It appears that the core is over-exposed, but this is not the case, as CCDops stretches the image.

the core of the galaxy! When the image mode was changed to RGB from greyscale, and the eyedropper tool placed over the brightest part of the core, the R, G and B numbers all gave the value 20. This is on the scale from 0 to 256, so the peak value of the galaxy that the CCD has measured is $20 \times 65,536/256$, or 5,120. This shows that a significantly longer exposure could have been made without over-exposing the core. As the CCD chip has a very linear response, with this 80-mm scope an exposure of 200 seconds would still not over-expose the core. However, the 80-mm refractor was mounted on an unguided Alt/Az mount and I wanted to keep exposures short so that the stars would not trail, so a sequence of light and dark frames each of 20 seconds was taken as determined by the tracking ability of the mount rather than the maximum exposure requirement. The SBIG software automatically subtracted the dark frame as each pair was taken.

For this first exercise, a flat field correction was not used, which did not cause any obvious problem. The ambient temperature at the time was 10 C and so the temperature set point was set at –20 C, which was easily achieved by the camera's Peltier cooler. These individual frames, exported as 16-bit TIFF images, would then in principle be stacked in 'Deep Sky Stacker' (DDS) but, without a reasonable number of stars to align on, DDS cannot work. So, before they were imported into the stacking program, each of the images was 'stretched' to make more stars visible. The objective of stretching the image is to bring the central core of the galaxy up to values close to ~50,000, towards peak white, but also to make the fainter parts of the galaxy more prominent relative to the core, so a non-linear stretch is required – the fainter parts of the image being boosted more than the core.

With software such as 'Maxim DL', there are functions to achieve this, as will a combination of the free programs 'IRIS' (to convert into FITS format if required) and 'FITS Liberator 3' (to apply the non-linear stretching), but I chose to achieve this by using the curves function in Photoshop. When the curves function is initiated (Image > Adjustments > Curves) a box appears, as shown in Figure 15.4. The image has had three applications of the curves function shown in the box. The histogram shows that virtually all of the image still resides at very low brightness levels, but a fair number of stars are now visible so that DSS would be able to align and stack all the images when similarly stretched.

Exactly the same process was applied to all 20 images, and these were imported into DSS and stacked. Further stretching with the same curve was applied to the stacked image used to bring up the fainter parts of the galaxy, and Figure 15.5 (top) shows the result. The noise in the darker parts of the image was very obvious, so the 'Levels' command (Image > Adjustments > Levels) is used to hide this by moving the 'black point' up to give the result in Figure 15.5 (bottom). The effects in the corners of Figure 15.5 (top) are due to the fact that an Alt/Az mount was used, so that the program had to correct for rotation and thus fewer images contributed to the corners – a very good reason for using an equatorial mount! This was not a particularly good image of M31 and its two daughter galaxies M32 and M101, but it was done as a tutorial exercise with a number of amateur astronomers observing the

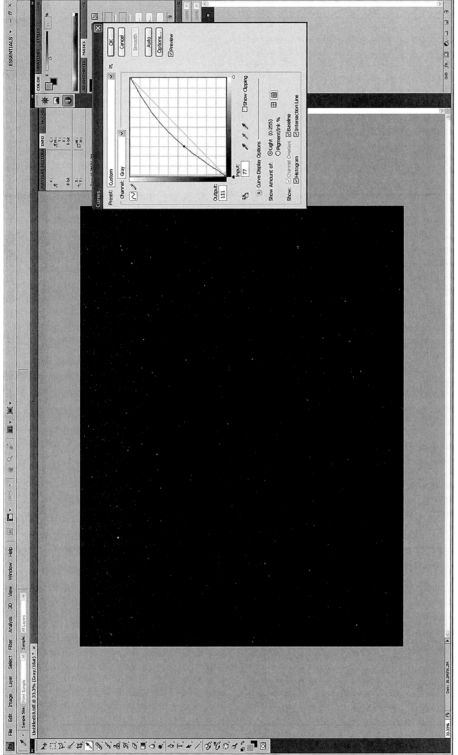

Figure 15.4 A single 20-second TIFF image having been stretched to bring up sufficient stars to enable Deep Sky Stacker to align and stack all 20 images.

Figure 15.5 The initial stacked and stretched image of M31 (*top*) and the image after the black level was lifted up to suppress the low-level noise (*bottom*).

imaging and analysis process. The total exposure time was only 400 seconds. To its credit, however, the image does not have an over-exposed core.

I was able to improve the image to give that shown in Figure 15.6. The aim was to smooth the nebula regions towards the edge of the galaxy which were very noisy but, at the same time, retain the faint stars visible in this region. This process is covered in detail in Chapter 18 with a supporting image (Figure 18.2).

Figure 15.6 The final monochrome image of M31 and its companion galaxies produced from a total exposure of 400 seconds after twenty 20-second exposures were stacked.

You may well be surprised by how much image manipulation is carried out following the initial production of an image. I promise you that this is usually the case and, quite often, images captured by some of the world's best astro-imagers are handed over to an expert in image enhancement to produce the image that is seen in print or on the Web. As the noise in the image will be reduced by less than half for each doubling of the exposure time, one can see that to obtain a really good image of M31 a considerably longer total exposure will be required.

One major enhancement would be to make it a colour image, and, having later acquired a filter wheel to incorporate a set of dichroic colour filters, I added colour to the monochrome version, as described later. (Happily, M31 does not change from year to year except when a supernova appears!)

15.5 LRGB Colour Imaging

This standard technique used by deep-sky imagers involves taking a monochrome image – called the luminance image – to which colour is added from an RGB image constructed from three individual monochrome images taken through red, green and blue filters. An infrared cut-off filter (the L filter) is used to form the luminance image in order to avoid any out-of-focus halos. Dichroic RGB filters also remove the infrared.

Adding Colour to the Monochrome Image of M31, the Andromeda Galaxy

To provide a first example, I decided to produce an RGB image and use this to give colour to the monochrome image described earlier. The camera was attached to the 80-mm telescope in the same orientation as when the luminance image is taken so that the luminance and colour images would approximately line up. Sequences of 20-second exposures were made to give a total of 5 minutes of exposure through each of the red, green and blue filters. A good initial procedure is to take some test exposures of a white sheet and vary the exposures until an RGB composite gives a white result. This will give the ratios of the exposures required for the three colours to get a good colour balance.

It is not necessary to have as high resolution for the colour image, as the detail will be provided by the luminance image, and therefore the pixels were binned in 2 × 2 mode so that each output pixel from the camera was made up from four adjacent pixels, thus improving the sensitivity and so requiring shorter total exposure times. The individual images taken through each filter were stretched in Photoshop exactly as for the luminance image and then imported into DDS to provide three monochrome images which contain the red, green and blue colour data.

Aligning the Individual Greyscale Images That Will Be Used to Build up an LRGB Image or to Add an H-Alpha Layer to an Image

There can be both lateral and (possibly) rotational difference between the images that will make up the RGB colour image, and the following method allows the three images to be precisely aligned. The three RGB greyscale images are opened up, and then a new image is opened in RGB mode having the same pixel dimensions. The 'Channels' window of the new image is opened and the individual red, green and blue greyscale images are then copied (Ctrl A, Ctrl C) and pasted (Ctrl V) into the appropriate channels. With all channels included (the eyes showing to their left) the colour image may well look pretty poor due to misalignment between the channels. The process is then to select the red and green channels (turning off the 'eye' of the blue channel by clicking on it), click on and highlight the red channel image (its surroundings go blue), activate the move tool and use the arrow keys to align the stars in the red channel with those in the green channel. The process is repeated with just the blue and green channels selected.

This will result in a well-aligned image unless there is any rotation between the fields. The process employed is exactly the same, but it will then be necessary to both rotate and move the red channel to align it with the green. This is done by highlighting the red channel and using Ctrl T to enable a free transform to be made. The cursor keys can be used to align the two channels as before, but if the mouse is clicked to the upper right of the channel image (not the exact corner) a rotation symbol appears 'and, by use of the mouse, the red channel can be rotated relative to the green and, in combination with lateral adjustments, the two channels can be aligned. In a similar manner, the blue image is rotated and aligned with the green so that all three are perfectly aligned and a sharp colour image results.

Figure 15.7 The smoothed colour image (*left*) and the result of adding colour to the monochrome image of Figure 15.6.

These colour images were taken with the same telescope and camera, but as the colour images were binned 2 × 2, the image scale will be halved. By means of Image > Image Size, the colour image was scaled by 200% to match that of the monochrome image. The colour image also had to be slightly rotated to match the monochrome image, as described earlier. In case there were still small errors, so that the stars would not exactly line up, the 'Dust and Scratches' filter was applied (Filter > Noise > Dust and Scratches) to remove the stars from the image, giving the result shown in Figure 15.7 (left) and Plate 15.7 (top). This image was selected (Ctrl A), copied (Ctrl C) and pasted over the monochrome image by opening that image and using Ctrl V. The blending mode was set to colour and its opacity adjusted before the two layers were flattened (Layers > Flatten Image) to give the final result, as shown in Figure 15.7 (right) and in Plate 15.7 (bottom). This image has no great merit, but the colour does add something to the image.

Figure 15.8 and Plate 15.8 show a superb image of M31 taken by Peter Shah using an Orion Optics (UK) AG8, f3.8 astrograph.

15.6 Remote Observing

There is an alternative approach to taking top-quality astro-photos, which is to buy time on a telescope system situated in a country where observing conditions will often be better than at home – particularly if one lives in the north of England. A number of companies and observatories offer this service. There are two possible modes of operation. One method is to take direct control of the telescope and its imaging systems using high-speed Internet access, whilst the second, similar to that often used by professional astronomers with large optical telescopes, is to provide an observing proposal that will be undertaken on your behalf by the robotic system. A very nice feature of the latter is that the observer buys an amount of imaging time – which is the actual length of time that the CCD camera is imaging

Figure 15.8 An image by Peter Shah of M31 taken with an 8-inch astrograph.

an object or objects. It does not include, for example, the time that a telescope might take to slew from one object to the next, so one knows exactly what will be provided. In some cases, the cost of telescope time may vary depending on the phase of the Moon, being least at full Moon and greatest around new Moon, as one might expect.

I felt that it would be well worth trying out this idea and obtained telescope time on three telescopes, one situated at Nerpio in Spain at a height of 1,680 m and two others at heights of ~1,500 m in Arizona and California. The former is operated by a company, MyAstroPic GmbH, based in Germany, which (at the time of writing) provides time on three telescopes: for wide-field imaging an 8-inch Newtonian with a field of view of 117 × 80 arc minutes and a Hyperbolic Newtonian with a field of view of 211 × 211 arc minutes, along with a 20-inch Cassegrain with a field of view of 30 × 30 arc minutes for galaxy hunting.

The Sierra Stars Observatory Network in the United States also provides time on three telescopes. The smallest, Rigel, is a 14.5-inch Cassegrain having a field of view of 25 × 25 arc minutes and is operated by the University of Iowa at a height of 1,524 m south of Tucson, Arizona. A larger telescope is the Sierra Stars 24-inch Cassegrain having a field of view of 21 × 21 arc minutes; it is situated on the California/Nevada border at a height of 1,545 m. The largest is the University of Arizona's 32-inch Richéy-Cretian situated at a height of 2,800 m on Mt Lemmon, 90 miles from Tucson; this has a field of view of 22.5 × 22.5 arc minutes.

Using the MyAstroPic 8-Inch Newtonian Situated at Nerpio in Spain

I elected to use the 8-inch Astrosysteme Austria (ASA) Newtonian astrograph referred to in Chapter 4, the cost of which was 39 euros per hour for imaging at new Moon. It is equipped with an Atic 383L+ 8.3-Mpixel camera (similar to my own SBIG ST-8300) and an Atic EFW2 filter wheel incorporating Astronomic Filters: broadband LRGB and narrowband H-alpha, OIII and SII. This would be an ideal telescope for doing some narrowband astro-imaging using the Hubble Palette mentioned earlier. Such imaging can be carried out even when the Moon is full and the cost of telescope time is lowest.

My project was to image the galaxy M33 in Triangulum and, with helpful advice from the imaging team, arranged for luminance and RGB images to be taken around new Moon, with a total observing time of 10,800 seconds. Ten sub-frames, each of 300 seconds, were taken for the luminance image, eight sub-frames of 300 seconds for the green and nine sub-frames of 300 seconds for the red and blue, making a total of 36 sub-frames. In addition, 20 bias frames and 20 dark frames of 300-second exposure were provided, along with 15 flats for each of the LRGB filters – making a total of 136 FITS images. These were placed in a 'Dropbox' folder so that I could download them into my own computer.

I used DDS to calibrate, align and stack the four LRGB images. For each image the light frames with their appropriate flats were loaded into DDS along with the bias and dark frames. All images were checked and, before too long, each calibrated and stacked image could be exported as a TIFF file for further processing in Photoshop.

The first step was to stretch each image. This was done using seven applications of a very gentle curve followed by the setting of the dark point to the point where the histogram rises sharply with the use of levels. At this point all four monochrome images looked broadly similar. The next step was to produce an RGB colour image. All three monochrome images were loaded into Photoshop, and 'New' image was opened up with the same dimensions and in RGB 16-bit colour mode. The appropriate image size may be set automatically but, if so, will be in greyscale mode and will have to be changed to RGB 16-bit colour mode. The 'Channels' box is opened up (Window > Channels), which will show three blank red, green and blue frames. Each monochrome image is selected (Ctrl A), copied (Ctrl C) and pasted (Ctrl V) into the appropriate channel box. A colour image will now appear. Unless the three RGB images are perfectly aligned, it may well look very messy. The three images may have to be aligned and perhaps rotated slightly, as already described.

One can then spend some time adjusting this image to give the best-looking result. There are several ways of doing this: one can select each channel in 'Levels' and adjust its black, middle and white points, or perhaps use the 'Colour Balance' controls. A final adjustment may well be to increase the saturation. (A couple of very good techniques to achieve this are given in Chapter 18.) This image may be rather noisy, as usually less time is spent imaging the red, green and blue filtered images and, due to the filters, the sensitivity of the CCD will be lower. One does not need such a high resolution for the colour image as for the luminance image and so, to reduce the noise, a Gaussian blur of a few pixels was applied.

The final procedure is to give colour to the luminance image. One can either open the luminance and RGB images, copy and paste the luminance image over the colour one and use the luminance mode to combine them or paste the colour image over the luminance image and use the colour mode. Most astro-imagers tend to use the former, but I cannot detect any obvious difference between them. The opacity slider can be adjusted to give the best result.

That's it! If the image is noisy, the 'Despeckle' filter in Photoshop may help somewhat or one can use a noise reduction program such as 'Picture Cooler'. The final result of this imaging exercise is shown in monochrome in Figure 15.9 and in colour in Plate 15.9. I have to say that I was quite pleased with it and have already booked an imaging session to use this telescope to image the Horsehead Nebula region when it is next in view on moonless nights.

Using the Sierra Stars Rigel 14.5-Inch Cassegrain in Arizona

This is the smallest, and hence least expensive, of the Sierra Stars network, costing around 50 dollars for an hour of observing time. It is equipped with a Finger Lakes Instrumentation ProLine camera – a state of the art research-grade CCD camera capable of maintaining stable temperatures to 65 degrees Celsius below the ambient temperature. The camera contains a Kodak KAF-16803 4,096 × 4,096 pixel CCD chip with 9-micron pixels, giving an image scale of 0.37 arc second/pixel over a field of view of 25 × 25 arc minutes. This is impressive!

It is worthwhile seeing how well a specified field size will encompass a particular object, and there is a very simple way of achieving this. This is to use the STScI Digitised Sky Survey (archive.stsci.edu/cgi-bin/dss_form). The object name is entered and the coordinates obtained by clicking on the 'Get Coordinates' box. Set the height and width of the image to the desired size (in this case 25 and 25 arc minutes),

Figure 15.9 The luminance image of M33.

select the GIF file format and click on 'Retrieve Image'. This field of view will nicely encompass both the galaxy M51, the Whirlpool Galaxy, a face-on spiral interacting with a smaller companion NGC 5195, and NGC 4565, the Needle Galaxy, each of which I was keen to image (Figure 15.10). Having registered with the site, I was able to schedule two sets of observations.

A problem that has been illustrated by my imaging of the Orion Nebula is that, due to the high brightness of the central region around the Trapezium stars, it is very easy for this to be burnt out. I thus wanted to stack a number of relatively short exposures to provide the luminance image, which would be coloured with a smoothed colour image derived from exposures taken through the red, green and blue filters. M51 has a number of bright HII regions and so, to add a little extra to the image, H-alpha exposures were also taken. Three sets of 60-second exposure images were thus programmed for luminance and RGB images, and 180-second exposures for the H-alpha image, giving a total exposure time of 1,080 seconds. This total time is far less than would be used by most astro-imagers, but I wanted to see just what could be achieved at relatively low cost.

A day or so later, an email alerted me to the fact that the data had been acquired, and the Web site gave directions as to how to 'ftp' the data from the Web site into my computer for analysis. The data was in the FITS format used by professional (and many amateur) astronomers but had already been calibrated with the appropriate bias, dark and flat frames. To carry out the analysis procedure I used DSS to align

Figure 15.10 The Sierra Stars Rigel 14.5-inch Cassegrain and the Digitised Sky Survey 25 × 25 arc second image centred on the galaxy NGC 4565.

and stack the FITS frames and exported them as 16-bit TIFFs for further processing in Photoshop.

There can be a problem with this process if there are not enough stars visible in the frame for DDS to carry out its alignment. In this case, each FITS frame can be individually imported into DSS and immediately exported as a TIFF file into Photoshop, and a series of gentle curve applications performed. It is important that the same curve profile be used with each application, so this should be made an action. Once the curve applications have been performed, the 'Levels' command is used to bring the black point up to the left edge of the histogram and the image saved. Hopefully a number of stars will be visible and the TIFF files can imported back into DSS and the frames stacked. In either case, there will be four, or in this case five, images to process in Photoshop.

First, the images were identically stretched using repeated applications of a gentle curve (Image > Adjustments > Curves). Then, in 'Levels' (Image > Adjustments > Levels) the black point of each was set to just to the left of the histogram. One has to decide how many curves applications are to be made. Each image had dimensions of 2,048 × 2,048 pixels. Because the images were of a relatively short total exposure they were, not surprisingly, somewhat noisy. I thus imported the 16-bit RGB TIFFs into the noise reduction program 'Picture Cooler' and applied some noise reduction to produce a smoother image.

Two further procedures were then applied to the luminance image. First, the 'Unsharp Mask' filter (Filter > Sharpen > Unsharp Mask) was applied with a radius of 250 pixels and an amount of 15% to provide some local contrast enhancement. This was followed by use of the 'Smart Sharpen' filter (Filter > Sharpen > Smart Sharpen)

Figure 15.11 Monochrome images of M51 (*left*) and NGC 4565 (*right*) imaged with the Rigel 14.5-inch Cassegrain.

with a radius of 9 pixels and an amount of 100% to give the final luminance image shown in Figure 15.11 (left).

As described earlier, a colour image was created from the three red, green and blue images. It is quite likely that the colour balance will not be correct. Using the 'info' dropper, the red, green and blue levels in the dark areas of the image can be measured to give an idea of how the relative levels need to be adjusted using the 'Levels' command with each of the colours individually selected. Some use of the colour balance adjustments (Image > Adjustments > Colour Balance) was also made. It has to be said that this is a pretty subjective process!

The colour image can be smoothed with a Gaussian blur of about 4 pixels and then selected (Ctrl A), copied (Ctrl C) and pasted (Ctrl V) onto the monochrome image, which must first be set to RGB 16-bit mode. The colour blending mode is then chosen to give colour to the image.

The final process was to add the H-alpha layer to the image. This was very noisy. The 'Levels' command was used to remove the low-level noise and isolate the point-like objects (HII regions) within the image. The image was duplicated and a suitable red colour, such as red = 231, green = 33 and blue = 11, was set as the foreground colour. This was painted over the duplicate layer, the blending mode set to colour and the two layers flattened. The resulting H-alpha image was copied, pasted and aligned onto the coloured whirlpool image and the 'colour dodge' mode selected to add the H-alpha regions into the image.

To produce the image of the Needle Galaxy, NGC 4565, the same three sets of 60-second LRGB exposures were made with an identical processing work flow. However, as the galaxy is edge-on, no HII regions were likely to be seen and so no H-alpha images were taken.

Figure 15.12 An image of the globular cluster M5 taken with the Sierra Stars 24-inch Cassegrain.

Using the Sierra Stars 24-Inch Cassegrain to Image the Globular Cluster M5

This telescope has a field of view of 21 × 21 arc minutes and, as indicated by the Digitised Sky Survey, it appeared that this would nicely encompass the globular cluster M5, which was well placed in the night sky. I thus scheduled a set of luminance observations totalling 700 seconds. It would have been nice to have had an answer for the question of how long an individual exposure could be whilst ensuring that the bright central core of the cluster would not be burnt out. I probably should have taken some single exposures to find out, but I made a guess that a single exposure of 70 seconds would be suitable and set the schedule so that the telescope would take ten 70-second exposures. Happily, my guess was right – but only just, and more, shorter exposures would have been a somewhat safer choice.

The zipped FITS files having been downloaded, they were imported into DDS and the resulting stacked version was exported as a TIFF for the final processing in Photoshop. Three things were done: in 'Levels', the black point was adjusted to darken the sky background, a mild curves function was used to stretch the image a little and a little sharpening was done using the 'smart sharpen' function. I was very pleased with the result, which had taken very little time to process (Figure 15.12).

I hope that this has shown that the use of a remote telescope does allow amateur astronomers the possibility of producing high-quality images which I believe they can call their own as much as a professional astronomer using a large telescope.

16
Auto-Guiding and Drift Scan Alignment

In several earlier examples it has been shown how one can give the effect of one long exposure by stacking a number of sub-exposures. Most telescope mounts can track sufficiently well to allow exposures of up to 60 seconds without elongating the star images, so why bother to auto-guide? The fundamental reason is that longer exposures have a greater signal-to-noise ratio than short ones so, in principle, one 60-minute exposure would be better than the stacking of sixty 1-minute exposures. Stacking a number of images *does* increase the signal-to-noise ratio but not quite enough to compensate. The maths relating to this is somewhat complex and will not be covered here, but it turns out that to get the equivalent to one 60-minute exposure would require perhaps eighty 1-minute exposures to be stacked.

But it would be totally stupid to attempt a single 60-minute exposure: firstly, any bright stars in the field would be grossly over-exposed, causing highly unattractive 'blooming' of their images; secondly, the chance of a single 60-minute exposure not having a satellite or aircraft trail is quite low; and thirdly, for best results, you are going to have to take a number of 60-minute dark frames. As a result, astro-imagers have to compromise. An optimum strategy might well be to take eight or nine 10-minute exposures to give a comparable result, but care must be taken not to over-expose an important part of the image, for example, the core of M31 discussed in the preceding chapter.

To achieve these longer exposures, the telescope's tracking must be corrected during the exposure to minimise the effects of periodic errors in the drive train or slight misalignment of the mount. The fundamental process is to simultaneously observe a reasonably bright star in, or close to, the field of view of the camera and, by measuring very tiny movements in its image, make minor corrections to the tracking so that star images remain circular and are not elongated. There are several ways to achieve this, but in all cases a secondary camera is employed. (Some CCD cameras incorporate a second small camera offset from the main sensor, but this is still in effect a second camera.) A continuous sequence of short exposures is analysed by a computer program to determine if the centroid of the selected guide star image is moving. Should such movement be found, the system issues appropriate drive commands to the telescope mount, so bringing the star image back to its nominal position. There obviously has to be some error in order for a correction to be applied but, somewhat amazingly, the software is able to detect and correct movements of less than an arc second, so an image in which stellar disks are typically at least 2 arc seconds in diameter will show no tracking errors. Magic!

16.1 Off-Axis Auto-Guiding

This is where some of the light at the edge of the field of view of the main imaging camera is diverted into a separate auto-guiding camera (Figure 16.1). There are two significant advantages to this approach. Firstly, no additional guide scope is required, reducing both the weight and complexity of the system, and, secondly, if the imaging telescope is a Schmidt-Cassegrain, it will correct for any slight movement of the primary mirror during the exposure. (The latest ACF and High-Edge Schmidt-Cassegrains from Meade and Celestron have 'mirror locks' to prevent this and, with earlier scopes, the effect can be minimised by always finalising the focusing with an anti-clockwise turn of the focuser.)

However, there can sometimes be a problem finding a suitable guide star, although the mirror attachments can be rotated so areas all round the main field of view can be searched. A secondary problem when a large CCD array is being used is that the images beyond the field edge may well suffer from coma and astigmatism, so giving somewhat misshapen stars that the software may have a problem analysing. There can be a further problem with a Newtonian imaging scope in that the off-axis guider attachment, placed in front of the CCD camera, may make it impossible to bring the CCD camera to focus. For these reasons, many employ a separate guide scope system, as shown in Figure 16.2.

16.2 Using a Secondary Guide Scope

The alternative is to use a secondary guide scope that is attached to the telescope mount. In some respects this is not as good as using an off-axis guider, as it cannot correct for any primary mirror movement and, unless the guide scope is very rigidly tied to the main scope, the guiding will not be as accurate. It does, however, give a far wider choice of guide star and, as the guide scope does not even have to look along the precise axis of the main scope, it is quite likely that a brighter guide star can be found than when an off-axis guider is used. This can make it easier for the system to provide accurate guiding corrections.

Short-focal-length 80-mm refractors are often used as guide cameras. Rather than using oversized tube rings with adjustment screws to adjust the alignment of the guide scope relative to the main scope – which is a pretty fiddly business and may not be too rigid – Sky-Watcher has produced a guide scope mount (Figure 16.2) that holds the scope in a Vixen dovetail and has adjustments to offset its optical axis in both directions. This allows a suitable guide star to be easily found. The 80-mm guide scope system used by a colleague is even equipped with its own finder to make this an easy process.

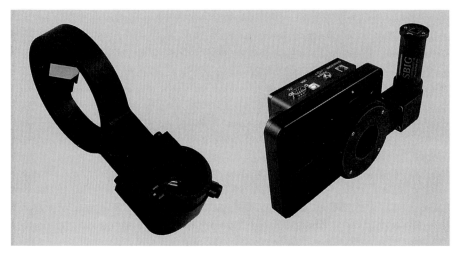

Figure 16.1 Off-axis guider for use with SBIG ST-8300 and STF-8300 CCD cameras and ST-i planet cam/guide camera. (Image: S-BIG)

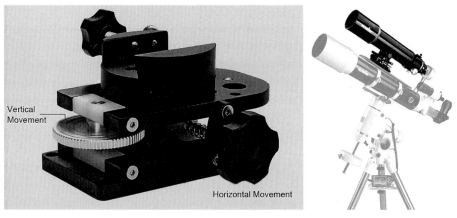

Figure 16.2 The Sky-Watcher guide scope mount and its use with an 80-mm guide scope. (Images: Sky-Watcher)

A much debated question concerns the specification of the guide scope. Many imaging systems have been using 60- or 80-mm refractors either mounted on or beside (using a dual dovetail system) the imaging scope. The situation has changed during the past few years as guide cameras have become available with pixel sizes of the order of 5.2 microns. These are comparable to or smaller than those used in the main camera and allow the guiding algorithms used in the software to work with smaller image scales. This means that guide scope focal lengths need not be so long. The rule used to be that the guide scope focal length should be no less than one-third the effective focal length of the imaging scope. With the latest cameras and software packages this ratio can be reduced to one-sixth or even less. Much imaging is done with focal lengths of 800 mm or less and would thus require a guide scope focal length of only 130 mm or so.

The guide scope has two relevant parameters: the first is its aperture – with larger objectives able to see fainter stars in a given time – and the second is the focal length. The latter will determine the field of view 'seen' by the guide camera and, obviously, the wider this is the more likely it is that at least one suitable guide star will be seen. So, to give wide fields of view, a short focal length is best. This, coupled with a large objective, implies that an objective with a low focal ratio would be ideal. The calculation of the optimum specification depends on the density of stars in the direction of view and there is no single answer. However, camera lenses tend to have shorter focal ratios than telescopes and, coupled with reasonably sized objectives giving focal ratios of f2.8 or even less, might well prove excellent for this task. By providing a wide field of view, it almost certain that a suitable guide star will be found without having to offset the guide scope from the main telescope's target. This tends to allow a more rigid mounting to be used so that the guide scope does not move relative to the imaging telescope.

This idea has been taken up by several manufacturers. Orion Optics (United States) sells a very neat f3.6 guide scope with a 50-mm objective and 180-mm focal length that is essentially a finder scope equipped with a 1.25-inch eyepiece socket. This uses the standard finder dovetail mount that many scopes are equipped with. If not, a dovetail is provided to fix to the imaging scope. Borg offers a guide scope with a 50-mm objective and a focal length of 250 mm and, even better, a 60-mm scope with a focal length of 228 mm. SBIG offers an exceedingly compact f2.8 guide scope with a focal length of 100 mm which is essentially a camera lens provided with a very small and solid mount. As the field of view when used with their ST-I camera is more than 2 degrees wide, it is virtually certain that a guide star can be found without having to offset its position relative to the imaging scope. This means that the mounting can be very solid, reducing the problem of differential deflection when the imaging and guide scopes do not remain perfectly aligned. SBIG states that even though a colour CCD sensor is used in the camera provided as part of the guide package, so reducing its sensitivity, 1-second exposures can detect 10th-magnitude stars. (These guide scopes are shown in Figure 16.3.)

I have followed this route, adapting a Zeiss 135-mm, f2.8 lens to directly fit a QHY, QHY6 CCD camera. With the few-second exposures used for guiding, Peltier cooling is not required.

16.3 On-Axis Guiding

An innovative product has recently come on the market which enables the main telescope to be used as the guide camera. It is called the Innovations Foresight On-Axis Guider (ONAG®) (Figure 16.4). The idea is very simple: the unit uses a 45-degree dichroic mirror in the telescope light path to reflect the *visible* light (including that from H-alpha emissions) sideways into the imaging camera, whilst the mirror passes the *near-infrared* light through and into the guide camera. (The guide camera must not

Figure 16.3 Orion 50-mm guide scope (*top left*), Mini Borg 50-mm guide scope (*bottom left*) and the S-BIG lens-based guide camera (*right*).

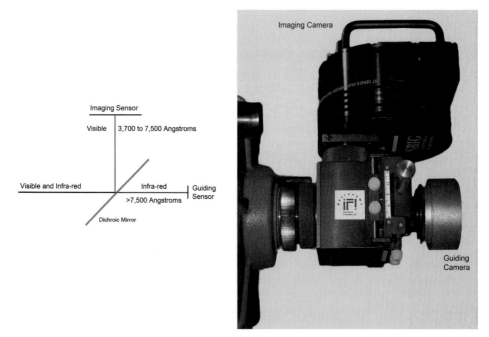

Figure 16.4 The Innovations Foresight On-Axis Guider.

have an infrared blocking filter in front of the CCD!) The mirror is of very high optical quality so as not to introduce any degradation of the captured image and reflects 99% of the visible light into the imaging camera whilst passing ~93% of the near-infrared into the guide camera. Two versions of the unit are available: the original ONAG® is suitable for sensors with diagonals up to 28 mm across (such as the Kodak KAF8300), with a larger version, the ONAG® XT, available for use with sensors having diagonals up to 50 mm across.

The units are equipped with a moveable stage on which to mount the guide camera. This allows a field of diameter 46 mm to be explored (covering more than 1 degree with a typical SCT) to allow a suitable guide star to be found. Though, with a large-aperture telescope – say an 11-inch SCT – the field covered by the guiding camera sensor will be small, the fact that the telescope *has* a large aperture means that far fainter stars can be detected. An 8-inch scope will have a gain of ~2 magnitudes over a typical 80-mm guide scope, and an 11-inch scope a gain of 3 magnitudes. This means that, no matter where the guide scope is placed in the allowable field, it is very likely that a guide star can be found and it will rarely be necessary to adjust the guide camera's position.

The fundamental accuracy of a guiding system is ultimately limited by the seeing. Longer wavelengths are less affected by the seeing, which is why I suggested that the red channel of an RGB image might be used as a luminance channel in what is called 'RRGB imaging' to give a somewhat sharper image than a simple RGB image. This, of course, holds true for infrared images, so the fact that this guiding system uses the infrared part of the spectrum for guiding should give a real improvement, with the effect of seeing being ~23% lower than that with visual wavelengths. All reviews that I have found indicate that this is a superb solution to the auto-guiding problem.

16.4 Driving the Mount

The computer carrying out the guiding must obviously be able to apply corrections to both the right ascension (RA) and declination (Dec) axes of the telescope. To this end many equatorial mounts are equipped with an auto-guiding port. This is a socket for an RJ-12 connector which is usually called an 'ST-4 port', as it employs the connections originally defined by SBIG for its ST-4 auto-guider system, and which has now become the de facto standard. To drive the mount, Dec+, Dec–, RA+ and RA– inputs are grounded as required. Some hardware is required to do this. It may be incorporated within the guiding camera, as discussed later, but otherwise an interface box, such as that supplied by 'Shoestring' in its GPUSB kit, can be used. As its name implies this links to the control computer software by means of a USB port and provided opto-isolated switches to control the mount to which it is linked using an RJ-12 cable.

Some mounts do not have an ST-4 guiding port but, instead, require a serial data control stream from the computer, which can also be used to remotely control the mount from a program such as 'Carts du Ciel' planetarium program. As virtually no laptop computers are now provided with a serial port, a USB-to-serial converter will be required.

16.5 Suitable Cameras

Virtually any small CCD camera could be used, but if one were to buy one, specifically what might one look for? It should be monochrome, as a colour version using

Figure 16.5 Guide cameras: Fishcamp Starfish, Starlight Express Lodestar and Orion Starshoot 52064.

the same chip will be one-third as sensitive. It should have a reasonably large size, so increasing the field of view with a given guide scope and thus providing more candidate guide stars. It should have a high quantum efficiency and be capable of integrating for a few seconds to provide low-noise images and thus allow more stars in a given field of view to be used. It should be lightweight to minimise any flex in its mounting to the guide scope and the overall weight of the guide system. Finally, it saves on additional cost and complication if it can include an ST-4 interface port. A nice feature, now common to most guide cameras, is that they are powered through the USB port that is used to control the camera and download its images. This reduces the number of cables that are in use, making for a tidier system.

One excellent guiding camera is the Starlight Express Lodestar, which uses a 6.4 × 4.75 mm Sony IXC429AL CCD chip having 752 × 580 ~8-micron-square pixels and an excellent peak quantum efficiency of 65%. It includes an ST-4 compatible interface and is highly regarded. Though not cooled, it can be used to do some useful deep-sky imaging as well. It has been the received wisdom that CMOS sensors are not as sensitive as CCD chips, but the Micron MT9M001 6.66 × 5.32 mm CMOS chip, which utilises 1,280 × 1,024 5.2-micron-square pixels, has a peak quantum efficiency of 56%. It is being used in a number of cameras which are ideally suited for auto-guiding purposes. The top flight of these CMOS-based cameras, but currently available for purchase only from the United States, is the Fishcamp Starfish camera in cooled or uncooled versions. This does include an ST-4 interface. At lower cost are the Opticstar AG-130M Coolair and Orion Starshoot 52064. The Opticstar incorporates a fan to help cool the CMOS chip and reduce hot pixels and dark current, whilst the Orion incorporates the USB ST-4 interface so an additional GPSUB interface box is not required. Both allow integrations of several seconds. The same chip is also used in the QHY5 video camera, which is also provided with an ST-4 port. (These guide cameras are illustrated in Figure 16.5.)

Another camera that can be used to good effect is the Watec 120N integrating video camera, described in Chapter 13, which employs a same-size (½-inch), high-sensitivity CCD chip but utilizes fewer, but larger (~8-micron) pixels in a 570 × 480 array. The Watec 120N does, however, utilise three cables (power, control and video) rather than a single USB cable, so making it more difficult to install neatly.

Most cameras are able to integrate for a few seconds, with the general consensus being that integrations of from 2 to 4 seconds are about right for auto-guiding: if too short exposures are used, the guiding software might try to 'chase' the seeing, which makes a star image move about quickly; if the exposures are too long, the guiding will not be as accurate. With a few seconds of exposure, the seeing is averaged out to produce a larger but uniform stellar disk ideal for the guiding software to work with. If video cameras that are capable of only short exposures are used, the auto-guiding software package can internally integrate the video images to allow fainter stars to be seen.

A somewhat more expensive option is to use one of the entry-level cooled CCD cameras such as the Atik Titan and QHY QHY6, both of which are equipped with an ST-4 guide port. This can give a CCD camera, initially used for imaging, a second lease on life. Cooling may well not be necessary with the short exposures needed for guiding, and so there will simply be a single USB2 cable linking the camera to the control computer and the ST-4 guide cable linking it to the mount guiding control port.

As already discussed, exposure times of a few seconds are best. This implies that the download time of the camera should not be longer than a second or so; otherwise this limits the rate at which guide commands can be given to the mount. A guide camera should thus have a USB2 rather than a USB1 serial interface. All modern cameras do use USB2 interfaces, which can give download speeds of up to 2 megapixel/second; thus the ~500 kilopixel of a typical guide camera can be downloaded in less than a second. However, older CCD cameras equipped with USB1 or even parallel interfaces will require significantly more time for downloading. This tends to make them less attractive as guide cameras even though they can often be bought cheaply on the second-hand market.

16.6 Auto-Guiding Tips

If the mount has a periodic error correction facility it is definitely well worth spending the time to train the mount (for my Losmandy GM8 this is 11 minutes). Then, perhaps surprisingly, it is actually best not to set up the telescope mount perfectly! The ideal situation is when that backlash is removed as far as possible and the guide commands to the mount are always in the same direction. To remove backlash from the right ascension axis, the telescope should not be perfectly balanced. It should be made slightly 'east heavy' instead, so that the worm gears are always working on the same side of the worm teeth. This must be done separately for both sides of the 'meridian flip'. The trick to improving declination tracking is to very slightly offset the polar axis from the North Celestial Pole. This causes a slow declination drift in one direction so that all the guide commands will be in the same direction – so again eliminating backlash. The polar alignment error can be quite small, so that there will be no problem with field rotation.

Focusing

It can be quite tricky if one first tries to focus the camera at night as, unless a very bright star is in the field of view, nothing will be seen until the focus is almost perfect. One solution is to first focus on a distant object in late evening when the light levels are low. Once the camera is focused, it can be worthwhile using an eyepiece to then reach focus without adjusting the focuser. This could well mean that the eyepiece barrel is not fully inserted into the focuser, and a piece of insulating tape or a 'par-focaling ring' could perhaps be employed to give a 'stop' to the eyepiece barrel. This eyepiece would then be 'par-focal' with the camera and could be used later to achieve an initial focus for the camera.

16.7 Guiding Software

The major CCD imaging packages such as 'Maxim DL' include auto-guiding facilities, but there are several free stand-alone guiding packages, such as 'Guidedog' and 'PHD' (which stands for 'Push Here Dummy'). The latter package is very highly regarded and is one that I employ because it is very simple to use. One first 'pushes' (i.e., clicks the mouse on) the camera icon to select the camera type being used, which has been connected through one USB port. One then 'pushes' the telescope icon to select the appropriate telescope interface – most often 'GPUSB' if the ST-4 interface is being used. The desired integration time is selected and, when one is prompted to cover the guide scope, four dark frames are taken. These having been taken and averaged, one is prompted to uncover the guide scope. One then 'pushes' a looping icon, which initiates the taking and displaying of a continuous sequence of images so that the guide scope can be accurately focused. A suitable star within the field should then be selected. It should not be so bright as to saturate the stellar image but not too faint either; otherwise noise can hinder the guiding software. When satisfied with the focus and star selection, one finally 'pushes' the PHD symbol to initiate the guiding process (Figure 16.6). It really is that simple! The process begins with an automatic calibration process in which the program sends drive commands to the mount and measures in what direction and by how much it moves for a given drive command. When the calibration process, taking perhaps 5 minutes, is complete, imaging can begin. As the way the mount responds to drive commands may vary in different parts of the sky, the calibration procedure should be ideally carried out before each sequence of images is taken.

Sub-Pixel Guiding

In the case of auto-guiding, it is actually better if the stars are not quite in focus, as the guiding software can then carry out what is called 'sub-pixel' guiding. The

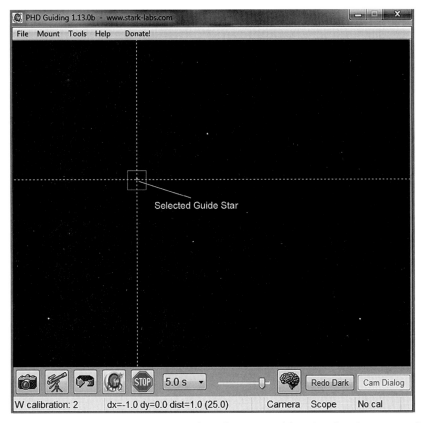

Figure 16.6 The PHD screen as the telescope calibration has just started.

requirement is to spread a star's light over several adjacent pixels on the camera's sensor. The auto-guiding software finds the centroid of the star's image and can then 'track' this to a precision of well within 1 pixel diameter! As an example, suppose a star's light is spread over a square grid of 4 pixels with the centre of its image at the precise centre of the grid. Each pixel would then have the same brightness. If the star's image were to move slightly away from this central position towards the right, the two right-hand pixels would receive more light and the two left-hand pixels less. The software, measuring these differences, can then calculate the very slight movement and so apply very precise guide commands to the mount.

16.8 Stand-Alone Systems

Astro-imagers using DSLR cameras would not need a laptop in the field for their imaging and might like a stand-alone guiding solution. Several manufacturers offer such systems. The Celestron Nexguide incorporates an LCD display and uses a small hand controller for setting up. This is very similar to the Sky-Watcher SynGuider, which uses the same Sony CCD chip. Baader Planetarium distributes the LVI Smart Guider, which uses a separate 2.5-inch display and control unit. These systems are shown in Figure 16.7.

Figure 16.7 The Celestron Nexguide system (*left*), the Sky-Watcher SynGuider (*top right*) and the LVI Smart Guider (*bottom right*).

16.9 Adaptive Optics Guiders

There is a sophisticated aid to achieving the best-quality stellar images that utilises a tilt-tip window in front of the main CCD camera to apply small corrections to the pointing without moving the mount. The tilt-tip window deviates the light through small angles to take out minor small-scale image movements arising from mount imperfections and atmospheric effects. That produced by Starlight Express can make up to 200 corrections per second, but in practice the rate is limited by the speed at which the guide camera can download images to the control computer. It is also equipped with an ST-4 output to allow the controlling software to correct for more major mount errors. SBIG also produces one for use with its CCD cameras. By reducing the effects of atmospheric turbulence and wind-induced vibrations, and eliminating the remaining periodic errors in telescope drives, these systems produce some of the best high-resolution images being made by amateur astronomers today.

16.10 Mounts with an Integrated Guide Camera System

Meade produces a German equatorial mount called the 'LX850' (Figure 16.8), which includes an integral 'Starlock' guiding system. The mount is constructed of machined stainless steel and aircraft-grade 6061-T6 aluminium and is equipped with roller bearings on both axes to support payloads of up to 90 pounds. The mount incorporates two imaging systems, first a wide-angle imaging system using Meade's Lightswitch technology to align the mount and find the target objects. Then a second camera allied to an 80-mm, f5 guide scope captures a field star down to 11th magnitude which is used to guide the telescope to a typical tracking precision of 1 arc second. Meade has also introduced auto-guiding versions of its fork-mounted Advanced Coma Free (ACF) telescopes in the LX600 series of 10-, 12- and 14-inch telescopes.

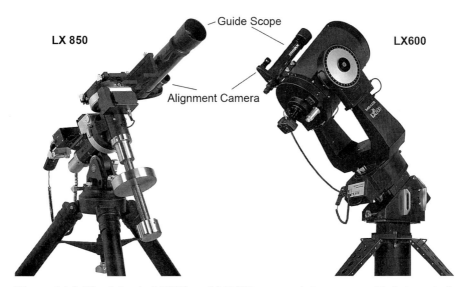

Figure 16.8 The Meade LX850 and LX600 equatorial mounts with integrated guide scopes. (Images: Meade Telescopes)

16.11 The 'Telescope Drive Master'

To greatly enhance a mount's tracking ability a device called the 'Telescope Drive Master' can be employed with many of the mounts used by amateur astro-imagers. Provided that the mount is accurately polar aligned and that it does not exhibit any flexure, it enables tracking accuracies down to the ±1 arc second level to be achieved by removing periodic and small random tracking errors. It is important that the mount is very accurately aligned and, if so, can allow unguided exposures of up to several minutes, with refraction in the atmosphere being the limiting factor. The latest version can accept external guide commands from an auto-guider system so that even longer exposures can be made. The system uses a very high accuracy encoder, which is attached directly to the RA shaft to monitor its rotation rate, and an electronics 'black box', which sends appropriate corrections to the RA drive system to keep the RA drive rate absolutely constant.

Coupling a good auto-guiding system with the Drive Master system should provide an astro-imager with the ultimate in tracking, with the auto-guider compensating for atmospheric effects (such as refraction) and declination errors and the Drive Master correcting all other sources of error. This innovative device enables good-quality mid-range mounts such as the Astro-Physics 1200, Celestron CGE, iOptron iEQ45, Losmandy G11, Meade LXD 75/LX200, Synta HEQ5/EQ6 and Viven GP-DX/GP-2 to provide tracking accuracies comparable to or better than those provided by mounts costing considerably more. Technical support is available to help equip other mounts with the system.

16.12 Drift Scan Alignment

Whether or not a guide scope is used, it would be sensible for any astro-photographer who wishes to take long-exposure CCD images to set up his or her equatorial mount as accurately as possible. Some techniques for getting a pretty close alignment were described in Chapter 8 on mounts, but to get the ultimate precision a technique called 'drift scan alignment' is used. The basic idea is to centre the telescope on a bright star, stop the RA drive and observe the drift across the eyepiece.

If one uses an eyepiece (which ideally has illuminated cross wires) perhaps in conjunction with a Barlow lens, a star that is close to zero declination due south is found and centred in the field of view. One then watches its position to see if the star is drifting either north or south and then adjusts the azimuth fine control screws to eliminate the drift. If the star is seen moving towards the top of the field of view, one rotates the polar axis in azimuth using the positioning screws so that the guide star moves to the right of the field of view. If the star is seen moving towards the bottom, one rotates the polar axis using the positioning screws so that the star moves towards the left. Having made an adjustment, one brings the star back to the centre of the field (using the fine slew controls) and repeats the process until no drift is seen. The faster the star is seen to drift, the greater the correction that will be needed.

The telescope must now be pointed to a star that lies exactly east or west and as low (perhaps 20 degrees of elevation) as possible. Observing whether the star drifts up or down in the field of view, one adjusts the altitude of the polar axis using the fine control screws until the drift is eliminated or very minimal. This often involves undoing one screw and doing up the opposite one. One must not adjust the azimuth screws. Although the two adjustments should be independent, it may well be worth repeating the whole exercise to get very precise alignment of the polar axis on the North Celestial Pole.

Using a CCD Imager to Observe the Telescope Drift

The process can be made a little easier if the main camera has a CCD or DSLR imager:

- Focus the camera on a star close to zero degrees declination and lying due south.
- Once the camera is focused, move the star to the west side of the image field.
- Select the lowest slew rate drive speed.
- Set the camera software to take a 105-second exposure and initiate the exposure.
- After the first 5 seconds have elapsed, press the 'West' button on the telescope keypad to cause the star to move towards the opposite side of the sensor.
- After 55 seconds reverse the telescope direction to drive east (i.e., after 50 seconds of driving).
- When exposure has finished, stop moving the telescope and observe the downloaded image.

The image should show a V shape; the more open the V the greater the misalignment in the azimuth position of the polar axis. If the bright spot marking the initial point on the track is lower than the final point of the track, the polar axis is pointing too far west, so make an adjustment to rotate the polar axis towards the east and vice versa. (Depending on the telescope type it might be the other way round, so if on the following exposure the V has opened out, make the corrections the other way round.) Repeat the process until the V has collapsed to a straight line.

The telescope is now slewed to a star at low elevation either due east or west. The procedure is carried out as before but, this time, it is the altitude of the polar axis that is adjusted.

As the accuracy in both axes improves, taking longer exposures will allow more precision, and the drift line can even go off the image sensor before the drive change brings it back on. As when the adjustment is carried out visually, the greatest accuracy will result when a second iteration of the two operations is performed.

I suspect that any astro-imagers who really need to use the drift scan technique will have an auto-guiding system using a CCD camera and may well be using PHD guiding. (PHD guiding can be downloaded for free if another program is used for guiding.) The process is very similar to that just described but produces a graphical display of the drift, allowing the appropriate corrections to be made very quickly.

Drift Alignment Using PHD Guiding

The following technique is illustrated in Figure 16.9.

- Mount the CCD camera on the guide telescope and align it so that the chip is orientated along the RA and Dec axes with the Dec axis up/down.
- To set the RA correction, choose a star close to declination zero that is near the meridian, due south, first calibrate the mount and then start guiding on it using PHD.
- Turn on the graph and select 'DX/DY' rather than 'RA/DEC' and turn off the Dec guiding.
- If your mount was perfectly aligned, the DY line (red) will be seen to track horizontally across the graph. Usually it will be seen to drift upwards or downwards. Then make fine adjustments (using the azimuth adjusters) to the azimuth position of the polar axis until the drift is eliminated. The effects of alterations will be immediately visible!
- Now slew the telescope to a star that lies near the horizon due east or west and recalibrate the guiding. Repeat the two previous steps, making sure that the Dec guiding is off, but this time adjust the elevation of the polar axis and the result of the changes will be seen immediately.

The use of any of these three techniques should result in a perfectly aligned mount but then, as already mentioned, the polar axis should be very slightly offset so that the auto-guiding declination corrections are always in the same direction to prevent backlash in the declination gears.

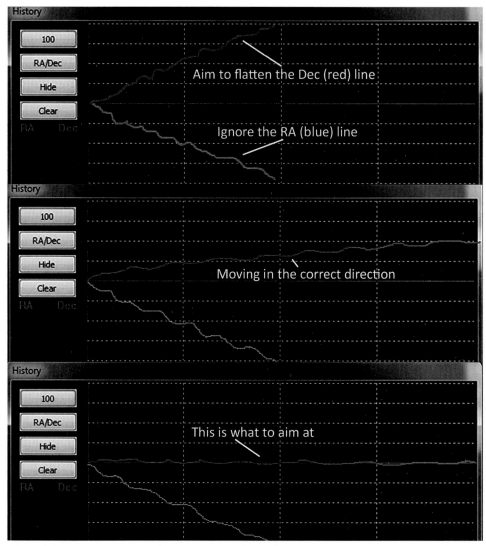

Figure 16.9 Using the graph plot of PHD guiding to carry out a drift scan polar alignment.

17
Spectral Studies

The majority of amateur astronomers have probably not even considered making any spectroscopic observations, but it is an immensely rewarding part of the hobby; the cost of the required hardware (a diffraction grating) is no more than that of a reasonable eyepiece, whilst software is readily available to process one's spectra. In the United States, Tom Field has produced an exceptionally user-friendly piece of software called 'RSpec', and his Web site includes a set of video tutorials to show one how to use it. It is possible to run a full specification trial version for 30 days. A freeware program, 'Visual Spectrum', is also available.

Almost any camera can be used to capture spectra, and the use of DSLRs, webcams and CCD cameras will be covered here. By simply screwing the grating into an eyepiece, one can even observe the spectra of some bright stars visually. I really do hope that this chapter encourages many more amateurs to have a try!

Much can be learnt about stars, galaxies and even quasars by studying the spectral details in the light that they emit. First one needs to learn a little about the three types of spectra that we can observe: continuum, emission and absorption.

17.1 Types of Spectra

It is worth understanding how these three types of spectra form and what their characteristics are.

Continuum Spectra

This refers to the overall electromagnetic radiation emitted by a body in, for the purposes of this account, the ultraviolet, visible and infrared parts of the spectrum due simply to the fact that it is warm. It has the property that both the emitted energy and the wavelength at which the electromagnetic radiation is a maximum are directly related to its temperature. Below about 700 K (430 C), bodies produce very little radiation at visible wavelengths and appear black to our eyes – though we could sense the infrared radiation emitted as this temperature is neared. Above this temperature, bodies emit enough radiation at visible wavelengths for us to see a colour which passes through red, orange, yellow and white to blue as the temperature increases. As the temperature increases further, the peak wavelength moves into the ultraviolet, but there is still considerable energy in the blue part of the spectrum so that the object appears blue. Thus, when we talk about the colour of a star – for example, referring to Rigel as appearing blue-white, Capella as yellow and Alderbaran as orange – we are, in fact, estimating its temperature.

Emission Line Spectra

If one sprinkles some salt into the flame of a Bunsen burner it turns a golden orange colour, and if this is observed through a spectroscope much of the light is seen to be concentrated at two close wavelengths in the yellow-orange part of the spectrum. These lines are called 'emission lines', as they are emitted when an element, in this case sodium, is excited. As we have seen in Chapter 14, in the section of astro-imaging, the pink-red colour in the North American Nebula, for example, arises from hydrogen which has been excited by ultraviolet light from a nearby star. The emission lines are formed when the outer electron of an atom drops from one excited energy level to another.

The Hydrogen Spectrum

Neutral hydrogen produces a distinctive set of spectral lines, called the Balmer series, that range in wavelength from 3,634 angstroms in the ultraviolet to 6,563 angstroms in the red (Figure 17.1). The most prominent is the 6,563-angstrom line, which is called the 'H-alpha line' and is seen in clouds of gas where the electrons have been lifted into excited states by incident ultraviolet radiation and then drop back down into lower energy states. To form the Balmer series of visible lines, the electrons drop down to the first excited state, level 2. The H-alpha line is caused by a transition from the second to the first excited state and thus from level 3 down to level 2. The green line, produced by the transition from level 4 down to level 2, is called the 'H-beta line', whilst the H-gamma line, in the mid-blue, is produced by the transition from level 5 down to level 2.

Absorption Line Spectra

In 1666 Isaac Newton, using a prism, observed the Sun's continuum spectrum and showed that sunlight is composed of all the colours of the spectrum. In 1804 William Wollaston observed that there appeared to be some gaps in the spectrum that looked like dark lines. Later, in 1814–17, Joseph Fraunhofer mapped many of these lines with reasonable accuracy. They have thus become known as Fraunhofer lines. They represent wavelengths where there is a lessening of the observed solar emission and are thus called 'absorption lines'. Subsequently, Gustav Kirchoff and Robert Bunsen found that the wavelengths of the absorption lines seen in the Sun corresponded

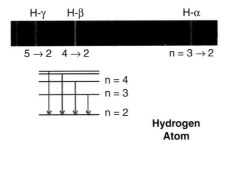

Figure 17.1 The spectral lines emitted by hydrogen.

Figure 17.2 The solar spectrum showing the Fraunhofer lines. The peak intensity is in the yellow-orange part of the spectrum close to the strong pair of sodium D lines in the centre of the spectrum.

to those of the emission lines observed when the atoms of a particular element are excited. Before long, Fraunhofer lines corresponding to all the known elements had been found in the Sun's spectrum (Figure 17.2). During a solar eclipse in 1868, Jules Janssen and, independently, Norman Lockyer detected an unknown yellow line in the Sun's spectrum. Lockyer proposed that it was due to an as yet unknown element and proposed that it be called 'helium' after *Helios*, the Greek name for the Sun.

How are these lines formed? The photosphere of a star emits a continuous spectrum. The photons then pass through the Sun's upper atmosphere, the chromosphere, where atoms can absorb photons that correspond to transitions between their energy levels. Thus the lines, called 'absorption lines', will be at just the same wavelengths as the emission lines that we can observe on Earth. The absorption spectrum that is observed is thus a mix of all these lines and depends strongly on temperature.

Stellar Spectral Types

In the latter part of the nineteenth century, the spectra of thousands of stars were photographed by astronomers at Harvard University and were used to classify the stars into their spectral types, but it was not until 1912 that the present nomenclature arose. For example, type A stars were those where the hydrogen Balmer lines were seen to be at their strongest. Stars where the hydrogen lines were weak but helium lines were seen were called type O. In all, the stars were split into seven spectral types: O, B, A,

F, G, K and M. Here they have been listed in decreasing order of temperature, O the hottest and M the coolest. Each type is split into tenths, so the hottest stars within a spectral type will be classified as, say, G0 and the coolest within that type G9. Our Sun is classified as a G2 star and is thus towards the hot end of the G-type stars.

17.2 Forming a Spectrum

There are two ways of dispersing the light to form the spectrum of a celestial object. The most obvious is to use a prism. It has the advantage of producing just one spectrum – which, of course, will contain all the light from the object – but has the disadvantage that the wavelength scale across the spectrum is not linear, so making calibration somewhat difficult. The second method is to use a diffraction grating, which can be either a reflection grating with closely spaced ruled lines etched onto a mirror surface (a CD can be used as a reflection grating) or a transmission grating with the lines etched onto a piece of plain glass through which the light is passed. Diffraction gratings have the advantage that the wavelength scale is essentially linear but the disadvantage that the light produces a number of spectra (called orders) on either side of the undiffracted image (called the zeroth order) so that the light that is to be analysed in a particular spectrum, normally that of first order, is reduced. However, there is a very neat trick called 'blazing', which, by carefully forming the etched lines on the glass, can direct much of the light into the first-order spectrum on one side of the undiffracted image. A well-blazed transmission grating may put just 15% of the light into the zeroth order (useful for calibration purposes) and 75% into one of the first-order spectra – leaving just 10% of the light that is wasted.

A relatively low cost blazed transmission grating is thus almost certainly the first type of spectroscope that an amateur should use. The simplest use is in what is called a 'slit-less spectrometer' as no slit is used. It will thus work only at its nominal resolution when star-like objects are being observed, and the resolution obtained will depend partly on the seeing conditions (which will affect the effective angular size of the star). This does not mean, however, that other objects cannot be observed and, by using a needle to give the effect of a thin slit, one can even observe the Fraunhofer lines in the Sun's spectrum – and that is certainly not a point source of light!

Diffraction gratings are specified as having a given number of lines per millimetre, and these typically range from 50 to 1,200 lines/mm. The finer the ruling, the greater the resolution in the resultant spectra, but as the spectra are more spread out more light will be required – hence the need for very large professional telescopes. For amateurs, a good starting choice would be the use of a 100 lines/mm blazed grating. The 100 lines/mm Paton Hawksley Star Analyser can be obtained both in the United Kingdom (for less than £100) and from Field Tested Systems (which produces the RSpec software described later) in the United States. Both companies can ship worldwide. Gratings with more lines per millimetre give potentially greater resolution but, in practice, due to atmospheric turbulence this may not be realised – in just the same way that going to higher magnification when one is using a telescope may not enable one to see more detail in an image. A good way to think about this is that

Figure 17.3 The spectral image of a single vertical bar of a low-energy light bulb using a blazed grating. It can be seen that the first-order spectrum on the right side (+1) is brighter than that on the other side (–1), as about 75% of the total energy is placed in the +1 spectral order.

the grating effectively produces a linear set of images of the star. The image at each wavelength is the size of the star image, so if this is large due to turbulence then the spectrum is smeared out, as the individual images will overlap. If one is using an f10 Schmidt-Cassegrain or other telescope of long focal length, it may well be best to use an f6.3 focal reducer to reduce the size of stellar images unless the seeing is very good.

So how is such a transmission grating used? The first thing to do is to simply hold up the grating towards a lamp. One will see a relatively faint undiffracted image of the lamp in the centre of the field with a very bright first-order spectrum on one side (this is due to the blazing as shown in Figure 17.3). There will be a fainter first-order spectrum on the other side with additional faint, higher-order spectra extending beyond on both sides.

A grating can be used in three ways: to observe the spectra of bright stars directly, to capture spectra of the Sun and brighter stars using a DSLR camera and lens and, perhaps the most common use, to include the grating in the light path of a telescope and camera to enable fainter objects to be observed. In all cases, as this is a slit-less spectroscope, the observed object must be star-like – but there is a trick to obtaining a solar spectrum.

17.3 Using a Grating to Make a Visual Observation

First light with a telescope could simply be visual, with the grating screwed into an eyepiece barrel (just as a 1.25-inch filter would be) and the telescope pointed at a suitable bright star. Bright orange-red stars such as 0.92-magnitude Antares or 0.5-magnitude Betelgeuse are very good, as they show easily seen broad spectrum lines, whilst stars such as 0.03-magnitude Vega or –1.96-magnitude Sirius, show dark narrow absorption lines due to the hydrogen in their chromospheres. The length of the spectra as seen when the grating is screwed directly into the front of the eyepiece is rather short. If instead, the grating is screwed into the front of a 1.25-inch star diagonal, its distance from the eyepiece is increased and thus the scale of the spectrum is greater. An intermediate extension can be used by incorporating a 1.25-inch extender between the grating and the eyepiece or by unscrewing the concave lens element of a Barlow lens and replacing it with the grating and placing this before the eyepiece.

There can, however, be a problem with some Newtonians, as there may not be sufficient back-focus when these are used.

A grating spectroscope can give a pretty good indication as to the correction of chromatic aberration in a refractor. If the telescope is focused so that the spectrum is made sharpest for green light – so this part of the spectrum is a very narrow band – then, if the correction is not good, the width of the spectrum will increase (rather like a fish tail) towards the extremes of the spectrum. If it remains narrow along its length, you have a pretty good apochromat!

17.4 Using a Grating with a DSLR Camera and Lens

This is perhaps the simplest first imaging project. The grating is mounted in front of the centre of the objective of the chosen lens. One could perhaps use a lens cap or ultraviolet filter as a basis for doing this but, failing that, Field Tested Systems sells a specially designed adapter which holds the grating and screws into a lens 55 mm in diameter. It is possible to buy step-up or step-down adapter rings if your chosen lens is not 55 mm in diameter. Two good features of this use of a grating are that the light passing through the grating has a plane wavefront (rather than a converging cone of light as when a grating is used in the light path of a telescope) and that the full width of the grating is employed.

Imaging the Sun's Spectrum to Observe the Fraunhofer Lines in the Spectrum

This seems like a very good first spectroscopy project and simply requires the Sun to shine. A telephoto lens of between 85- and 200-mm focal length will work quite well. At this point one might well ask how this can be done, as the Sun is certainly not point-like. You might know that John Dobson used an insulator on a distant telegraph pole to produce an artificial star formed from reflected sunlight. The convex surface produced a tiny image of the Sun just as would a ball bearing. In this case, rather than a ball bearing, a needle is used – its cylindrical shape works in much the same way when reflecting sunlight and produces the effect of a very thin strip of sunlight just as would a very thin slit such as that used in professional spectroscopy.

Note that, when the needle is reflecting direct sunlight, the effective width of the 'needle slit' is considerably thinner than the actual thickness of the needle. A nice colour spectrum will be observed when one is photographing a needle under overcast skies, but then it is the full width of the needle that forms the effective slit and the Fraunhofer absorption lines in the spectrum will be smeared out. The image of the needle should be vertical (by aligning the needle and camera) and the spectrum should be horizontal (by rotating the grating). This reduces the need for any later rotation of the image.

Figure 17.4 (top) shows the experimental set-up that was used. One end of the needle was pinned into a matt-black netbook cover and the other end affixed with black tape. This was set up roughly at 70 degrees (towards a right angle) to the Sun's position and angled upwards so that one could see the Sun's specular reflection in the needle at a height of about 1.6 m from a distance of ~2 m. The camera, located at this point, was a Nikon D7000 coupled with a prime 200-mm lens. The lens was left at full aperture, as the effective aperture of ~f5.6 is determined by the diameter of the grating placed in front of the lens. With an ISO of 100, exposures in the range 1/1,000 to 1/2,000 second were required. To get the highest image quality, raw data capture was used and the images were converted into 16-bit TIFFs. (However, in this application JPEGs will give virtually identical results, so it may well not be worth taking raw images.) The grating was aligned so that the blazed first-order spectrum was to the left of the first-order 'needle' image rather than the right. (The image is easily flipped horizontally in Photoshop to provide the conventional orientation for processing.) This was so that a good shadow could be arranged to fall on the area of the matt-black backing material against which the spectrum would appear in the image. As a result, the areas of the image adjacent to the spectrum were effectively pure black.

The images produced can be immediately imported into RSpec for analysis, and this is required if quantitative results are necessary. For purely visual purposes Photoshop can be used to provide an attractive image of the observed spectra.

Using Photoshop to Produce the Spectrum Image

Photoshop or the equivalent is first used to rotate the spectrum so that it is accurately horizontal – that is, any absorption lines seen within it are vertical. The image is then cropped so that it shows just the strip, including the needle and spectrum, and is thus perhaps a few hundred pixels high. In the 'Image Size' box, the tick in the 'Constrain Proportions' box is removed and the image resized to reduce the height to just 1 pixel. I know this sounds pretty odd, but it essentially makes each pixel the average of its vertical column. (This is why it is important to have the absorption lines vertical as, if they are not, they will be smeared out.) Finally, the image is resized to increase the height to, say, 150 pixels. This produces a nice clean spectrum where the lines should show up well. The captured spectrum is shown in Figure 17.4 (bottom, a), and the 'cleaned-up' spectrum is shown in Figure 17.4 (bottom, b). (The result is pretty amazing the first time you try this!) The 'Smart Sharpen' filter can be used to sharpen up the spectral features. The histogram of the data can be viewed, and you can use the levels sliders to improve the image by bringing the left slider up to the left edge of the histogram to darken the background and bringing the right slider down to brighten the image if necessary. Be careful not to blow out the brighter parts of the spectrum and so hide the Fraunhofer lines. The full strip is required to calibrate the spectrum, but to show the spectrum better just crop the spectrum section of the image.

17.5 Calibrating a Spectrum

The next step is to produce a plot of the spectrum and then calibrate it to convert from pixel number to wavelength. The procedure is very simple if RSpec is used (Figure 17.5). The image of the spectrum is imported and, if necessary, the rotate tool is used to make it horizontal. A bounding box is then placed over the spectrum, which is then automatically analysed to give a plot. If the 'slit' (zeroth-order) image is very

Figure 17.4 The set-up to image the solar spectrum (*top*), with the raw spectrum (*bottom, a*) and the cleaned spectrum (*bottom, b*) made as described in the text.

bright relative to the spectrum, this will tend to squash the height of the spectrum plot. To overcome this, you can reduce its brightness in Photoshop prior to importing it into the analysis programs: the region of the spectrum surrounding the slit is selected and the brightness control used to make it less prominent. Its position in the overall spectrum will still be visible, but the spectrum will then extend to a greater height in the plot. The 'Calibrate' box is clicked and the cursor moved to the vertical, zeroth-order star strip. Its pixel number is shown and this is put in the upper box. As this corresponds to zero dispersion, '0' is put into the wavelength plot. In the far-red part of the spectrum you should see an absorption line due to H-alpha. You can expand the appropriate area by using the mouse's roller wheel, or, using the control key and the left mouse button, you can box in an area and so accurately find the pixel number of this absorption band. This and the wavelength, 6,563 angstroms, are entered into the second box and the spectrum calibrated. The x-axis markers are now in wavelengths rather than pixel numbers. Simply by moving the mouse to each obvious absorption line, you can read off its wavelength. Just into the orange part of the spectrum are seen two very closely spaced lines, the sodium D lines at wavelengths of 5,890 and 5,896 angstroms. The plot produced by RSpec showed that the two lines *were* separated, with their wavelengths given as 5,889 and 5,895 angstroms – each just 1 angstrom below the true value. I think that this was pretty impressive for such a very simple set-up!

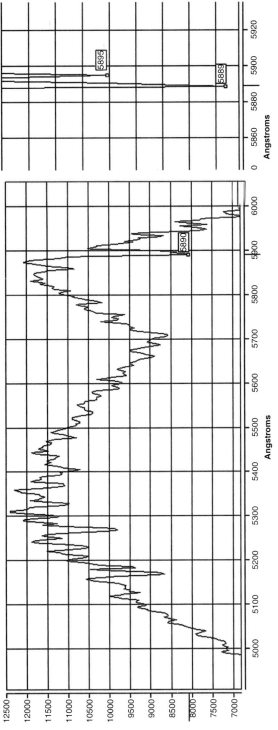

Figure 17.5 The wavelength-calibrated spectral plot produced by RSpec of the solar spectrum (*left*) described earlier in the text with (*right*) an expansion of the region of the two sodium D lines.

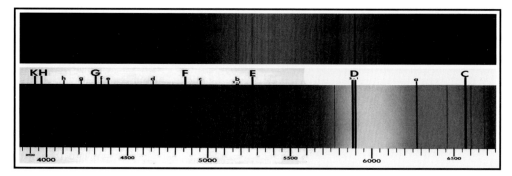

Figure 17.6 The solar spectrum taken using a needle to give the effect of a very narrow slit, along with a comparison spectrum.

The Spectral Resolution, or 'Resolvance', of a Grating

I have to say that I was pretty amazed that this simple set-up was able to split the sodium D lines, just 5 angstroms apart, and tried to understand why. There are three limitations to the possible resolving power. The first is that the spectrum is sufficiently well sampled by the sensor elements: if a single pixel corresponds to, say, 10 angstroms in spectral width, there is no way the resolution can be better than this. In fact, from the Nyquist sampling theorem there should be at least 2 pixels covering the spectral width that is determined by the two other factors that determine the spectral resolution. Using the Nikon D7000 with a 200-mm lens gives 2 angstroms per pixel, so limiting the effective resolution to 4 angstroms, which, as will be shown, gives an excellent match to the fundamental resolving power of the grating.

The second limitation is determined by the effective size of the grating. This is not the actual size, but the width of the grating that the light from the object is passing through. The smaller this is, the more the spectrum detail is limited, just as the fundamental resolution of a telescope is limited by its aperture. It turns out that the fundamental resolving power, called 'resolvance', which is short for 'chromatic resolving power', is determined by the number of lines of the grating that the light passes through. In the case of the solar spectrum the sunlight is passing through the full width of the grating, ~12 mm in diameter, so covering 1,200 grating lines, as the grating has 100 lines/mm. The resolvance is then simply the wavelength within the spectrum divided by this number, so for the sodium D lines at 5,893 angstroms, this is 5,830/1,200, or 4.85 angstroms. Thus the D lines could just be split, provided that the final limitation does not limit the spectral resolution.

The third limitation to the spectral resolution is given by the width of the spectrograph 'slit'. In this case the apparent width of the needle might have limited the possible resolution, but examining the image in detail showed that this was not a limiting factor and that the resolving power was determined by the grating size. Theoretically, then, the sodium D lines could be resolved in the solar spectrum shown in Figure 17.6.

I hope this shows that the set-up used to analyse the solar spectrum was essentially perfect and could be improved only by the use of a larger grating, assuming that the effective slit width does not become the limiting factor.

The way to capture and process the spectra of relatively bright stars is identical to that used to capture and analyse the solar spectrum.

17.6 Using a DSLR and Lens to Capture Stellar Spectra

The same camera/lens set-up that was used to obtain the solar spectrum can be used for observing some of the brighter stars when exposures of perhaps 30 seconds are needed. One can simply mount the camera on a fixed tripod and adjust its orientation so that any drift during the exposure simply spreads out the spectrum, with the star's image moving at right angles to the spectrum, so giving a straight line rather than a dot for the zeroth-order 'star' image. This is then processed exactly like the solar spectrum described earlier. Alternatively, the camera can be used with an AstroTrac or mounted on a scope with a driven mount.

For one's first attempt at capturing a stellar spectra, a type A star (such as Vega) is ideal, as it has a readily identifiable H-beta line that makes it easy to carry out the initial calibration of the instrumental set-up and derive the dispersion constant in angstroms per pixel. Once this is known, it can then be applied to any spectrum where the zeroth order is shown along with the dispersed first-order spectrum of the object being observed.

Imaging Vega's Spectrum with a DSLR and 200-mm Lens

The spectrum of the star Vega was captured in this way with both my Nikon D7000 and modified (to increase the sensitivity at the H-alpha wavelength) Canon EOS 1100D camera. Vega was centred in the telescope field of view and test exposures made which, if the camera alignment was perfect, would give an image of Vega in the centre of the frame with parts of the first-order spectrum on either side. The pointing was then off-set from Vega, so that the full length of the brighter, blazed spectrum was covered by the frame and exposures taken with a variety of ISO values and exposure times.

The resultant spectra were processed in just the same way as for the solar spectrum, with the plots derived from both cameras produced using RSpec, as shown in Figure 17.7. These observations showed how much more sensitive the modified Canon camera was at the wavelength of H-alpha. The H-alpha absorption line is not visible in the Nikon plot but is clearly visible in the Canon plot, with H-alpha, -beta and -gamma absorption lines all well seen.

17.7 Imaging Stellar Spectra with a Telescope and a DSLR, CCD Camera or Webcam

This will allow the spectra of fainter stars to be obtained and is achieved by mounting the grating in front of the camera sensor at the rear of the telescope. Some ingenuity

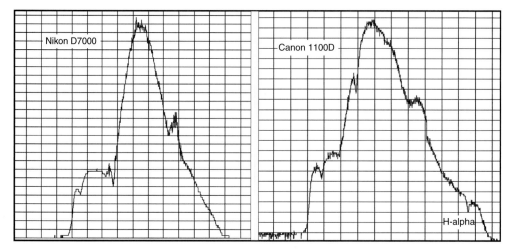

Figure 17.7 Vega spectrum taken using a 200-mm lens with both a standard Nikon D7000 and modified Canon EOS 1100D. The difference in sensitivity at the wavelength of H-alpha is very obvious.

may be needed to do this, as the grating is a 1.25-inch screw fit, and the mount used for coupling the DSLR or CCD camera will normally be a 2-inch fitment. Field Tested Systems sells an AD-T2 adapter to mount a 1.25-inch grating into a 42-mm DSLR T-mount adapter. Both monochrome and colour cooled CCD cameras can be used, allowing longer exposures to be made and so making it possible to observe the spectra of fainter objects. A monochrome CCD camera is ideal as, having no Bayer filter matrix, it will be more sensitive and have twice the effective resolution of a colour camera since each pixel is sensitive to all colours. For use with a CCD camera the grating can be placed either between the 2-inch barrel and CCD camera or, to gain a greater dispersion, at some point within the 2-inch barrel. In this case, the grating has to be mounted within a tube that can slide within the barrel.

Transmission gratings will also work well with webcams, and the grating can then be screwed directly into the front of the 1.25-inch barrel. With my Imaging Source webcam, the supplied 1.25-inch barrel was, at 37-mm length, a touch too long to fit the spectrum horizontally across the sensor but, as will be discussed, the spectrum would fit diagonally across the chip.

The camera should first be focused on a bright star without the grating in place. Once the grating is incorporated, the star should still be visible, but fainter (as only ~10% of its light will be in this zeroth-order image) and should be brought back into focus. On one side of the star image a bright spectrum will be visible. Rotate the camera to bring this to the right side of the star image and offset the pointing so that the star image is to the left of the field; its spectrum is thus spread out on its right, with increasing wavelength (from blue to red) running from left to right. The exposure may well have to be increased as the star's light is spread out. Webcams will make an AVI file which can either be directly imported into RSpec (which can average up to 100 images) or converted into a single image using Registax.

Nominally one would arrange to have the spectrum along the major axis of the CCD chip, but if it is too long to fit, one can rotate the grating so that it lies along the diagonal. (Image processing programs or RSpec can easily rotate it to the horizontal position.) The scale of the spectrum will be reduced if the separation between the grating and sensor is reduced or, though it makes the calibration more difficult, the zeroth-order star image can be moved out of the field. The length of the spectrum can also be increased by the addition of spacers to increase the grating-to-chip distance. There is bound to be a slight shift in focus due to the increasing angle of the light beam from the blue to the red end of the spectrum, so a compromise focus position may be required.

The great advantage over the telescope/camera set-up is that far more light is collected. However, the effective size of the grating will be smaller and this will reduce the spectral resolution. The light falling onto the camera is a narrowing cone that reaches near zero size at the sensor surface. Let's assume that a 200-mm-aperture, f4 scope is used. When used with a webcam or small CCD sensor, the grating could well be just 37 mm in front of the sensor, so the cone of light passing through the grating is $200 \times 37/800$, or 9.25, mm in diameter. As already described, the effective resolution will thus be reduced somewhat. If, however, a 200-mm, f10 Schmidt-Cassegrain is used, the effective size of the grating is only 3.3 mm, so the fundamental resolution is significantly reduced. This why using an f6.3 focal reducer will help to sharpen the spectral features, as the effective diameter of the grating increases to 5.2 mm. The situation is obviously better when a CCD or DSLR with a larger sensor is used, as the grating can be placed further from the sensor. In this case, as the light is spread out over a larger area of the sensor, longer exposures will be needed and this is where a large-sensor cooled CCD camera would be ideal.

A Grism

To obtain higher spectral resolution, one might well use a higher dispersion grating, but one problem is that the zeroth-order spectrum (the undiffracted star image) will lie in the centre of the imaged field with the blazed first-order spectrum at some distance away and possibly outside the imaging sensor's area. This may be overcome by the placement of a thin wedge prism adjacent to the grating to deflect the light so that the blazed first-order spectrum lies in the centre of the field. The combination of grating and prism is, not surprisingly, called a 'grism'.

17.8 Calculating the Dispersion Constant Produced by a Specific Instrumental Set-Up

The dispersion in angstroms per pixel is given by

10,000 * (pixel size in microns) / (grating lines/mm) * (grating-to-CCD distance in millimetres)

As an example, suppose I was using my Imaging Source webcam, which has a pixel spacing of 5.6 microns and a distance between the grating and the CCD sensor of 33 mm. The dispersion is then given by

$$(10{,}000 \times 5.6) \,/\, (100 \times 33) = {\sim}17 \text{ angstroms /pixel}$$

So measuring the number of pixels from the zeroth-order star image and multiplying by 17 will give the wavelength.

A more accurate way is to measure, in pixels, the distance from the star zeroth-order image to a known spectral feature – it might even be an absorption line due to oxygen in our atmosphere – and divide the wavelength in angstroms by this distance to calculate the dispersion constant. Of course, the software used to analyse the spectrum does all this maths for you.

17.9 Calibrating the Instrument Response Curve and Applying It to the Spectra

The spectra of the star Vega shown earlier do not relate very well to the actual spectrum of the star, as it has been modified by a number of factors. Two of these are that the sensitivity of the CCD or DSLR camera will fall off towards the blue and red ends of the spectrum and, if a blazed grating has been used, it will be a little more efficient in the yellow-green part of the spectrum and less so towards the blue and red ends of the spectrum. The uncalibrated spectrum will thus peak in the yellow-green part of the spectrum. If an unmodified DSLR were used, the sensitivity towards the red end of the spectrum would be significantly reduced, as illustrated by the two Vega spectra of Figure 17.7. However, it should be pointed out that what makes a spectrum interesting are the emission and absorption features that are seen within it rather than its overall shape, so this is not something to get too carried away with.

To derive the true stellar spectrum one thus needs to calibrate out these instrumental effects. This is a fairly complicated procedure. If you are using RSpec, there is an excellent video tutorial on the RSpec Web site that can be played in one window and paused as each step of the process is carried out using the program in another. (The beautifully presented set of video tutorials provided by RSpec makes it very easy to learn how to use the software package.) Visual Spectrum provides details in its extensive 'Help' PDF document. Both software packages provide a database of calibrated spectra covering all spectral types that are used to make the calibration. Vega is a very good star with which to make a first calibrated spectrum and is the example used in the RSpec video. The calibrated spectrum of Vega is shown in Figure 17.8 and illustrates how there is significantly more radiation towards the blue end of the spectrum than seen in the uncalibrated spectrum.

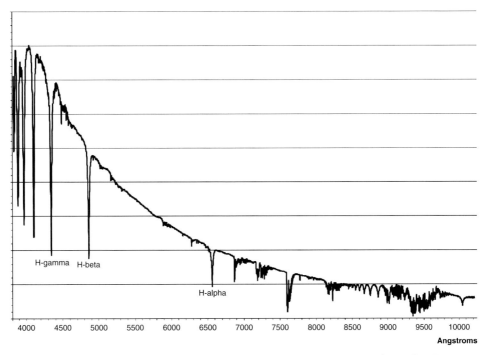

Figure 17.8 The calibrated spectrum of the star Vega. (Image: Wikimedia Commons)

17.10 Higher-Resolution Spectrographs

If you get captivated by what can be achieved with a simple slit-less grating spectroscope, you may like to take spectroscopy to the next step by building or purchasing a spectroscope based on a grism or reflection grating. Designs that enable one to build such a spectroscope can be found on the Internet, and they can be bought from a number of companies.

The Elliott Instruments CCDSPEC Medium-Resolution Spectrometer

In the UK, Elliott Instruments produces a spectrometer costing ~£1,300 (~$2,000) that uses a grism to provide a resolution of ~15 angstroms. It is built into a single block of aluminium to make a very stable system. Very nicely, it includes a built-in guiding port: the slit is electroformed in a metal plate which reflects the light not passing through it towards the guiding port, where it can be viewed through an eyepiece. The field size is 5 mm across, which allows the object to be imaged within the field of most guiding cameras. The IRIS acquisition software includes an algorithm to allow guiding on the remnant light of the object – this will be the outer parts of the object's image as, when properly aligned, the central light from it will pass through the slit and so will not be reflected. It can be used with both DSLRs and cooled CCD cameras. Figure 17.9 shows the optical design along with a picture of the CCDSPEC mounted on

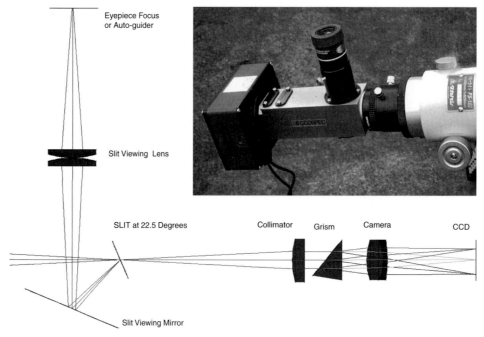

Figure 17.9 The optical configuration of the grism-based CCDSPEC spectrometer with (*inset*) the spectrometer mounted on an FS102 refractor and coupled to an SBIG ST-8300 cooled CCD camera. (Images: Elliot Instruments and Ian Morison)

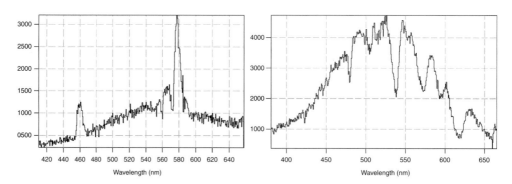

Figure 17.10 Two spectra taken with the CCDSPEC spectrometer allied to an 8-inch Schmidt-Cassegrain with (*left*) the spectra of a Wolf-Rayet star showing wide emission lines, indicating a massive stellar wind, and (*right*), the spectrum of Uranus, which shows deep absorption bands due to methane in its atmosphere. (Images: Elliot Instruments)

my FS102 refractor with an SBIG ST-8300 cooled CCD camera to image the spectrum. The spectroscope is provided with an analysis program called 'PC Spectra'.

A medium-resolution spectrometer such as the CCDSPEC can make many interesting observations, such as those shown in Figure 17.10.

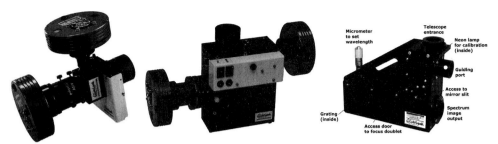

Figure 17.11 The three Shelyak spectrographs: (*left to right*) ALPY 600 with guiding and calibration modules, LISA with calibration module attached and Lhires III. (Images: Shelyak Instruments)

The Shelyak Instruments ALPY 600, LISA and Lhires III Spectrographs

The French company Shelyak Instruments produces a number of spectrographs, including a new modular medium-resolution spectrograph, ALPY 600. This uses a 600 lines/mm transmission grating and prism (grism) to give a resolution of ~10 angstroms and can be equipped with optional guiding and calibration modules. The LISA low-resolution spectrograph is specially designed for faint and extended objects, whilst the Lhires III spectrograph has been specifically designed for use with telescopes such as 8-inch Schmidt-Cassegrains (Figure 17.11). This uses a 2,400 lines/mm reflection grating as standard but with a number of alternative gratings as options and has a typical resolution of 0.12 angstrom. The slit width is adjustable from 15 to 35 microns and the grating is rotated to bring different parts of the high-resolution spectrum onto the imaging sensor.

17.11 A Spectroscopic Challenge: The Red Shift of 3C 273

In the early 1960s radio astronomers at the Jodrell Bank Observatory discovered a number of radio sources which had angular sizes of less than 1 arc second. They would thus appear as stars on a photographic plate – hence the name 'quasar', short for quasi-stellar object, meaning an object that looks like a star. An object known as 3C 273 was the first of these to have its precise position measured, and this enabled Maarten Schmidt to take its spectrum using the Palomar Observatory's 200-inch (5.1-m) Hale Telescope (Figure 17.12). Somewhat to his amazement, he discovered that the hydrogen emission lines in its spectrum were red-shifted by ~16%, implying that it lay at a distance of 2.4 billion light years.

The fact that even at such a great distance it has an apparent magnitude of 12.9 means that it must be producing vast amounts of energy, being more than 4 trillion times brighter than our Sun. It is powered by matter falling in towards a super-massive black hole at the heart of a giant galaxy, a process which is far more efficient than the nuclear fusion process that powers our Sun.

Amateur astronomers who have a 100 lines/mm transmission grating and a telescope with an aperture of 200 mm or greater can re-create his observation with less

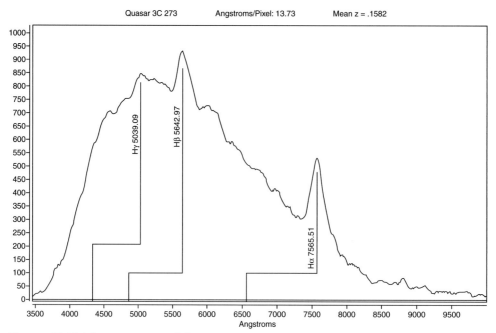

Figure 17.12 The spectrum of the quasar 3C 273 taken by William Wiethoff using a Celestron 14-inch Schmidt-Cassegrain and a 100 lines/mm transmission grating. The spectrum was captured with a Meade DSI Pro II CCD with a total exposure time of 300 seconds and processed in RSpec.

than 15 minutes' exposure, and a spectrum has even been obtained using a TEC 140-mm refractor with a total of 20 minutes' exposure. It is thus not that difficult to undertake one of the most interesting spectroscopic challenges open to amateurs.

18

Improving and Enhancing Images in Photoshop

In the imaging examples given throughout the book, some of the processes that can be used to improve the quality of the resulting images have been described. In this chapter, these will be summarised along with other techniques that may be well be useful in producing the best results from your captured images.

Use Raw Whenever Possible but Preferably Raw Plus JPEG

RAW data is that which has been captured by the camera sensor without any processing and will be digitised to 12 or 14 bits per channel. In contrast, a JPEG will be digitised to only 8 bits – just 256 levels as opposed to 4,096 or even 16,384 – per channel. In addition, the raw conversion software used to provide the image in your computer may well be more sophisticated than that used in your camera. To keep all the inherent quality in the image, the raw files should be converted into 16-bit TIFF files. Not only will a JPEG have less depth, the compression will cause some artefacts within the image. One result of using raw and TIFF files is that they will be fairly large both when one is capturing the image and when processing them later. I use an external USB hard drive to store the images when imaging with my laptop. Very compact drives that are powered from the USB connection are available with capacities of 500 gigabytes or more.

Why save JPEGs as well? Suppose one has taken 20 or more images to be stacked; it is very useful to have a quick look at each one to see if any have misshapen stars or if, perhaps, a satellite has passed through the image. In Windows, clicking on the first of a sequence of JPEGs will open a viewing window and allow one to view each image and note any that need to be rejected before the (preferably raw or TIFF) files are loaded into a program such as Deep Sky Stacker. It might well be worth stacking the JPEGs as well to view the result. It is possible that you might find, as I once did, that the JPEG-derived image actually looks better. Though the raw files will have a greater bit depth (12 or 14 bits rather than 8) the addition of a number of 8-bit frames increases the effective bit depth in the image so, in this case, the use of TIFF files may not produce an image with greater dynamic range and it is possible that the raw-to-JPEG conversion within the camera may produce a more pleasing result. It costs nothing to have a try!

Removing the Sky Glow from an Image

When one is observing from an urban location, the images will usually have a reddish colouration due to the sky glow, and this may vary across the image – usually being brighter at lower elevations. This is very easy to remove if the imaged object is not too large in the frame (e.g., M31, the Andromeda Galaxy). Make a duplicate copy of the image: Layer > Duplicate Layer. If there is a bright region within the image, such as a globular cluster, the region around the Trapezium in the Orion Nebula, or a small galaxy, this should be cloned over using parts of the image that were at the same elevation in the sky. There will thus be just a star field left. The stars can be removed by using a noise filter (Filter > Noise > Dust and Scratches) with the radius set to, say, 20 pixels, so that the stars disappear. One can then smooth the layer using Filter > Blur > Gaussian Blur, again with ~20 pixels radius. The result is effectively an image of the sky glow. There are then two ways of removing it: the simplest is to simply set the blending mode in the 'Layers' box to 'Difference', whilst a slightly more complex method (but one that may well give a nicer result) is to invert the sky glow layer (Image > Adjustments > Invert) and then set the blending mode to 'Colour Burn'. In either case use Layer > Flatten Image to produce an image with the sky glow removed.

In the case where the object of interest fills much of the frame (e.g., the Andromeda Galaxy), one could take an image of an adjacent part of the sky at the same elevation. It can have a relatively short exposure, as the two filters, 'Dust and Scratches' and 'Gaussian Blur', used as just described, will smooth the image. The same technique will, of course, remove the sky light from an image of the Moon taken during twilight. In this case one can take an image of the sky to the right or left of the Moon. In either case the sky glow/twilight image is selected (Select > All or Ctrl A, Edit > Copy or Ctrl C) and, by means of Edit > Paste or Ctrl V, pasted over the lunar image. Then, as before, the difference mode is used to remove the sky glow.

Stretching the Image to Make the Fainter Details Visible

If you have followed my advice when imaging astronomical objects that have a very wide dynamic range and exposed the image so that the brightest parts of the image are not over-exposed (it does not matter so much if some very bright stars in the image are over-exposed), the result of combining the sub-exposures in, say, Deep Sky Stacker will be that only the very brightest parts of the image are shown. The stacked image must then be 'stretched' so that the brightest regions remain at their original brightness, but with the fainter regions 'lifted' so that they become visible. This advice is particularly relevant to the central region of the Andromeda Galaxy, the region around the Trapezium stars in the Orion Nebula, and the cores of globular clusters which are far brighter than their surrounding regions.

Two techniques can be applied. The first, and simplest, is to use the levels control: Select Image > Adjustments > Levels. This brings up the 'Levels' box displaying a histogram of the data which will show that virtually all the image is of very low brightness. Leaving the right-hand marker fixed so that the brightness of the brightest part of the image is not increased, one moves the middle marker to the left. The fainter parts of the image will begin to appear. It is best not to do this all at one go, combining several lesser applications of the levels tool. Often, when the best image has been obtained, one can move the left-hand curser a little to the right to set the black level and so reduce the background noise level.

The more sophisticated approach is to use the curves control: Select Image > Adjustments > Curves. This brings up the 'Curves" box. To the right of the words 'Show Amount of' is seen 'Light (0–255)', which is selected if not the default mode. A histogram of the image data (well to the left of the plot) overlain with a straight line from bottom left to top right then appears. The cursor is clicked on the line about one-third of the way from the left and used to 'pull' the curve upwards, as shown in Figure 18.1 (left). Clicking on 'OK' applies the adjustment. Fainter detail in the image should now be apparent. Entering the 'Curves' dialogue box again one, can see that the left-hand end of the histogram has been stretched to the right. Again, it is best to apply a small amount of stretching several times rather than bring up the image in one go. It can be worth making the curves sequence into an 'Action' so, once the first correction has been made, following ones require only one stroke of a function key. The 'Levels' dialogue box may then be used to move the left-hand curser a little to the right to reduce the background noise level, as previously described.

By setting more points on the curve one can tailor the result. One common 'curve' that is used is shown in Figure 18.1(right). This retains the straight line over the right-hand quarter of the curve, so ensuring that the highlights will not be blown out. It helps to experiment a bit, as different images may respond better to specific curve shapes. It would be a good idea to build up some actions with differing curve shapes so that you can apply them easily to find which suits an image best.

Correcting Star Images That Have Been Slightly Trailed

This is quite a common problem! What should be a circular star looks like a very small sausage. If the length of the trail is not too long, there is a simple way of correcting the problem. A duplicate layer (Layer > Duplicate Layer) is made. In the 'Layers' box, the combining mode is set to 'Darken'. One will initially see no difference. The 'magnifying glass' is then used to expand the image scale so that the star images in a small part of the frame are shown at ~500%. The move tool is selected and the effect of pressing the four arrow cursor keys observed. Each press will move the top layer 1 pixel relative to the bottom. As the darken mode is selected, the result will be that the darkest pixels of the two layers will be selected and you will see that the star images are reduced in size. Dependent on the direction of the star trails, an appropriate use of the arrow keys should make the star images round.

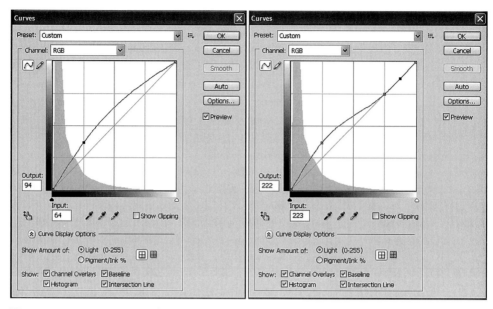

Figure 18.1 Two 'curves' that can be used for stretching an image.

There is a second method if the trailing is very slight, which is to use Filter > Sharpen > Smart Sharpen. The 'motion blur' mode rather than the default Gaussian mode is selected and the correction axis rotated using the mouse until the stars become rounded.

Three Colour Correction Tools

The renowned astro-imager Rogelio Bernal Andreo has produced three Photoshop plug-in filters which may be downloaded for free from www.deepskycolors. com/tools.html. These can be used to remove colour casts from an astronomical image.

DSLR images tend to have a green cast, and the technique for removing sky glow previously described may also have an effect on star colours, making some distinctly green. This can be corrected by means of the 'Hasta la Vista, Green!' filter, which is usually used in the strong mode. It works best when the image is in RGB colour mode.

'WhiteCal' is a colour calibration tool that allows one to select a small area of an image which one would expect to be white, such as a G2V star or a galaxy core. The tool will then do an accurate RGB colour balance across the whole image, assuming that the average colour of the area is white or neutral.

'RGBC' is a colour balancing plug-in that allows one to easily and very precisely modify the weights of the red, green and blue channels in an RGB image. Rather than adjusting the curves or levels of the individual colours, one can use RGBC to enter a coefficient for each channel to adjust their relative weightings.

Enhancing Star Colours

The central parts of bright star images are often over-exposed, so that their colour is lost. However, towards the edges of their images, which are not as bright, some colour is often seen. The aim of the following technique is to transfer this colour to the whole star image. The steps are as follows. First the stars are selected within the image and made into a new layer; then their saturation is increased and a Gaussian blur is applied to 'spread' the colour over a larger area. Then the colour blending mode is selected and the two layers flattened. Let's take this step by step.

There are two approaches to selecting the stars. In the first, one selects everything that is not a star and then inverts the selection. The 'magic wand' tool is used with a tolerance of about 10 and only the 'Anti-Alias' box ticked. The wand is clicked on a part of black sky. All the stars are then seen to be ringed. To select them and not the sky, one inverts the selection using Select, Inverse. It may be that one wants only to select the brighter stars and, if so, a second method can be used. In this case, the 'Select' pull-down menu is entered and 'Colour Range' chosen. The mouse then takes the form of the colour sampler tool (which looks just like the eyedropper tool), which is moved over the centre of one of the brightest stars. By adjusting the fuzziness control one can determine the range of stars that are selected. The selection is made by clicking 'OK', and the selected stars will be seen to be ringed by 'marching ants'. The value will need to be found for each image by trial and error – one tries a fussiness value and sees how many stars are selected when one presses 'OK' and then steps backwards to try again as necessary. There is a preview window, but I have not found this to be too helpful. When the brighter stars have been selected, it is necessary to expand the selection so that the outer parts of the star – where the colour is found – are included. Again, the 'Select' menu is selected and 'Modify' is clicked on, followed by 'Expand'; perhaps 4 pixels are chosen (this will depend on the scale of the image) so that outer parts of the stars are included.

If there is a bright part of a nebula that has been selected, this must be removed from the selection. At the bottom of the tools vertical bar below the 'Black/White Colour' boxes is a rectangle containing a circle. A right click on this will bring up 'Edit in Standard Mode'. If this is clicked on, all *but* the stars along with the bright parts of the nebula appear red. Then, with the paint brush the nebula region can be painted over with red paint. Left clicking on the rectangular box will then return one to the 'ringed' stars but without the nebula selected.

A new layer, containing just the stars, is then created above the original layer by the use of Ctrl A and then Ctrl C. It will be called 'Layer 1'. The layer rectangle appearing as a box with a checkerboard pattern is clicked to activate this layer. The saturation (Image > Adjustments > Hue/Saturation) of this layer is then increased by moving the saturation slider to the right to about +95%. This produces a very pretty effect! The star images are then blurred with a radius of ~5 pixels by means of Filter > Blur > Gaussian Blur so that the brighter stars appear to have colour and will also appear somewhat larger. The blurring reduces the stars' brightness somewhat, so one brighten them if desired by using the levels control and bringing the middle slider a little to the left.

Finally the two layers are merged with a blending mode. A mode often used is 'colour', but I have found that the brightest stars do not really show their colour as well as I would like, and it might be worthwhile trying the saturation mode with an opacity of ~50%. The image is then flattened with Layer > Flatten Image.

Improving 'Bloated' Stars

An image of a faint galaxy – hence requiring long exposures – may well have a number of over-exposed and hence overly large star images within the field that will detract from the image. The following techniques will help to reduce their impact.

First the brightest stars are selected by means of the 'select colour range' process described in the preceding section. Then the 'Minimum' filter is selected by means of Filter > Other > Minimum. The radius in the dialogue box should be adjusted as one observes the effect – it may well be left at 1 pixel. In the edit menu, 'Fade Minimum' (Edit > Fade Minimum) is selected and the effect reduced to somewhere between 50 and 70%.

A second and better way to reduce their size in later versions of Photoshop is to use the 'Liquefy filter' (Filter > Liquefy). This brings up a new window. Moving the mouse over the image shows a cross surrounded by a circle. The image is expanded using the magnifying glass in the left tools column and the size of the circle adjusted until it just surrounds the star image. The bloat tool (an ellipse with four arrows) is then selected from the tools column and the cross moved over the centre of the star. With the Alt key pressed, left clicks of the mouse will reduce the size of the star. Clicking on the 'Show Mesh' option in the 'View Options' box will show how the region including the star has been distorted. The liquefy tool can then be moved over all the stars that one would like reduced in size and the same procedure carried out.

A third method is to select the star with the elliptical marquee tool with the feather set to 5–10 pixels and use the 'Spherize' filter (Filter > Distort > Spherize). Moving the slider to a negative value will reduce the size of the star.

Restoring Damaged Stars

Sometimes, perhaps due to a slight tracking error, the brighter stars will have a small extension to one side. Any attempt to clone this out directly from the surrounding area will rarely leave the star profile looking good. One way to clone out the extension and give the right profile is to save a copy (with another name) of the image and bring back the original into the work area. The copy is then rotated by 180 degrees (Image > Image Rotation > 180). Once the two images showing the stars are increased up to ~200%, the correct side of the rotated star is cloned over the bad side of the misshapen star.

If a star has an irregular shape it can be cleaned up with the 'Radial Blur' filter set to spin. The elliptical marquee tool is used to select the star with the 'Feather' box set to ~4 pixels. The filter is entered (Filter > Blur > Radial Blur) and in the dialogue box the blur method is set to spin and the quality set to best. The amount is set to around

40–60% followed by a click on 'OK'. If there is still any obvious pattern, one can use the 'Gaussian Blur' filter (Filter > Blur > Gaussian Blur) to smooth the star's image.

Enhancing a Digital Sky Image to Appear More Film-Like

As discussed in Chapter 15, digital constellation images do not tend to look as good as images taken by film, and the following techniques may be used to improve the result by making the brighter stars larger and so more prominent.

Two copies of the original images, called 'mid-stars' and 'bright stars', are made. The mid-stars image is selected and the colour range selection tool selected (Select > Colour Range) with the picker over the centre of a star. The 'fuzziness' is set to ~80 followed by a click on 'OK'. Marching ants will be seen around the brighter stars. The fuzziness can be adjusted to control the number of stars selected. For the mid-stars image this has to encompass a fair number of stars. One first removes all the other stars by inverting the selection (Select > Inverse), selecting levels (Image > Adjustments > Levels) and bringing the black slider up to the extreme right. The selection is inverted again and the circle surrounding the stars expanded by ~4 pixels (Select > Modify > Expand). A Gaussian blur (Filter > Blur > Gaussian Blur) of ~4 pixels is now applied and the selected stars will appear larger but fainter. Entering the 'Levels' adjustment again, one moves the centre slider to the left to bring up their brightness and then increases their saturation (Image > Adjustments > Hue/Saturation) to some extent.

The same process is applied to the bright-stars image but with a lower 'fuzziness' level so that only the very brightest stars are selected. To make them even larger in size, the selection is enlarged to perhaps 8 pixels. The values that should be used will depend on the resolution of the DSLR sensor with the amount of Gaussian blur applied. One then brightens the selected stars using levels (Image > Adjustments > Levels) while applying judgement and some experimentation. The final process is to overlay the mid- and bright-star images over the original. The original and mid-stars images are opened and the mid-stars image selected and copied (Ctrl A, then Ctrl C). The original image is opened and the mid-stars image pasted over it (Ctrl V). The Screen blending mode in Layers is selected (Window > Layers) and the opacity adjusted to suit before flattening the image (Layer > Flatten Image). All the brighter stars will now appear brighter and more colourful. Apply the same process to the resultant image and the bright-stars image to increase their size even more.

An attractive result can be obtained by adding a soft halo around the very brightest stars. Once can achieve this by using a soft-focus filter or stretched black stocking in front of the lens when one is taking the image. A similar result can be obtained from an enhanced digital image such as that in Figure 14.1 by selecting the regions around the stars in the image to which one wishes to apply the effect, adding a large-radius Gaussian blur (perhaps 30 pixels) to these stars, brightening the blurred images to some extent and then overlaying them over the original image. The idea is to leave these stars with a bright central core surrounded by a fainter halo. Figure 18.2 shows the result of adding halos to the brightest stars of the plough image shown in

Figure 18.2 The result of adding soft halos to the brightest stars of Figure 14.1.

Figure 14.1 and Plate 14.1. With this technique, one can achieve results that go some-
way towards the wonderful (film-based) images made by Akira Fujii!

Overall Image Contrast Enhancement

There are a number of ways in which the contrast can be enhanced. The simplest, but not
necessarily the best, is to open up the image in Photoshop, then open the 'Brightness/
Contrast' box (Image > Adjustments > Brightness/Contrast) and use the sliders to
improve the image. A second method uses the 'Curves' box as described in the section
on 'stretching' the image. The 'curve' will initially be a straight line from bottom left to
top right. To increase the contrast one turns the straight line into an 'S' curve, pulling
the lower half down and lifting the upper part up as one observes the result.

Local Contrast Enhancement

A technique that can 'lift' an image is called 'local contrast enhancement'. Perhaps
surprisingly, this is implemented using the 'Unsharp Mask' dialogue box (Filter >
Sharpen > Unsharp Mask), which provides one (probably not the best) way to sharpen
an image. In this application, the radius is set to the maximum of 250 and the amount
to a value between, say, 10 and 20 – the image being observed with the preview
box checked to see the resulting effect. (This works very well on lunar images, provided that the initial image is a little under-exposed, as the technique will brighten
the brightest parts of the image.)

It is also possible to achieve this effect by opening the image in Adobe Camera Raw by first bringing the image into the browser window of Adobe Bridge, right clicking on it and then opening it in Camera Raw, the fourth entry in the list of opening possibilities. This brings up the image in the 'Camera Raw' window. As when processing a raw image there are a set of sliders allowing one to enhance the image by adjusting the exposure and contrast in a similar way to Photoshop. Camera Raw does, however, provide a slider called 'Clarity'. This controls two separate contrast enhancing techniques. The first is that just described using the 'Unsharp Mask' process which was originally devised by Thomas Knoll. The second is a rather complex mid-tone contrast enhancement technique that was originally devised by Mac Holbert. Both have been combined in one simple control slider. As well as having a very positive effect on the perceived contrast of the image it also helps to bring out detail in the image.

Enhancing the Overall Colour in an Image

Perhaps the most obvious method is to simply increase the saturation (Image > Adjustments > Hue/Saturation). Another technique is to use the 'Shadows/Highlights' dialogue box (Image > Adjustments > Shadows/Highlights). This, by default, lifts up the darker parts with the Shadows > Amount slider (at the top) at 50%; this is not wanted but, nevertheless, it should be left there as, if the slider is returned to zero, no colour adjustment can be made. It may be necessary to tick the 'Show More Options' box to reveal the colour corrections slider. This can then be moved to the right to perhaps +50 to nicely enhance the colour. Once 'OK' is clicked, the levels dialogue box is entered with Image > Adjustments > Levels and the middle and perhaps the left pointer (to set the black point) adjusted to bring the overall brightness back to what is desired.

A technique which works very well is to convert the image into 'Lab Color': Image > Mode > Lab Color (US spelling). In RGB mode, when one is adjusting the saturation, the brightness of each pixel has to adjust as well and this may introduce some chrominance noise into the image. However, in 'Lab Color', the luminance contrast and colour are separated and so the saturation can be increased without altering the brightness. There are three channels: Lightness, 'a' and 'b'. The image is loaded, converted into 'Lab Color' and saved with a '_sat' added to the file name (this is to make sure that an inadvertent key stroke will not replace your original image). The 'Channels' window (Windows > Channels) is opened and the 'a' channel selected, which will appear grey with differing shades depending on the image. The 'Levels' command (Image > Adjustments > Levels) is entered and a histogram peaked around the centre will be seen. Now the input levels are adjusted with the two outer sliders to give 55 and 200 (rather than 0 and 255). The 'b' channel is selected and the same process carried out. Selecting the top ('Lab') image in the channels box will show a highly saturated image which is converted back to RGB colour mode (Image > Mode > RGB). The original image is opened and the saturated version copied (Ctrl A and Ctrl C) and pasted (Ctrl V) on top in a second layer. In the 'Layers' window, the two images will be one above the other. Finally the opacity slider is adjusted to give the required amount of saturation before the image is flattened (Layer > Flatten Image) and saved. This sounds like a rather involved process, but after a couple of tries does not take too long and, I promise you, the result is far better than applying a simple saturation adjustment.

One problem with these techniques is that they will increase the saturation of all areas of the image, even those which are well saturated already. Photoshop CS4 and above and Adobe Camera Raw have a vibrance slider (Image > Adjustments > Vibrance), which boosts the saturation only of lesser saturated colours so that it will not over-saturate the colours which are already bright. With earlier versions of Photoshop, one can bring the image into Camera Raw as described in the preceding section relating to the clarity function and adjust the vibrance slider as desired.

Enhancing the Colour of an Image Using the 'Match Color' Tool in Photoshop

There is a very quick and easy way to enhance a colour image in later versions of Photoshop which I have found to give superb results. The image is opened and a duplicate layer made (Layers > Duplicate Layer) with the opacity left at 100% so that the copy is seen. This is simply so that the effect of the tool can be easily seen. The tool is entered by using Image > Adjustments > Match Color (US spelling). In the box that opens up, the 'color intensity' slider is moved to the right and adjusted (probably in the range 130–50) to get the required effect. By switching on and off the adjusted layer by clicking in its 'eye', one can easily compare the before and after images. With the enhanced image visible at 100%, flattening the image (Layers > Flatten Image) will give the enhanced image.

It may well be that one wishes to enhance the colour only in a particular area of the image. This can be done by adding an adjustment mask to the enhanced image layer. The layer is highlighted by clicking in the image box and masked using Layer > Layer Mask > Hide All. The layer now appears black and only the original image is seen when both are selected. The paintbrush colour is now set to white and the area that is to be enhanced is painted, revealing the enhanced colour in the painted area. (Black conceals and white reveals.) The transition from the original to the enhanced region is likely to be too obvious, and this transition can be 'smoothed' by the application of a Gaussian blur of perhaps 50 pixels to the mask (Filter > Blur > Gaussian Blur). The two layers are then flattened as earlier to give the final image.

Sharpening an Image

Sometimes an image may not be quite as sharp as one would like, and a little judicious sharpening may well improve it. The emphasis here should be on the word 'judicious' as over-sharpened images are not very attractive!

Photoshop provides three methods of sharpening an image. The first is the 'Unsharp Mask' filter (Filter > Sharpen > Unsharp Mask). In the dialogue box one sets a radius and an amount. The radius should be no more than a few pixels, and then the amount is adjusted to give a suitable result. Some trial and error will be required to get a good compromise of radius and amount. This, however, is probably the least satisfactory method of sharpening an image.

The second method is called high-pass sharpening. A duplicate layer is made (Layer > Duplicate Layer) and the high-pass filter applied to it (Filter > Other > High Pass). The radius is set to a few pixels and the mouse clicked on 'OK'. The layer now looks grey but with some detail visible. In the 'Layers' box the blending mode is set to 'Overlay' and the sharpened image become visible. The opacity slider can be pulled back to lessen the sharpening. High-pass sharpening produces fewer artefacts than the Unsharp Mask filter.

Photoshop also provides a 'Smart Sharpen' filter which does give rather more control and, some believe, may actually incorporate a technique called deconvolution (described later) in that it has the capability of removing or diminishing various kinds of blur: Gaussian blur, lens blur and motion blur – it is best to select lens blur for astro-images. It is entered by using Filter > Sharpen > Smart Sharpen. As in the previous sharpening methods, a radius of perhaps 1–5 pixels is selected and, with the image viewed at 100% the amount is adjusted to get the desired result. At the bottom of the window is the 'More Accurate' checkbox. Since Smart Sharpening is reasonably CPU intensive, it may be best to leave it unchecked while adjusting the other controls – but check it before clicking 'OK'. Clicking on the 'Advanced' button, gives additional 'Shadow' and 'Highlight' boxes, allowing the sharpening to be faded (reduced) in the shadow or highlight regions of the image. In many astro-images it would be useful to fade the sharpening in the darker areas to avoid accentuating the noise that is often found there.

By following any of these methods with a further step, one can remove any colour fringing that might be produced. The 'Fade Unsharp Mask' dialogue box is entered with Edit > Fade Unsharp Mask. The opacity is left at 100%, the luminosity mode selected and the mouse clicked on 'OK'.

Both these methods use edge enhancement to give a sharper perceived image. A far more sophisticated approach makes use of a technique called 'deconvolution', by means of which one tries to determine what the image would have looked like had it not been blurred by the seeing, resolution of the telescope and so on. One key require-ment is a knowledge of what is called the 'point spread function', which is what the imaging system (including the atmosphere) would give for a point source. For a refractor under perfect seeing conditions this would be the Airy pattern. The pro-grams attempt to estimate this by analysing features in the image. The Richardson-Lucy (RL) algorithm is one that is applied in many of the sharpening programs that are available.

The free program 'Raw Therapee' provides an RL deconvolution method for sharp-ening which assumes that a Gaussian blur (like applying a Gaussian filter) has been produced by the imaging system. A radius (of a few pixels) is selected and an amount (too high will give a harsh result) set along with a damping control which avoids sharpening the noise in smooth areas. Finally, the number of iterations used in the process is defined. A value of 30 will be fine for most uses. The number of iterations increases the time taken by the process and there is also a danger of introducing halo artefacts if too many are used.

Noise Reduction

As has been pointed out, if a number of noisy images are stacked in Registax or Deep Sky Stacker, the noise level will be greatly reduced, so this is always a useful technique to produce images that are as free of noise as possible. But sometimes more noise reduction may be required.

Often the noise in an image is below the brightness of the faintest stars that are easily visible, and the obvious way to improve the image is to lift up the black level. The image is opened in Photoshop; the 'Levels' box (Image > Adjustments > Levels) is entered and the left marker moved to the right until the low-level noise disappears.

I have never been overly impressed by the noise reduction algorithms used by Photoshop, but I do find the 'Despeckle' filter quite useful (Filter > Noise > Despeckle). It can be used more than once but if used too often will lose fine detail in the image. I have found it to be very useful in cleaning up noise in the lunar maria.

There are many noise reduction programs available to purchase, and one that I have found both easy to use and very effective is 'Picture Cooler'. It can open both JPEG and 16-bit TIFF images provided that they are in RGB colour mode. To noise-filter a monochrome image, it has to be first set to RGB mode by means of Image > Mode > RGB and saved as a colour version. It can be returned to mono mode after the noise reduction has been applied. This program works very well indeed on most astro-images and is very simple to use.

A Novel Approach to Noise Reduction, Particularly in Large Nebulae such as M31 and M42

The example image of M31, the Andromeda Galaxy, which was the 'first light' image taken with an SBIG ST-8300 camera and described in Chapter 15 section 4, showed a lot of noise in the fainter, outer parts of the galaxy. It seemed almost impossible to smooth the faint nebulosity without eliminating the fainter stars in the image, but I finally realised that it was possible – and actually very easy. The idea is first to remove all the stars from the image into a 'stars only image', so just leaving the nebulosity in the original image. This process does smooth the nebulosity, and further smoothing can be applied if necessary by the use of an appropriate amount of Gaussian blur. The stars are then put back. Simple!

The image is brought into Photoshop and a duplicate layer made (Layer > Duplicate Layer). The 'Dust and Scratches' noise filter (Filter > Noise > Dust and Scratches) is applied to the image, with a sufficient pixel radius selected (~12) to just remove all but the very brightest stars. In the 'Layers' box, the difference mode is selected and an image showing just the fainter stars will appear. The two layers are flattened (Layers > Flatten Image) and saved as a new image such as Stars.tif). One then steps backwards (Edit > Step Backwards) until just the smoothed nebula image is seen after

Figure 18.3 Smoothing the nebulosity in an image of M31 as described in the text.

the 'Noise and Scratches' filter is applied. One might then wish to add some further smoothing using Gaussian blur. The star's image is brought back and copied (Ctrl A, Ctrl C). The smoothed nebula image is opened and the stars image pasted over it by means of Ctrl V to form a stars layer over the nebula layer. The layers blending mode is changed to 'Screen' and the stars will have returned to the image, which is then flattened (Layers > Flatten Image) and saved.

This may well not be an original technique but I have not seen it described anywhere. I do, however, think that it could be a very useful weapon in an astro-imager's armoury. Figure 18.3 shows a 100% section of the original Andromeda image, the stars image produced from this and the result of adding back the stars to the smoothed nebula image after this technique had been applied. The faint stars in the image have remained.

Software for Astro-Imaging

This comes in two forms. The first, for those with Photoshop, is sets of 'actions' that will carry out specific tasks such as for enhancing the colours of stars and removing a sky glow gradient from an image. The best-known set has been produced by Neil Carboni and is available from Pro*Digital* Software. It contains 34 actions, one of which will even add diffraction spikes to the brighter stars of an image – I have no idea as to how Neil has been able to do that in Photoshop! The cost is very reasonable.

The second class of software provides a complete astro-imaging platform into which one imports images and applies a variety of processes such as noise reduction and contrast enhancement. A relatively inexpensive package is called 'StarTools' (startools.org/drupal/home). It is quite easy to use, and one can download and use the package for free to see how it can be used – one can just not save the resultant images. The license to fully activate it can then be bought for a reasonable cost.

A somewhat more sophisticated package, and with a rather steeper learning curve, is called PixInsight. In this case one can request a trial licence which will allow the program to be fully used for 45 days before one needs to buy a full licence. As one might expect, this fully professional software package is somewhat more expensive – but it costs no more than a premium eyepiece.

Index